SILICON PHOTONICS
FOR TELECOMMUNICATIONS AND BIOMEDICINE

SILICON PHOTONICS

FOR TELECOMMUNICATIONS AND BIOMEDICINE

EDITED BY

SASAN FATHPOUR
BAHRAM JALALI

CRC Press
Taylor & Francis Group
Boca Raton London New York

CRC Press is an imprint of the
Taylor & Francis Group, an **informa** business

CRC Press
Taylor & Francis Group
6000 Broken Sound Parkway NW, Suite 300
Boca Raton, FL 33487-2742

First issued in paperback 2019

© 2012 by Taylor & Francis Group, LLC
CRC Press is an imprint of Taylor & Francis Group, an Informa business

No claim to original U.S. Government works

ISBN-13: 978-1-4398-0637-1 (hbk)
ISBN-13: 978-0-367-38204-9 (pbk)

Visit the Taylor & Francis Web site at
http://www.taylorandfrancis.com

and the CRC Press Web site at
http://www.crcpress.com

To Our Wives: Haleh and Mojgan

Contents

Preface

Today, silicon photonics, the technology for building low-cost and complex optics on a chip, is a thriving community, and a blossoming business. The roots of this promising new technology date back to the late 1980s and early 1990s to the work of Soref, Peterman, and others. There were three early findings that paved the path for much of the subsequent progress. First, it was recognized that micrometer-size waveguides, compatible with the CMOS technology of the time, could be realized despite the large refractive index difference between silicon and silicon dioxide (SiO_2). Previously, this large refractive index was thought to result in multimode waveguides that are undesirable for building useful interferometric devices such as directional couplers, Mach–Zehnder modulators, and so on. Although today's submicron (nanophotonic) waveguides are routinely realized and desired for their more efficient use of wafer real estate, the advance fabrication capability needed to fabricate such structures was not widely available to photonic device researchers. Second, it was proposed by Soref that by modulating the free-carrier density, which can be done easily with a diode or a transistor, electro-optic switching can be achieved through the resulting electroabsorption and electrorefraction effects. Third, it was shown that infrared photodectors operating in the telecommunication band centered at 1550 nm can be monolithically integrated onto silicon chips using strained layer GeSi (and eventually Ge) grown directly on silicon.

The potential for creating low-cost photonics using the silicon CMOS chip manufacturing infrastructure was gradually recognized by the photonics research and business community in the late 1990s and early 2000s. Fueling this development was the concurrent commercial emergence of silicon-on-insulator (SOI) CMOS as the platform of choice for high-performance complementary metal-oxide semiconductor (CMOS) processing. SOI also offers an ideal platform for creating planar optical circuits by providing an optically confining layer below the waveguide core. Also, the strong optical confinement offered by the large refractive index contrast between silicon and SiO_2 makes it possible to scale down the size of photonic circuits. Such lateral and vertical dimensions are required for economic compatibility with integrated circuits (IC) processing. In addition, a large nonlinear optical index in silicon plus a high optical intensity arising from the large index contrast between Si and SiO$_2$ make it possible to create nonlinear optical devices in chip-scale devices such as those based on Raman and Kerr nonlinearities. Optical amplifiers, optically pumped lasers, and wavelength converters—functions that were traditionally considered to be beyond the reach of silicon—were created.

The potential for silicon photonics extends beyond low-cost data communication products. The compatibility with CMOS notwithstanding, silicon has excellent material properties on its own. These include high thermal conductivity (about 10 times higher than gallium arsenide [GaAs]), high optical damage threshold (about 100 times higher than GaAs), and high third-order optical nonlinearities (about 100 times higher than silica optical fiber). Silicon is highly transparent in the wavelength range of 1.1 μm to nearly 7 μm. Furthermore, the absence of two-photon absorption at wavelengths larger than 2.25 μm renders silicon an excellent nonlinear optical material in the mid-wave infrared spectrum, where there are numerous important applications in remote sensing and biomedical applications.

Presently, it is believed that the highest impact of silicon photonics may be in optical interconnection between digital electronic chips. This technology addresses the communication bottleneck in very-large-scale integrated (VLSI) electronics. However, the benefits of integrated optics and electronics extend beyond the realm of computers. For example, in next-generation ultrasound medical imaging systems, the rate for signals generated by the array of transducers will exceed 100 GBps, once digitized, and will continue to increase as radiologists demand better image resolution. The size and power dissipation of conventional optical transceivers prevent them from being used in the imaging probe. Silicon integrated circuits with on-chip optical interfaces can potentially solve this problem.

Applications beyond telecommunications are being pursued for silicon photonics. For example, silicon photonics may be able to produce disposable mass-produced biosensors. One likely application is the so-called lab-on-a-chip in which both reaction and analysis are performed on a single device. Such sensors, along with integrated intelligence and wireless communication circuitry, may form nodes of an intelligent sensor network or environmental monitoring. High-power photonic and biomedical applications of silicon at mid-wave infrared are other possibilities on the horizon.

This book is meant to complement, rather than replace, previous books on silicon photonics. Indeed, there are the excellent books edited by L. Pavesi and D. J. Lockwood (*Silicon Photonics*, 2004); G. T. Reed and A. P. Knights (*Silicon Photonics: An Introduction*, 2004); G. T. Reed (*Silicon Photonics: The State of the Art*, 2008); and L. Khriachtchev (*Silicon Nanophotonics*, 2009). The topics covered in the present book are advanced, as familiarity with integrated photonics, in general, and with basics of silicon photonics, in particular, is assumed. Readers interested in more fundamental topics may refer to the three books mentioned above.

We attempt to offer a balance between theory and experiment on one hand, and current and forthcoming industrial trends on the other. An introductory chapter reviews the present state of the art and future trends and technological challenges. Following are selected topics on two major applications of silicon photonics—namely, telecommunications (Chapters 2 to 5) and high-power photonics and biomedicine (Chapters 6 and 7). The next four chapters are devoted to technological challenges that must still be overcome if silicon photonics is to fulfill its destiny. Chapters 8 and 9 cover the challenge of hybridization of III-V compound semiconductors on silicon in order to achieve monolithic light sources. Economic compatibility and the heat dissipation problems in CMOS chips—important challenges that are often neglected by the research community but reign supreme—are discussed in Chapters 9 and 10, respectively. The issues in the design of electronic-photonics ICs and the need for standardization in computer-aided design of industrial chips are addressed in the final chapter of this book.

Last but not least, we would like to thank the authors of each chapter for making this book possible.

Editors

Sasan Fathpour is an assistant professor at the College of Optics and Photonics (CREOL) at the University of Central Florida (UCF). He also holds a joint appointment at UCF's Department of Electrical Engineering and Computer Science. He received a PhD in electrical engineering from the University of Michigan–Ann Arbor in 2005. He then joined the University of California–Los Angeles (UCLA) as a postdoctoral fellow. He won the 2007 UCLA Chancellor's Award for postdoctoral research for his work on energy harvesting in silicon photonics. Dr. Fathpour is a coauthor of over 70 journal and conference papers and book chapters.

Bahram Jalali is a professor of electrical engineering at UCLA, a fellow of IEEE and of the Optical Society of America, and recipient of the R.W. Wood Prize from the Optical Society of America. In 2005, he was elected to the *Scientific American* Top 50, and received the BrideGate 20 Award in 2001 for his contributions to the Southern California economy. Dr. Jalali serves on the board of trustees of the California Science Center and the board of Columbia University School of Engineering and Applied Sciences. He has published over 300 journal and conference papers and holds 8 patents.

Contributors

Pallab Bhattacharya
Department of Electrical
Engineering and Computer
Science
University of Michigan
Ann Arbor, Michigan

Ozdal Boyraz
Department of Electrical
Engineering and Computer
Science
University of California–Irvine
Irvine, Califoria

Ernst Brinkmeyer
Technische Universität
Hamburg–Harburg
Hamburg, Germany

Walter Buchwald
Air Force Research Laboratory
Hanscom AFB
Bedford, Massachusetts

Sang-Yeon Cho
Department of Electrical and
Computer Enginecring
New Mexico State University
Las Cruces, New Mexico

Justin Cleary
Department of Physics
University of Central Florida
Orlando, Florida

Sasan Fathpour
Department of Electrical
Engineering and Computer
Science
College of Optics and Photonics
University of Central Florida
Orlando, Florida

Frederic Y. Gardes
Department of Electronic
Engineering
University of Surrey
Guildford, United Kingdom

William R. Headley
Department of Electronic
Engineering
University of Surrey
Guildford, United Kingdom

Diana L. Huffaker
Electrical Engineering
Department
University of California–Los
Angeles
Los Angeles, California

Tejaswi Indukuri
Intel Corporation
Santa Clara, California

Bahram Jalali
Electrical Engineering
Department
University of California–Los
Angeles
Los Angeles, California

Prakash Koonath
Tanner Research Inc.
Monrovia, California

Michael Krause
Technische Universität
  Hamburg–Harburg
Hamburg, Germany

Daniel Kucharski
Luxtera Inc.
Carlsbad, California

Jenifer L. Lawrie
Department of Electrical
  Engineering and Computer
  Science
Vanderbilt University
Nashville, Tennessee

Goran Z. Mashanovich
Department of Electronic
  Engineering
University of Surrey
Guildford, United Kingdom

Gianlorenzo Masini
Luxtera Inc.
Carlsbad, California

Attila Mekis
Luxtera Inc.
Carlsbad, California

Zetian Mi
Department of Electrical and
  Computer Engineering
McGill University
Montreal, Quebec, Canada

Milan M. Milosevic
Department of Electronic
  Engineering
University of Surrey
Guildford, United Kingdom

Robert E. Peale
Department of Physics
University of Central Florida
Orlando, Florida

Thierry Pinguet
Luxtera Inc.
Carlsbad, California

Varun Raghunathan
Department of Chemistry
University of California–Irvine
Irvine, California

Graham T. Reed
Department of Electronic
  Engineering
University of Surrey
Guildford, United Kingdom

Hagen Renner
Technische Universität
  Hamburg–Harburg
Hamburg, Germany

Richard Soref
Air Force Research Laboratory
Hanscom AFB
Bedford, Massachusetts

Jun Tatebayashi
Electrical Engineering
  Department
University of California–Los
  Angeles
Los Angeles, California

David J. Thomson
Department of Electronic
    Engineering
University of Surrey
Guildford, United Kingdom

Kevin K. Tsia
Department of Electrical and
    Electronic Engineering
University of Hong Kong
Hong Kong

Sharon M. Weiss
Department of Electrical
    Engineering and Computer
    Science
Vanderbilt University
Nashville, Tennessee

Jun Yang
Department of Electrical
    Engineering and Computer
    Science
University of Michigan
Ann Arbor, Michigan

CHAPTER 1

Silicon Photonics—The Evolution of Integration

Graham T. Reed, William R. Headley,
Goran Z. Mashanovich, Frederic Y. Gardes,
David J. Thomson, and Milan M. Milosevic

Contents

1.1 Introduction

Few can argue that the incredible advancements the microelectronics industry has made over the past 50 years are primarily due to integration. Integration allows the integrated system or device to be more advantageous than the sum of its individual components. Jack Kilby's ingenuity, vision, and effort in the late 1950s to develop integrated transistor circuits brought integration to the field of electronics, which helped spawn the microelectronics era [1]. Over the last 50 years, microelectronic devices have evolved dramatically as efforts to increase the number of components on a substrate have continued. This progress is captured by the much-heralded Moore's law, the trend for the number of transistors on a single CMOS substrate to double every two years [2] (revised

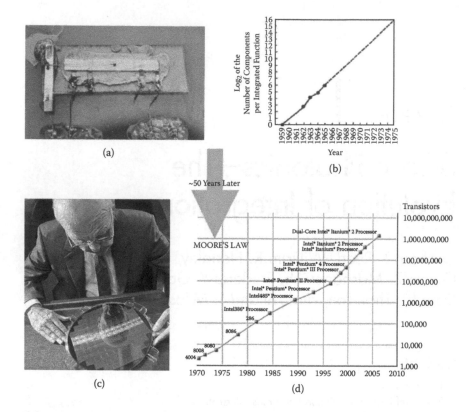

FIGURE 1.1 (a) The first integrated circuit (from Texas Instruments, First Integrated Circuit, http://www.ti.com/corp/graphics/press/image/on_line/co1034.jpg, 2009). (b) The original version of Moore's Law (from Moore, G. E., *Proc. IEEE*, **86** (1), pp. 82–85, 1998. With permission). (c) Jack Kilby inspecting a 300 mm wafer containing dozens of chips each with millions of transistors (from Texas Instruments, Jack Kilby Examines 300 mm Wafer, http://www.ti.com/corp/docs/kilbyctr/downloadphotos.shtml, 2009). (d) A more recent version of Moore's Law demonstrating its continuing validity (from T. Ghani, Challenges and Innovations in Nano-CMOS Transistor Scaling, http://www. intel.com/technology/silicon/Neikei_Presentation_2009_Tahir_Ghani.ppt, 2009. With permission).

by Gordon Moore as to the number of transistors approximately doubling every year [3]). To emphasize the impact of integration on the microelectronics industry, a figurative comparison is made between the state of the art in the late 1950s/early 1960s and today in Figure 1.1. The first integrated circuit (Figure 1.1a) had an approximate area of the size of a fingernail [1]. Today, individual transistors have dimensions of the order of tens of nanometers, and in the near future, a single chip will include approximately 2 billion transistors [7]. Likewise, Moore's law has remained nearly steadfast in its prediction of the number of transistors on a chip as a function of time. Perhaps even more impressive than this is the fact that this trend is expected to continue in the near future with the use of improved dielectric materials and innovations such as strained silicon transistors [8].

There are several key advantages to be obtained from integration of electronic components. These are primarily: device reliability as all components are fabricated at the same time under the same controlled conditions, control of stray impedance typically associated with discrete electrical interconnects, and huge cost savings due to the benefit of mass production. The resulting decrease in transistor size and control of stray impedances results in electronic circuits that can operate at higher speeds. Such integrated devices also improve opportunities for dimensional scaling and increased complexity, particularly where a constrained set of design rules can be implemented. Such opportunities have been seized by the semiconductor industry, resulting in modern lifestyles that are utterly dependent on electronic devices.

However, the benefits of integration do not come without consequences. One in particular is that power consumption increases as transistor density increases (see Figure 1.2), which can in turn lead to lifetime and performance degradation due to heating [9]. It can also lead to power-hungry devices, which may create a more global issue as technology becomes more ubiquitous, and as more

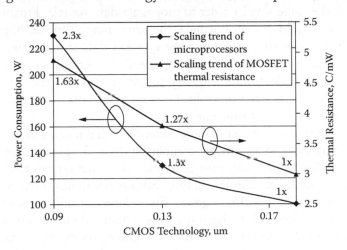

FIGURE 1.2 A plot demonstrating that as the gate width of a transistor decreases, (CMOS technology) power consumption and thermal resistance increase. (From O. Semenov et al., *IEEE Trans. Device and Materials Reliability*, **6**, pp. 17–27, 2006. With permission.).

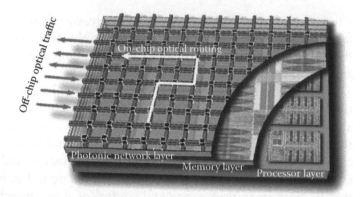

FIGURE 1.3 IBM's vision of optical intra-chip and inter-chip connections. (From International Business Machines, Silicon Integrated Nanophotonics, http://domino. research.ibm.com./comm/research_projects.nsf/pages/photonics.index.html, 2009).

users embrace it. One candidate solution to the heating and power (and other) problems is the use of photonics to transfer information to various sections of the chip (intrachip) and between the chips (interchip), thereby minimizing the use of metallic pathways as much as possible. It is these pathways that introduce many of the deleterious effects discussed previously. Figure 1.3 illustrates how, in the near future, a photonics solution might be utilized within a multilayer integrated circuit (IC) [10]. Optical pathways do not suffer from joule heating and hence do not dissipate power in the same way in order to propagate data signals. Furthermore, these pathways do not have the same magnitude of impedance limitations that may affect the speed of data signals, and so data can be transmitted at nearly the speed of light. Hence, in order for the microelectronics industry to continue along the integration path, it may need to evolve by embracing optical components in addition to electronic ones.

One issue with a photonics solution is that it can add substantial complexity to an already complex system. As greater complexity implies higher cost, then photonics as a solution must also demonstrate its economic viability. Silicon photonics offers a unique opportunity in this regard. The microelectronics industry has demonstrated that silicon is a superb material for mass production of complex devices. Photonic integration, on the other hand, has been modest to date, partly because traditional photonic materials are more difficult to use effectively in mass production than silicon, and partly because low-cost photonic materials other than silicon (such as silica) do not offer sufficient photonic functionality.

Furthermore, silicon offers the tantalizing opportunity of integrating both electronics and photonics into a single substrate. The technical difficulty of such integration should not be underestimated given the huge differences in the current state of development of silicon electronics and silicon photonics. Nevertheless the rewards for such integration are potentially huge, and can perhaps facilitate another revolution in communications and computing technology.

1.2 A Brief History of Silicon Photonics Integration

The very existence of this text is an indication of the fact that silicon photonics is a topical subject at present. However, the subject has existed as a research field for a surprisingly long time. The early work was published in the mid-1980s, focusing primarily on waveguides, the fundamental building block of the technology [11–13], but also discussed mechanisms related to switching and modulation. Perhaps a little surprisingly, integration was also a topic of discussion in the earliest papers [14,15], even before the most basic building blocks of the technology had been successfully demonstrated. On the other hand, integration is such an obvious target for silicon photonics that it is paradoxically also unsurprising that integration was an early subject of discussion, particularly given the success of integrated electronics in silicon, the very backbone of modern life. Soref was the earliest architect of integration in silicon photonics, in the early papers [13–16], but most specifically in his proposal of an Opto Electronic Integrated Circuit (OEIC) based on silicon in 1993 [17]. His vision is shown in Figure 1.4, and it effectively identifies the required building blocks of silicon photonics.

Included in the schematic are the following key components: optical waveguides, modulators, switches, photodiodes, light sources, and amplifiers, and perhaps most significantly, electronics, together with the purely photonic components. Of the photonic components, only one arguably essential function is missing, a multiplexer/demultiplexer system. In a philosophical discussion of technology requirements, the inclusion of electronics as well as photonics in a near-monolithic architecture immediately changes the landscape of the OEIC, because it effectively confines the contribution of purely passive technologies to history, and introduces the requirement of a semiconductor technology. Therefore, the question of "which semiconductor?" is introduced. At the outset of the silicon photonics work in the 1980s, the answer to

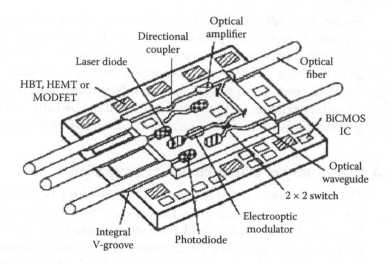

FIGURE 1.4 Soref's schematic of an OEIC superchip published in 1993. (From R. A. Soref, Silicon-based optoelectronics, *Proc. IEEE*, **81**, pp. 1687–1706, 1993.)

this question was far from obvious, but if pressed, most observers would probably have suggested that some sort of III-V compound would emerge. However, the changing face of silicon photonics, together with emerging applications, has transformed the landscape such that most observers would now identify silicon as the leading candidate, and many would say it is the *only* candidate that can deliver acceptable performance at sufficiently low cost and high yield.

So, what are the individual building blocks that comprise the OEIC? The most critical components for integration are the waveguide, the modulator, the photodetector, and perhaps the light source, although opinion is divided about whether the latter needs to be part of the integration process. Recent consensus appears to be leaning toward an external "optical power supply" as a more effective solution. The challenge of light sources and their integration will be discussed in the next section.

To provide the reader with some insight into the current state of the art in the field, it would be useful to start with some figures of merit for each component. In 2008, Kimerling proposed a series of performance metrics for success [18]. While the absolute values of these metrics can be debated, they offer a benchmark against which to measure the progress made in the field in less than two short years. First, he identified a high index contrast between core and cladding to enable a high packing density of >10^6 devices per chip. The Silicon on Insulator (SOI) platform meets this requirement better than any other technology, with a core cladding index ratio of approximately 3.5:1.5. Next, he identified a waveguide loss requirement of less than 0.1 dB/cm. This is a demanding target, arguably a little too demanding. Silicon can easily meet such a target in large waveguides (>1 μm), but in the submicron waveguides required for high packing density (~200–400 nm), this remains a challenge, but is very close to routine fulfillment. For modulators, Kimerling proposed a 3 dB bandwidth of >25 GHz, with a switching power <150 pJ, and an extinction ratio >5. Although demanding, these targets are actually within reach. Dramatic progress has been made in this area in recent years, and the bandwidth and extinction requirements are close to being met (they have been met in separate devices, but not simultaneously). Coupling these requirements with low switching power is now the primary challenge. For the photodetector, Kimerling proposed a bandwidth-external quantum efficiency product of approximately 75 GHz, However, this value has already been surpassed by a monolithically grown germanium/silicon avalanche photodiode with a gain-bandwidth product of 340 GHz [19].

Of course, the performance of these devices has primarily been measured in isolation rather than in an integrated form. There is no doubt that the integration of these devices while retaining or even improving performance is nontrivial. This is particularly true when also integrating the necessary electronics, as the different processing requirements of current electronic and photonic devices are usually incompatible. Addressing these challenges is still in its infancy, although some have made significant progress [18], and examples will be discussed later. Thus, we are again facing a discussion of the current state of integration. We will consider this in more detail later, when we will discuss the topic more directly. However, before doing so, a brief history of the field is provided so as to emphasize the similarities between the development of the microelectronics industry and silicon photonics.

As mentioned previously, integration was envisioned from the outset of silicon photonics as compared to the revolution brought about by Kilby's vision of the future of electronics. This is particularly interesting as the field is rather broad in the applications that it potentially encompasses. For instance, biology can benefit from silicon photonics in that lab-on-a-chip and biosensors can be realized in silicon and the particular biomaterial of interest can be interrogated optically. However, the field with the longest historical interaction with silicon photonics is telecommunications. This is probably due to the fact that silicon is transparent at the standard telecommunication wavelengths of 1310 and 1550 nm and yet is an inexpensive and well-understood semiconducting material when it comes to manufacturing. Hence, in hindsight, it would be logical that silicon photonics would follow a similar path to that of the microelectronics industry. This logic was far from obvious in the late 1980s when the concept of using silicon as a photonic material first materialized. Similar to Kilby's vision, it was but a small handful of individuals who believed that silicon could be used as more than just a material for electronic devices. Indeed, it was Soref's version (Figure 1.4) [17] of Arbistreiter's vision of the "superchip" [20] that started the field on the integration path. What is important about this illustration, besides the photonic components previously discussed, is the fact that the necessary drive electronics were included within the integrated optoelectronic circuit. This is important because two functionally and fundamentally different characteristics (i.e., light and electronics) had now been brought together onto a single platform. The vision presented in this diagram is similar to the aim of much of the research carried out today in the field of silicon photonics, with one notable difference: today the entire circuit is often (but not always) viewed as a monolithic integrated circuit, with all components fabricated from silicon or perhaps silicon compounds, as opposed to the hybrid approach originally proposed by Soref.

The field of silicon photonics began to take shape during the early to mid-1990s, with much of the work focusing on discrete, passive component development. The most important component studied during this time was the waveguide, as all other components are based on this fundamental building block. Some of the first ever SOI waveguides were planar in nature with losses of the order of several tens of dBs per centimeter [21]. However, as efforts to study waveguides continued, these losses have been dramatically reduced. It has been determined that the largest cause of propagation loss in submicron waveguides is due to surface and sidewall roughness [22]. However, oxidizing the waveguide after it has been etched can lead to significant improvements in loss values. In fact, it has been reported that strip waveguides with submicron cross-sectional areas have been fabricated with loss values that have decreased from more than 30 dB/cm to less than 1 dB/cm (including the process of oxidizing the waveguide) [22]. More recently, strip waveguides with cross-sectional dimensions of 226 nm (height) by 510 nm (width) have been fabricated with propagation losses of 1.7 dB/cm and a bend loss of a 1 μm bend of less than 0.1 dB/cm for TE-polarized light [23]. While this propagation loss value is greater than the aforementioned 1 dB/cm, it has been provided here as a demonstration of how small waveguide dimensions have become as fabrication tools and techniques are continuously improved. It therefore demonstrates the kinds of dimensions that can be achieved with photonic devices made from silicon. This implies that very high packing densities are achievable.

Waveguide development also benefited from efforts to improve on the modeling of these devices. In 1991, Soref et al. provided a preliminary design guideline that would yield single-mode rib waveguides. Their guideline is dependent on the waveguide height, width, and etch depth [13]. Since that time, and with the help of more advanced computer simulation packages and methods, the accuracy of this original model has been improved upon for rib [24] and strip [18] waveguides. Furthermore, this modeling now considers the zero birefringence condition (ZBC) of the waveguide in addition to its modal properties for both oxide-clad and unclad waveguides. An even more recent modeling effort has incorporated all of the previous effects while also considering the stress induced by the oxide cladding as well as the angle of the rib sidewalls [25]. Figure 1.5 demonstrates how much impact all of

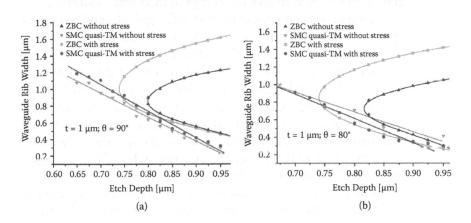

FIGURE 1.5 Demonstration of the effects of stress and sidewall angle on the single-mode condition (SMC) and ZBC (zero birefringence condition) of oxide-clad silicon waveguides for oxide thickness (t) of 1 μm and sidewall angle (θ) of (a) 90°, (b) 80° (where 90° is a vertical sidewall). (From M. M. Milosevic et al., *J. Lightwave Technol.*, **26**, pp. 1840–1846, 2008.)

these parameters can have on the optical properties of the waveguide. As more complex optical devices are realized using these fundamental building blocks, the importance of this work cannot be overestimated.

Waveguide technology is still a topic of interest as researchers continue to try shrink waveguide dimensions while maintaining low propagation loss values. Plasmonics, discussed later, is one way of possibly realizing very short waveguides and devices. Photonic crystal waveguides is another. While losses as low as 2 dB/cm have been reported for specific designs [138], propagation loss is often an issue for these waveguides, particularly at large enhancement factors, with many loss values reported to be more than 10 dB/cm [26]. However, providing loss values in units of loss per centimeter is somewhat misleading as photonic circuits utilizing photonic crystal waveguides can be very short in comparison to conventional waveguides, and so the millimeter length is a more common and practical measure, and arguably the insertion loss is a yet more valuable parameter than the propagation loss. Thus, when considering these dramatically smaller device lengths, the insertion loss values stated in the literature are equivalent to those of larger, micron-sized conventional waveguides. These loss values may continue to improve as lithographic resolution continues to improve. Therefore, it is possible that waveguides and devices utilizing photonic crystal technology will play a significant role in future integrated devices so as to take advantage of their small footprint.

An important component necessary for optical transceivers and optoelectronic integrated circuits is the modulator. Significant bandwidth advancement of these devices occurred only relatively recently with the demonstration by Liu et al. of the first experimental GHz modulator, which was based on a metal oxide semiconductor (MOS) capacitor-like structure [27]. In this device, a layer of oxide was grown horizontally in the center of a waveguide such that free carriers could be accumulated on either side of the oxide layer to actively change the effective index of the propagating mode. This modulator, based essentially on majority carriers, was primarily limited in bandwidth by the resistance-capacitance (RC) time constant and the rate at which carriers recombine when the modulator is switched from an "on" to an "off" state. This device was comparable in performance (other than bandwidth) and functionality to other injection-type modulators studied at the time [28–38]. However, despite the bandwidth limitations of injection devices, researchers have demonstrated the possibility of extending their speed by using a preemphasis driving signal [39–43]. Such a driving signal consists of an initial relatively large forward bias that quickly injects majority carriers into the region of interest. Once the required carrier concentration has been reached, the voltage is quickly dropped to a level that maintains that carrier concentration in the waveguide. Then, at the point where the modulator is switched off, a reverse bias is applied in order to quickly sweep out the carriers that were initially injected into the waveguide.

In 2005, an alternative submicrometer modulator based on the depletion of a *p-n* junction was proposed by Gardes et al. that was the first of its kind at the time of publication [44]. Similar to the MOS capacitor, the depletion-type phase shifter is not limited by the minority carrier recombination lifetime and is based on the principle of removing carriers from the junction when a reverse bias is applied. The main advantage of using depletion to adjust the index of refraction in the waveguide is the fast response time, simulated in this work to be 7 ps. This design also provided a simplified means of fabrication as compared to the capacitive device, as no gate oxide layer is required. Figure 1.6 shows the cross-section schematic of a four-terminal asymmetric *p-n* structure, where the concentration of *n*-type doping is much higher than the concentration of *p*-type doping. The reason for this structure is, first, to minimize the optical losses induced by the *n*-type doping

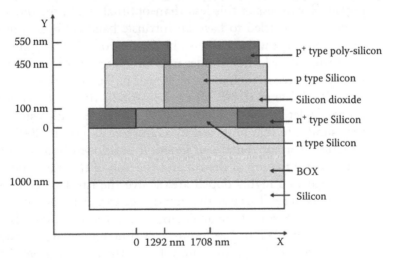

FIGURE 1.6 Schematic of a four-terminal depletion type modulator. (From F. Y. Gardes, G. T. Reed, N. G. Emerson, and C. E. Png, *Opt. Express*, **13**, pp. 8845–8854, 2005. With permission.)

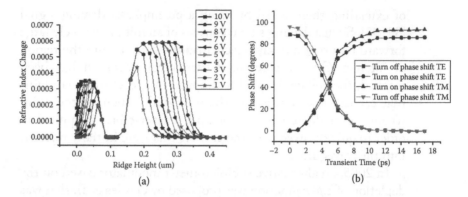

(a)   (b)

FIGURE 1.7 (a) Variation of the refractive index in the waveguide, (b) rise and fall time for TE and TM polarizations. (From F. Y. Gardes, G. T. Reed, N. G. Emerson, and C. E. Png, *Opt. Express*, **13**, pp. 8845–8854, 2005. With permission.)

and, second, to enhance the overlap between the optical mode and the *p*-type depletion region, which provides a better phase shift to length ratio.

The carrier concentration variation in this device is not uniform, however. As shown in Figure 1.7, the predicted refractive index changes across the vertical axis of the waveguide arises on both sides of the junction over a width of around 200 nm. It is believed that the device could be better optimized by increasing the overlap between the optical mode and the *p*-type depleted region. Regardless of this less-than-optimal result, the proposed device was modeled to have an intrinsic bandwidth of approximately 50 GHz when a reverse bias swing of 5 V was applied to both arms of a Mach–Zehnder interferometer (MZI) (a 2.5-mm-long active area in each arm), in a push-pull configuration.

In 2007, Liu et al. [45,93] experimentally demonstrated a *p-n* carrier depletion-based silicon optical modulator with a structure very similar to that proposed by Gardes et al. [44] in 2005. The structure is a horizontal *p-n* junction positioned in a rib waveguide where the top of the rib and the slab of the waveguide are connected to a highly doped area above the rib that serves as a resistive contact. This was the first device to experimentally demonstrate the possibility of achieving a bandwidth of 30 GHz and a data rate of 40 Gbps (see Figure 1.8) although having a modulation depth of only approximately 1 dB. The active region of this device was 1 mm long.

In 2008, Liu et al. reported a demultiplexer (DEMUX) configuration of 8 MZIs based on the same depletion-type modulator

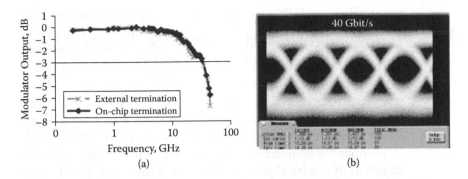

FIGURE 1.8 (a) Frequency response of the modulator tested by Liu et al., (b) the optical eye diagram of the modulator. (From L. Liao et al., *Electron. Lett.*, **43**, pp. 1196–1197, 2007.)

[46]. The slightly longer 1.5 mm modulators operated at 25 Gbps with an extinction ratio of 2 dB. The amplified single-ended output of 3.2 V_{pp} (6.4 V_{pp} differential) was combined with 2 V_{dc} using a bias tee to ensure reverse bias operation for the entire AC voltage swing. The DEMUX shown in Figure 1.9 demonstrates the possibility of transmitting an aggregate data rate of 200 Gbps.

The research trend of moving toward smaller waveguides to reduce power consumption and device real estate led to devices such as the one recently reported by Park et al. in 2009 [47]. This device is an MZI-based modulator that utilizes carrier depletion to obtain a change in the index of refraction. Similar in cross-sectional design to that proposed by Gardes et al. [48], the modulator is realized by using regions of high *p*- and *n*-type doping concentrations positioned vertically in a 220-nm-high, 500-nm-wide rib waveguide to achieve a large modal overlap with the depletion

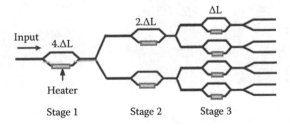

FIGURE 1.9 Schematic of the 8-channel DEMUX based on asymmetric MZIs; 4·ΔL = 97.2 μm for Δλ = 3.2 nm around 1550 nm, where *ΔL* and *Δλ* are the length differences between two arms of the asymmetric MZIs and optical channel spacing, respectively. (From L. Ling et al., *IEEE Electron Devices Meeting 2008*, pp. 1–4, 2008. With permission.)

region. Park et al. showed an eye diagram demonstrating a data rate of 12.5 Gbps. A $V_\pi L_\pi$ (the voltage-length product required to achieve a π-phase shift in the modulator) efficiency of 2 Vcm and a loss of approximately 4 dB/mm was also reported. A 3 dB electro-optic bandwidth of 7.1 GHz was reported, and data transmission was demonstrated at 12.5 Gbps for a 700-μm-long device, with an extinction ratio of 3 dB. Similarly, 4 Gbps was reported for a 2-mm-long device with an extinction ratio of approximately 7 dB.

To improve compactness further, a modulator based on the depletion of a vertical *p-n* junction was demonstrated by Gardes et al. [48] in 2009. The proposed ring resonator modulator was based on a 300-nm-wide, 150-nm-etch depth, and 200-nm-high rib waveguide. This structure is capable of providing single-mode propagation. As shown in Figure 1.10, the *p-n* junction was asymmetrical in size and in doping concentration in order to maximize the depletion area that overlaps with the optical mode. The *n*- and *p*-type regions were 75 nm and 225 nm wide, respectively, and the net doping concentration of this particular junction varies between 6×10^{17} cm^{-3} and 2×10^{17} cm^{-3} for *n*- and *p*-type regions, respectively. The device was fabricated in a Europe-wide collaboration [48].

The modulator, based on a 40 micron radius ring resonator, exhibited a DC on/off ratio of 5 dB at −10 V, and a 3 dB bandwidth of 19 GHz (Figure 1.11). The principal issues regarding the fabrication of ring resonator modulators are the difficulties encountered in finding the right coupler configuration to obtain high resonance contrast for a wide range of wavelengths. These issues are exacerbated when the fabrication variations of the

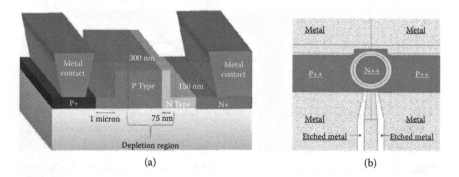

(a)   (b)

FIGURE 1.10   (a) Cross-section schematic of the rib waveguide modulator, (b) plan-view schematic of the positioning of the doped area and electrodes around the ring. (From F. Y. Gardes et al., *Opt. Express*, **17**, pp. 21986–21991, 2009. With permission.)

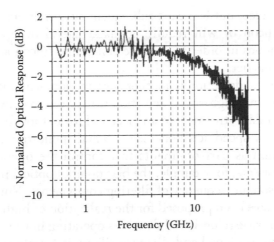

FIGURE 1.11 Normalized optical response as a function of frequency. (From F. Y. Gardes et al., *Opt. Express,* **17**, pp. 21986–21991, 2009. With permission.)

coupling area and the waveguide losses for a particular fabrication process are unknown. One way to counteract this uncertainty was proposed and demonstrated by Gill et al. [49] in 2009. The interesting feature of this device was the ability to control the resonance of the ring by replacing the coupler with an MZI, hence enabling full control of the extinction ratio of the ring. The ring resonator modulator fabricated here was based on SOI with a silicon overlayer of about 230 nm in height. The modulator was also based on a vertical *p-n* junction inserted into a rib waveguide, where the ring had an RF/optical interaction length of 400 μm, and the demonstrated electro-optic bandwidth was in excess of 35 GHz.

Moving away from *p-n* junction modulators, it is interesting to note that a recent modulator design based on accumulation of carriers on both sides of an oxide region has been reported by Lightwire, Inc. [50,51]. This device is based on the same principle used in [27], but they report a much higher efficiency, with a $V_\pi L_\pi$ of 2 Vmm, which results in a compact device (800 μm × 15 μm). The eye diagram associated with this device indicates data transmission at 10 Gbps with an extinction ratio of almost 9 dB. One of the possible drawbacks of this device and all reported devices based on carrier accumulation is the relative complexity of the fabrication process required to produce the MOS capacitor-like structure where an oxide layer has to be located in the waveguiding region. This is not to say that the oxide layer itself is complex to produce, but it means subsequent silicon layers will either be polysilicon or will require epitaxial lateral overgrowth (ELO) techniques in order to produce crystalline silicon adjacent to the oxide barrier.

Optical detection is another key function of an integrated optical circuit. High-speed detection is required to convert incoming optical data streams into an electrical format for further processing. Low-speed detectors may also find application in power level detection as part of circuits that control the operation of other components, for example, the bias point of optical modulators, or the resonant frequency of filtering components. Numerous structures have been proposed for the realization of high-performance silicon waveguide-based detectors operating in the 1550 nm telecommunications band. Because silicon is largely transparent for wavelengths greater than 1100 nm, these structures require either modification to be made to the silicon material itself or additional materials need to be utilized. The methods that show the most promise are germanium-on-silicon, III-V materials integrated onto a silicon platform, and silicon-based detectors incorporating defects.

While the absorption of light in silicon reduces dramatically as the wavelength is increased above ~1100 nm, germanium absorption extends beyond the 1550 nm telecommunication bands and is a viable material for detection at this wavelength. Furthermore, detectors formed by the integration of germanium onto silicon substrates have already been demonstrated. Germanium-on-silicon based detectors generally consist of a metal-semiconductor-metal (MSM) or p-i-n junction where light either couples evanescently into a region of germanium that is epitaxially grown onto the waveguide surface, or is end-fire coupled into a region of germanium that is grown into an etched trench in the waveguide. Using the butt-coupled approach, a p-i-n junction based germanium on silicon photodiode has been demonstrated with a bandwidth of 42 GHz, a responsivity of approximately 1 A/W, and dark current density of 60 mA/cm^2 [52]. A schematic and cross-sectional SEM image of the device is shown in Figure 1.12.

While not strictly a detector that can be integrated in its present form, the recent avalanche diode detector developed by Kang et al. is important to the field [19]. The photodetector comprises monolithically grown germanium/silicon layers. A cross-sectional schematic of the device is shown in Figure 1.13. The importance of this device is that it has an (external quantum) efficiency-bandwidth

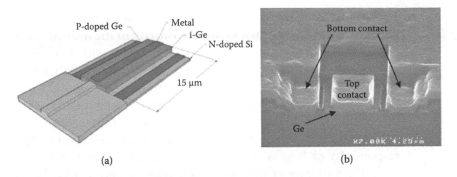

FIGURE 1.12 (a) Diagram of butt-coupled germanium on silicon photodetector, (b) SEM cross-sectional image of PIN diode region. (From V. Laurent et al., *Opt. Express*, **17**, pp. 6252–6257, 2009. With permission.)

product of 340 GHz and a sensitivity of –28 dBm at 10 Gbps for 1300 nm light [19]. Very recently, the IBM group also reported very impressive results: a gain-bandwidth product of 300 GHz with an operational speed exceeding 30 GHz and a bias voltage of only 1.5 V [139]. These devices demonstrate that silicon-based detectors can be realized that are comparable in functionality to those of the previously superior III-V-based detectors. Thus, low-cost, high-volume CMOS fabrication techniques can be used to fabricate these devices. Their CMOS compatibility may also allow them to be utilized in the interconnection of silicon-based electronic and optoelectronic devices.

The integration of III-V materials with silicon waveguides can offer solutions for light emission, modulation, and detection.

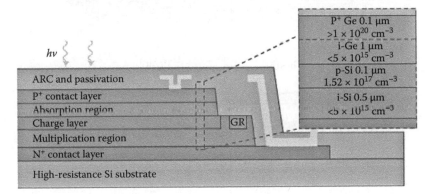

FIGURE 1.13 Cross-sectional schematic of a monolithically grown silicon/germanium avalanche photodiode with an efficiency-bandwidth product of 340 GHz (ARC is Anti-Reflection Coating, GR is Guard Ring). (From Y. Kang et al., *Nat. Photonics*, **3**, pp. 59–63, 2008.)

InGaAs-based detection devices offer superiority in terms of low dark current, high speed, and high sensitivity in the near-infrared region [53], but the incompatibility of InGaAs with CMOS is an obvious issue. III-V materials can be integrated with a silicon-on-insulator based waveguide system using a variety of techniques, for example, epitaxial growth, wafer bonding, or flip-chip integration [53]. Using a III-V die-to-wafer bonding technique, MSM structure photo detectors have been demonstrated with a responsivity of 1 A/W at 1550 nm and a dark current of 4.5 nA [53]. The structure of this device is shown in Figure 1.14. Light propagating along the silicon wire waveguide evanescently couples to the InGaAs layer, which performs the detection function.

Further results have also been obtained using bonded AlGaInAs quantum wells and a *p–i–n* diode structure as is shown in Figure 1.15 [54]. A responsivity of 1.1 A/W is reported with a dark current of under 100 nA.

Silicon-based photodiodes employing defect engineering of the waveguide region have also been demonstrated. A cross-sectional diagram of one example device is shown in Figure 1.16. Silicon ions are implanted into the intrinsic waveguide region of the

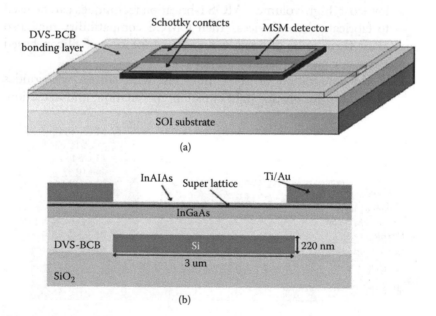

(a)

(b)

FIGURE 1.14 (a) Top-down and (b) cross-sectional diagrams of die-to-wafer bonded III-V MSM photodetector. (From J. Brouckaert et al., *IEEE Photon. Technol. Lett.*, **19**, pp. 1484–1486, 2007. With permission.)

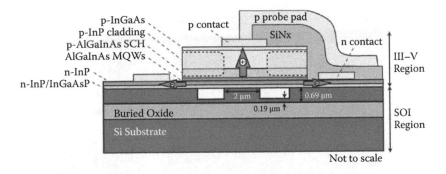

FIGURE 1.15 Cross-sectional schematic of a bonded III-V PIN photodetector. (From P. Hyundai et al., *Opt. Express*, **15**, pp. 6044–6052, 2007. With permission.)

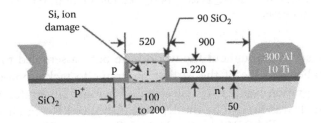

FIGURE 1.16 Cross-sectional schematic of a defect-engineered photodiode cross-section (all measurements are in nanometers). (From M. W. Geis et al., *IEEE Photon. Technol. Lett.*, **19**, pp. 152–154, 2007. With permission.)

device to create defects in the crystal lattice, which consequently enhances the photo response over the wavelength band from 1270 to 1740 nm. A 3 dB bandwidth of between 10 and 20 GHz was reported with a responsivity of 0.8 A/W at 1550 nm [55].

It is hoped that the reader will appreciate that while the foregoing discussion of the individual components has been necessarily selective and brief, it is intended to convey the rapid progress of the field in a very short time. In such a short review, it is impossible to capture the enormous worldwide effort. However, even with considerable worldwide effort, relatively little true integration of photonic components has emerged and even less integration with the necessary electronics. One of the reasons is the substantial resources required to undertake such integration. This resource demand may not be purely material in nature but rather may require a pooling of talent in order to provide the knowledge and experience to overcome many of the more significant obstacles that are faced with integration. As a result, a number of

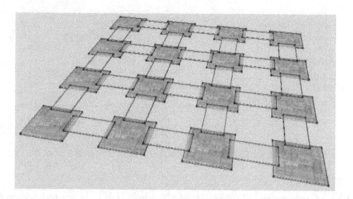

FIGURE 1.17 UNIC's vision of interconnected multiprocessors utilizing photonic interconnects. (From M. Feldman, *HPC Wire*, March 28, 2008, available at: http://www.hpcwire.com/features/DARPA_Places_44M_Bet_On_Optical_Interconnects.html, 2008. With permission.)

consortia throughout the world have been assembled to face this challenge. One of the earliest programs to tackle optoelectronic integration was EPIC (Electronic Photonic Integrated Circuits), which was funded by the Defense Advanced Research Programs Agency (DARPA) in the United States (see Figure 1.17) [56]. This program was established in the middle of the last decade and has worked to create CMOS-compatible nanophotonic circuits that are integrated with the necessary drive electronics via funding several consortia to work on these challenges. A more recent DARPA project, the Ultraperformance Nanophotonics Intrachip Communications (UNIC) project, funds a consortium of U.S. industrial and academic institutions [57]. The goal of this project is to develop high-bandwidth, low-latency optical interconnects for multiprocessor communications.

In Europe, there are currently several projects under way to investigate silicon photonic integration, too numerous to highlight individually. HELIOS is one of the largest, and is a collaboration between industrial and academic partners that aims to develop CMOS photonic components that are readily accessible to a larger group of users [59]. As mentioned previously, they have already realized a 40 Gbps germanium detector integrated with a silicon waveguide and ultimately seek to integrate a photonics layer with standard CMOS electronics, similar in vision to the IBM concept depicted in Figure 1.3. UK Silicon Photonics [60] is a more recently established consortium that involves some of the leading research groups in the United Kingdom, whose focus is to develop high-speed integrated optoelectronic circuits in silicon. Thus far, they have made good progress and have already contributed the design of a 19 GHz modulator in silicon, as was mentioned previously [49]. Terahertz applications utilizing silicon components are also being investigated by a consortium of European academic institutions [61].

What is interesting about these consortia is that they have a good mix of industrial and academic resources. The industrial partners are sometimes smaller companies focused on a near-term approach to selling products containing silicon photonic components. Lightwire, whose modulator was discussed earlier, has a 10 Gbps transceiver currently on the market with the expectation of providing 40 and 100 Gbps transceivers in the near future [62]. Luxtera has also released a 10 Gbps transceiver, however, this variant can be mounted directly onto a printed circuit board [63], as well as a 40 Gbps Active Optical Cable (AOC) that uses silicon photonic components [64]. Kotura has been selling variable optical attenuators (VOAs) for several years and recently has become an application-specific photonic component provider [65].

Larger companies are also essential to the development of silicon photonics. For example, Sun Microsystems is investigating optical interconnect technology using optical proximity communications [66] and is involved in the UNIC project (with Luxtera and Kotura). Alcatel-Lucent is actively involved in the HELIOS project and is also working on silicon photonic modulators [49], and was among the early collaborative EPIC team, together with BAE Systems and MIT. The Intel Corporation has made significant contributions to the field of silicon photonics with their detector [19], modulator [45], and integrated light source [67,68] work. HP Labs is working on optical interconnects with 1.5 μm radius silicon microring resonator modulators for their Corona architecture, a multiprocessor architecture that utilizes photonic interconnects to link the processors together [69]. While the smaller companies tend to focus on near-term solutions for telecommunication devices, it appears that the larger companies are setting their sights on longer-term projects such as dealing with the interconnection of microelectronic components using photonics, in a manner similar to that depicted in the vision provided by IBM in Figure 1.3.

This section has attempted to give a brief history of the field of silicon photonics with respect to the field of telecommunication component integration. Huge progress has been made in a few short years due in large part to the substantial contribution made by academic and industrial partners throughout the world, similar to the initial development of the microelectronics industry. However, even with the significant advancement that the field has made in the past few years, there are still some readily identifiable challenges to contend with. These are discussed in the next section.

1.3 Current Challenges

In this section, we identify several key issues that present challenges to the field of silicon photonics. The first such challenge is the light source. Silicon is a poor light emitter at the telecommunication wavelengths (1310 and 1550 nm) due to its indirect bandgap, which in the minds of some make it an inferior material to III-V material compounds, where practically all of the necessary active components for an optical transceiver can be realized. However, much effort has been made by researchers around the world to try and find a solution to silicon's light emission deficiencies. A good discussion of much of the work on light emission from silicon can be found in [70], ranging from the use of silicon nanocrystals to utilizing the Raman effect. The most common issue with proposed silicon light source solutions is that they are either less than ideal for integrated devices (e.g., a high-powered external laser is required for use in a Raman laser) or that they are inefficient, particularly when compared to direct bandgap III-V materials. Furthermore, these inefficiencies may have detrimental consequences. Power consumption, for instance, can be considerable for an inefficient light source. The consequence is that the inefficient power utilization can result in joule heating. In turn, this heating can have deleterious effects on the optical components in close proximity to the light source, because silicon has a large thermo-optic coefficient [18]. Consequently, opinion is divided among researchers on whether the light source should be integrated onto the chip, or whether it should be off-chip [71].

This should not imply that silicon-based light emitter research should be abandoned. In fact, some recent results indicate that an efficient light emitter, and ultimately an electrically pumped laser comprising group-IV materials (e.g., silicon or germanium), could be realized in the near future [72]. Researchers taking part in a MURI (Multi University Research Initiative) collaboration investigating a silicon laser [73] have recently published results reporting a germanium layer epitaxially grown on a silicon substrate that has a 0.25% tensile strain that demonstrates a net gain of 500 cm^{-1} [72]. Xu et al. claim that an n-type doping of approximately 7×10^{19} cm^{-3} is necessary to achieve this direct-gap net gain. They also investigate the properties of a p$^+$Si/n$^+$Ge/n$^+$Si heterojunction diode structure to allow for electrical pumping of the gain medium. They state that by doping the germanium area with n-type material they can achieve an external quantum efficiency of approximately 10%. These results show promise that a fully CMOS-compatible light source may be realized. If the efficiency of the device can be improved, then it may be possible to achieve an integrated optical transceiver where all components are based on group-IV materials.

An alternative to a group-IV light emitter that has been gaining ground recently is the hybrid approach. The idea is to incorporate III-V materials with silicon where the silicon acts as the waveguiding or optical cavity material. Chapters 8 and 9 are devoted to this important topic in this book. Chapter 8 covers novel III-V on silicon epitaxial growth techniques, while Chapter 9 particularly discusses epitaxially grown III-V lasers on silicon. An alternative technique is wafer-bonding of III-V active regions on silicon. A recent example of this method is discussed in the following. Fang et al. demonstrated such a device in 2006, and a cross-sectional schematic of the device is shown in Figure 1.18

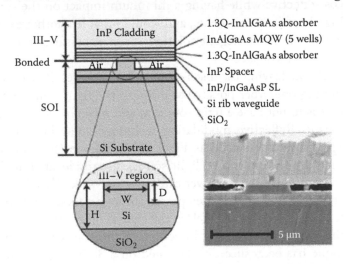

FIGURE 1.18 Cross-sectional schematic and SEM image of the hybrid laser fabricated by Fang et al. (From A. W. Fang et al., *IEEE Photon. Technol. Lett.*, **18**, pp. 1143–1145, 2006. With permission.)

[67]. The device works by electrically pumping the III-V active region that has been bonded to the top of a silicon rib waveguide using low-temperature oxygen plasma-assisted wafer bonding. The light is confined primarily in the silicon waveguide. The evanescent field from the light in the waveguide is then amplified by the III-V region above the waveguide. Fang et al. state that with this configuration they are able to realize a continuous wave lasing at 1568 nm with a threshold optical power of 23 mW. The drawback of this device is that it requires back-end processing in order to add the active region to the photonic circuit, which is not CMOS compatible and also requires extra fabrication steps. This results in a more complex fabrication of the laser, which in turn may be more costly to fabricate. However, if these devices can be incorporated into mass-produced circuits on a wafer scale, then this approach may be more effective in realizing a low-cost, fully integrated optical transceiver consisting primarily of silicon.

The best option for the light source for silicon photonic circuits remains to be seen. An elegant solution of a fully group-IV, electrically pumped light emitter may be on the horizon. However, the hybrid approach was realized several years ago, and so methods to overcome cost, thermal, and efficiency issues are more advanced for this technology option. Thus, the hybrid method may ultimately become the dominant technology if it is cost-effective while having a minimum impact on the surrounding photonic components, and would provide another example of where the most elegant solution is not necessarily the one that emerges most successful.

Before leaving the discussion on light sources, it is worth mentioning one other possible solution to the light source issue, and that is to not have a light source at all on the integrated photonic circuit. Reflective modulation techniques may be employed in optical networks such as Fiber to the Home (FTTH) [74]. In such a network, a CW light source, usually located at the central office, is used to power the entire network. This light, once it arrives at a receiver, is then passed through a modulator that sends the modulated signal back to the central office, thereby removing the requirement of a light source at the transceiver. This technique has been successfully tested on a 30 km optical link with a 1.25 Gbps data rate utilizing a lithium niobate Mach–Zehnder modulator [74]. Based on the progress made thus far in modulators and detectors in silicon photonics, it may be that lightless optical transceivers could be realized that would provide low-cost components for FTTH networks.

While power consumption is an issue for light sources, it is also a challenge for creating and transferring data across chips and boards using standard metallic interconnects. In fact, electrical trace density and data rates are closely following the trend of computing power increase. However, interconnections using electrical traces are facing increasing difficulties [75–80]. The problem lies with the buses between boards, but also between chips as well as within the chips themselves. For example, the power requirement at the interconnect level on a chip demands that approximately 50% of microprocessor power is consumed by the interconnects, in 130 nm gate technology [81]. This level of consumption is expected to rise to about 80% in the future [82]. Hence, to obtain a high data rate link with lower power consumption is of the utmost importance. The metric used to compare systems is the energy required to generate each bit of data, sometimes referred to as the "power per bit." State-of-the-art electronic transceivers have an energy per bit parameter of 2.8 to 6.5 pJ/bit for board or backplane interconnects [83] and approximately 2 pJ/bit [84] for moderate-length chip-to-chip interconnects. Recent work has demonstrated that for an ideal electrical channel, a figure of 1 pJ/bit can be achieved at approximately 10 Gbps [85,86]. This value is a useful benchmark for future optical transceivers as they will need to match, or more realistically, outperform, the best electrical transceivers.

As mentioned in the previous section, silicon photonics modulators currently predominantly use one of three possible high-speed means of changing the index of refraction in a waveguide; the injection, accumulation, or depletion of carriers. For accumulation devices, there are currently two types of modulators, based on Mach–Zehnder interferometers or resonators such as the ring resonator, each of which have demonstrated data rates of at least 10 Gbps (see Table 1.1). While the data rates are comparable to the state-of-the-art electrical transceivers, they do suffer from high energy consumption because of the relatively large charge (and hence current) that needs to be manipulated within the waveguide. This implies that substantial energy is required to create a single bit in an optical modulator.

One can see a trade-off emerging from the data in Table 1.1 between power consumption, optical spectrum, and device

TABLE 1.1 Comparison of Various Optical Modulator Metrics

Modulation principle	Depletion of a horizontal p-n junction [87,93]	Forward-biased diode [42]	Forward-biased diode [40,41]	Reverse-biased p-n junction [91]	Reverse-biased p-n junction [92]
Structure	MZI	MZI	Ring	Disk	Ring
Device footprint (μm²)	≈1 × 10⁴	≈1 × 10³	≈1 × 10²	20	≈1 × 10³
Speed achieved (Gbps)	30	10	>12.5	10	10
Energy per bit (fJ/bit)	≈3 × 10⁴	≈5 × 10³	≈300	85	50
Modulation voltage (V)	6.5	7.6 (preemphasis)	3.5 (preemphasis)	3.5	2
DC Modulation depth and insertion loss (dB)	>20, ≈7	6–10, 12	>10, <0.5	8, 1.5	6.5, 2
RF Modulation depth/ speed (Gbps)	1dB/30, 1dB/40	—	8dB/16, 3dB/18	—	8dB/10
Working spectrum (nm)	>20	—	≈0.1	≈0.1	≈0.1

footprint. Resonators have power consumption close to the expected target required on the order of 10 fJ per bit, as well as a very small footprint. However, they suffer from a narrow-spectrum bandwidth and tight fabrication tolerances. Furthermore, the thermal tuning power required to stabilize them at a predefined frequency of operation must be included in estimating the total system power requirement. Interferometers, however, have much more relaxed fabrication tolerances, wider operating spectra, and are less dependent on temperature variations. Nevertheless, they suffer from a larger footprint that increases their power consumption.

To obtain the best of both device types, the future of highly integrated optical modulation in silicon may lie with electroabsorption modulators using effects such as the Franz–Keldysh effect, seen in bulk semiconductors, and the Quantum Confined Stark effect (QCSE) observed in quantum-confined quantum well layers. QCSE modulators have the advantage of exhibiting the strongest high-speed electroabsorption effect. This would allow modulators with active areas of only a few microns in length, even without the use of resonators.

Packaging is another identifiable challenge currently facing silicon photonics. In fact, as much as 70% of costs typically arise from the packaging and testing of optical components [71]. Thus, the commercialization of photonic integrated circuits introduces the need for a suitable packaging solution. Bringing optical signals onto the chip needs to be performed with minimal coupling loss. However, the large mismatch in mode size between a standard single-mode fiber and an SOI-based waveguide creates a significant coupling challenge. As well as being low loss, it is desirable that the coupling process is insensitive to variations in alignment and polarization. Broadband coupling is also required for applications that utilize wavelength division multiplexing (WDM).

The trend to reduce the dimensions of SOI waveguides, which has been driven by component performance enhancements and device footprint miniaturization, has further increased the mismatch in mode size. Many coupling techniques have been proposed in the literature, which can be generalized into two categories, in-plane or out-of-plane coupling. There are arguments supporting both approaches. For example, out-of-plane coupling allows for wafer-scale testing of devices, requires a relatively simple fabrication and preparation technique, and generally has slightly larger alignment tolerances. In-plane coupling, on the other hand, tends to operate over a broader wavelength range and typically allows for a higher coupling efficiency. Too many examples of each approach exist to provide an exhaustive account here, so a few of the more mature techniques are briefly discussed.

Inverse tapers, which have a narrowing of waveguide width toward the edges of the chip (Figure 1.19), have been demonstrated in SOI with a coupling loss of 0.5–3 dB from a lensed fiber [94,140–142]. As light propagates along the taper (from left to right in the diagram), the mode expands as a result of the decreasing confinement of the narrowing waveguide. The mode will continue to expand until it reaches the end of the taper, and

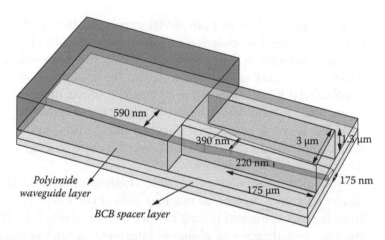

FIGURE 1.19 Diagram of inverse taper coupler. (From G. Roelkens et al., *IEEE Photon. Technol. Lett.*, **17**, pp. 2613–2615 2005. With permission.)

therefore the chip, so that upon exiting the taper the mode has a similar cross section to the optical fiber mode. The result is an improved mode overlap between the waveguide and the fiber. A polyimide waveguide layer has been used on top of the layer so as to condition the optical mode so that it has the approximate cross-sectional area to match that of the fiber [94], but other materials have also been used. The polyimide also has a similar index to that of the fiber, and so losses due to the index mismatch between the fiber and the waveguide can be minimized. A potential drawback to this method is that the device loss relies on a high-quality, high-resolution fabrication technique that can provide smooth sidewalls while also providing a clear and well-defined tip that is of the order of a few hundred nanometers in width.

Two-dimensional taper structures have been reported [143] that provide a gradual mode transformation in both vertical and lateral directions. Fabrication of the tapering in the vertical direction, however, is more complex, requiring, for example, gray-scale lithography where the photoresist that is used to protect the waveguide during the silicon etch has graded amounts of light applied to it, which in turn modifies how "developed" and therefore how protective the resist is to the waveguide during the etch process. An alternative method is shown in Figure 1.20, where a series of discrete etched steps may be employed to taper the waveguide in the vertical and horizontal directions [95]. This approach has yielded an improvement in coupling efficiency over an equivalent one-dimensional taper.

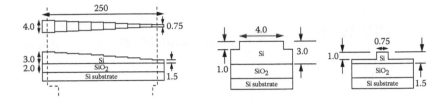

FIGURE 1.20 Diagram of a two-dimensional taper consisting of discrete etched steps. (From A. Choudhury et al., *IEEE Photon. Technol. Lett.*, **17**, pp. 1881–1883, 2005. With permission.)

Another type of in-plane coupler has been proposed that utilizes gratings to couple light to and from the waveguide. The Dual Grating Assisted Directional Coupler [96], with the cross-sectional schematic shown in Figure 1.21, uses two gratings to overcome the mode mismatch between a butt-coupled optical fiber and the silicon waveguide. Two gratings are required due to the large mismatch between the size and refractive indices of silica fibers and submicron silicon waveguides. Theoretical coupling efficiencies of up to 90% have been predicted, and up to 59% have been measured; however, they come at the cost of a narrow bandwidth and complex fabrication.

The most successful out-of-plane coupling structures have consisted of periodic structures. Shallow etched surface gratings have been demonstrated with coupling efficiencies between 30% and 70% [97,144,145]. The gratings are normally positioned in widened regions as shown in Figure 1.22 and employ some lateral

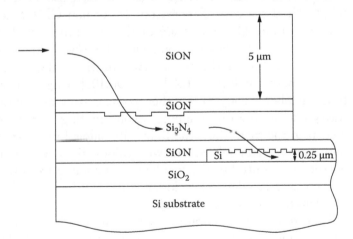

FIGURE 1.21 Diagram of the dual grating assisted coupler. (From G. Z. Masanovic et al., *Opt. Express*, **13**, pp. 7374–7379, 2005.)

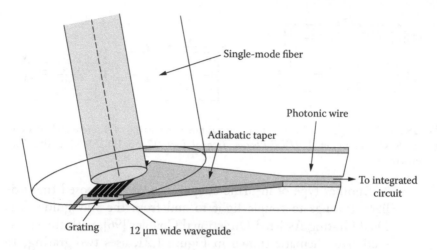

FIGURE 1.22 Diagram of out-of-plane grating coupler. (From D. Taillaert et al., *Jpn. J. Appl. Phys. Part 1*, **45**, pp. 6071–6077, 2006. With permission.)

tapering of the of the waveguide to allow for the grating area to be more closely matched to the core size of a single-mode fiber. The grating is detuned so that the fiber approaches slightly off-normal to the sample surface in order to avoid second-order reflections coming from the grating to pass back into the fiber. The difficulty with this method is that precise alignment of the fiber is required not only in the x- and y-axes, but also in terms of the angle that the fiber makes with respect to the grating surface.

Slightly more elaborate structures have been also demonstrated [88,98,99] at the cost of more complex or non-CMOS compatible fabrication. One downside of the one-dimensional grating coupler is its polarization dependence. However, this issue may be overcome through the use of a two-dimensional grating structure as shown in Figure 1.23b [89,100,101]. Light transmitted from the fiber is coupled to either to one grating and subsequent waveguide or to an overlapping second grating that is perpendicular to the first. Thus, the light will couple into either one or both gratings and subsequent waveguides, thereby relaxing the polarization control of the incident light coming from the fiber. The polarization-dependent loss can be kept below 1 dB over a wavelength range of 15 nm [90].

Linked with the coupling issue is the challenge of reliably aligning and mounting the fiber coming from the outside world to the photonic integrated circuit within the package. Techniques have been proposed and demonstrated for both in-plane and

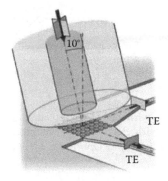 

FIGURE 1.23 (a) Schematic of the out-of-plane coupling scheme with polarization diversity and a (b) SEM image of the two-dimensional grating. (From D. Taillaert et al., *IEEE Photon. Technol. Lett.*, **15**, pp. 1249–1251, 2003; F. Van Laere et al., *IEEE Photon. Technol. Lett.*, **20**, pp. 318–320, 2008. With permission.)

out-of-plane coupling techniques. The use of etched V-grooves in the chip substrate to support the end of the fiber is a popular choice for both schemes. Silicon-based in-plane coupling using V-grooves and inverted tapers have been demonstrated as shown in Figure 1.24.

The larger alignment tolerances of grating couplers has meant that many component designers regard them as the most promising prospect for injecting light into nanoscale silicon photonic circuits. The technical issues of mechanical stability and compactness need to be overcome in order to realize a suitable packaging solution. One proposed method involves the bonding of a commercially available fiber array connector onto the upper surface of the photonic circuit [103], as shown in Figure 1.25. The end of the V-groove and glass block can be polished to ensure the fibers approach the chip surface at the correct angle. This method of

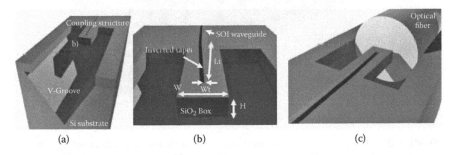

FIGURE 1.24 Diagram of in-plane coupling with V grooves and inverted taper structures. (From F. Van Laere et al., *IEEE Photon. Technol. Lett.*, **20**, pp. 318–320, 2008.)

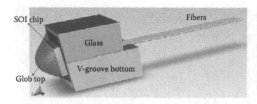

FIGURE 1.25 Diagram of a V-groove fiber mounting with out-of-plane coupling scheme for (a) an all-optical solution and (b) a solution with electrical input/outputs. (From L. Zimmermann et al., *Proc. European Conference Integrated Optics*, pp. 33–36, 2008. With permission.)

packaging provides all-optical coupling solutions (Figure 1.25a) as well as a means of coupling optical and electrical input/outputs (Figure 1.25b) to a photonic chip. Luxtera [104] has demonstrated a packaged fully functional silicon-based photonic WDM transceiver based on this approach, a photograph of which is shown in Figure 1.26.

An additional consideration for device packaging is the choice of the optical fiber itself. As demonstrated by the variety of coupling methods discussed earlier, the choice of optical fiber is important as it affects the coupling efficiency and hence overall output efficiency. Tapered fibers, for example, can be used so that the end of the fiber is tapered down so that it has a similar, or at least closer, cross-sectional mode size as the waveguide to which it is being coupled. Intermediate structures such as a microlens may be employed between a standard single-mode fiber and the

FIGURE 1.26 Photograph of packaged transceiver with out-of-plane coupling scheme. (From P. De Dobbelaere, Demonstration of First WDM CMOS Photonics Transceiver with. Monolithically Integrated Photo-Detectors, Proc. European Conference on Optical Communication, pp. Tu.3.C.1, 2008. With permission.)

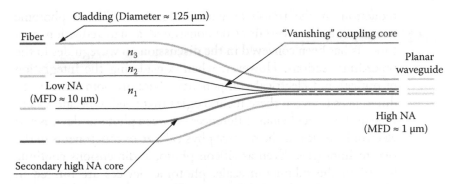

FIGURE 1.27 Cross-sectional diagram of Helica tapered coupler. (From Chiral Photonics, Inc., http://www.chiralphotonics.com/Web/default.html, 2008. With permission.)

on-chip waveguide in order to provide some form of mode matching. Novel structures such as the commercially available Helica tapered coupler [105], depicted in Figure 1.27, may be a favorable substitute to discrete structures such as lenses. This fiber structure has two concentric cores at its input, the inner of which matches the size of a standard single-mode fiber and the outer core acts as the cladding. Its diameter then undergoes a tapered reduction such that the inner core is too thin to support propagation and, as a result, the light propagates in the outer core, which is size-matched to the waveguide of the integrated photonic circuit, thus providing suitable coupling between the fiber and a waveguide with an approximate cross-sectional area of a square micron.

The packaging solutions discussed here were provided to give the reader a sense of not only how much work has been dedicated to overcoming the packaging/coupling issues surrounding photonic circuits, but also to demonstrate the significance of the issues themselves. Power/energy consumption is an issue that is faced by the light source and the modulator, but the efficiencies obtained by continued research into these devices can be diminished by requiring extra components to boost signals that have been reduced due to inefficient coupling between the circuit and the outside world. Hence, packaging remains a challenge, as not only are efficient optical/electrical connections required but they must also be cost-effective so as not to diminish the benefits of large-volume integration.

Let us now return to the challenge of integration. Many of the performance enhancements obtained by the microelectronics industry in terms of integrated circuits came from the continuous reduction in the transistor gate dimensions. Silicon photonic components have certainly demonstrated a size reduction over time, as has been conveyed in the discussion on waveguides in the preceding sections. However, when considering the integration of optical and electrical components, there are some important differences between these technologies that need to be addressed. One of the most obvious differences between photonic devices and electronic devices is the relative physical size of components in the two technologies. Even as silicon photonic dimensions continue to fall to the submicron scale, photonic devices are limited in ultimate size by optical diffraction. Photonic devices and circuits also tend to be relatively long compared to their width, as optical functions are cascaded sequentially (even as part of a parallel processing architecture). While researchers strive for ever-smaller and more efficient devices, the question arises as to whether alternative technologies can have a significant impact on the physical size of photonic components. The use of photonic crystals, and the emergence of the phenomenon of "slow light" [106,107], may impact the length of photonic devices, but will probably do little to reduce lateral dimensions. Plasmonic waveguides, on the other hand, inherently offer an opportunity for dimensional reductions measured in orders of magnitude [108], making these devices comparable in dimensions to electronic devices. The term *plasmonics* refers to the propagation of light via metallic nanostructures [109], and the associated term *active plasmonics* [110] refers to the possibility of manipulating the light to facilitate active functions such as modulation, at extremely high frequencies up to the terahertz regime [111]. In plasmonics, surface plasmon polaritons (SPPs), which are bound oscillations of electrons and light, propagate at metal–dielectric interfaces. Modulation is achieved by switching the SPPs via an electrical or optical signal, such that the local properties alter the propagation in some reversible way that facilitates modulation or switching. The extremely high speeds attainable, together with the very small volumes involved, and hence potentially low power, are very attractive. The fact that metal is used in the transmission medium also invites the interaction with an electrical drive signal to control the optical propagation. Furthermore, it is also intuitive that the small scales involved are attractive for integration with electronic circuitry.

Given the inherent and potential advantages of the plasmonic field, it is perhaps easy to believe that plasmonics offers solutions to many of the problems that larger-scale photonics is seeking to solve. However, there are still significant hurdles to overcome. It is difficult to fabricate efficient and low-loss plasmonic waveguides, the conversion efficiency from other optical waveguides to plasmonic waveguide modes needs to be improved dramatically, effective interaction with electrical drive signals needs much more development to impact the advancement of plasmonic modulators, and the integration issues await significant attention. Plasmonic waveguides are dealt with in more detail in Chapter 2 of this text, but for the purposes of this introduction, suffice to say that plasmonics offers significant promise, if key advances can be made. Probably a pragmatic solution will be the integration of aspects of silicon photonics with aspects of plasmonics, to enhance the overall performance of a given circuit, while mitigating some of the difficulties with either isolated technology.

1.4 The Future of Integration

What does the future hold for silicon photonics in terms of integration? Large commercial institutions appear to believe that it will solve the interconnect problem, while smaller companies expect the cost advantages of the technology to allow them to provide solutions for some of today's telecommunications challenges. It is likely that the field will continue along the path of integrating photonic and electronic components so that the distinction between the two technologies will fade. It is also likely that the more "exotic" technologies of today such as photonic crystal and plasmonic devices will also become integrated with mainstream components in order to facilitate the continual shrinkage of photonic circuits to gain from the economies of scale in terms of performance and fabrication costs. However, it may be that the current successes of silicon photonics may actually pave the way for integration of other technologies with silicon electronic and photonic components. Wireless communication components may become a part of the integration path, for example [112].

Radio over Fiber (RoF) networks are a specific example, which seeks to combine the flexibility of wireless mobile networks with the substantial bandwidth provided by optical networks [113]. The idea is to produce a network consisting of a multitude of simplified wireless base stations connected to a single (or minimal number of) central office hubs via fiber-optic links. Wireless signals received by a base station are converted to an optical signal that is then transmitted to the base station for coding/decoding.

Similarly, coded wireless data emanating from the central office is converted to an optical signal that is then transported along a fiber to the base station, where it is then converted to a wireless one, which is then transmitted by the base station.

One of the benefits of this network configuration is that practically all of the complex wireless signal coding, analysis, and switching is accomplished at the central office. This minimizes the capital expense of the network compared to having these capabilities at each base station. Furthermore, upgrade of the network as improved wireless communication protocols become available is simplified by having to only upgrade the equipment at the central office, assuming that similar optical modulation frequencies are used in the upgraded network [113].

RoF networks are ultimately limited by the latencies induced by the optical to electrical (O/E) and electrical to optical (E/O) conversions that occur in the transformation between wireless and optical signal conversions, and vice versa [114]. Until recently, the microwave photonic components that enable such a conversion have been made of lithium niobate or III-V materials as these components require GHz modulation and detection speeds [115]. However, silicon photonic integration of the optical, electrical, and wireless components may provide a low-cost, highly efficient solution for RoF optical-to-wireless transceivers. Indeed, work has already begun to investigate silicon as a solution for such devices. For example, Lecoy and Delacressonniere have investigated an optically controlled oscillator for RoF applications [116]. Their device is a silicon-germanium heterojunction bipolar phototransistor that directly transforms an optical modulation into RF at 5.2 GHz, thus eliminating the need for the intermediate step of optical-to-electronic and then electronic-to-wireless conversions and their associated latencies, power consumption, and semiconductor die space.

Vacondio et al. have also been investigating RoF components in silicon [117]. They have developed a high-speed modulator capable of working at a frequency of approximately 5 GHz. Their device is a depletion-type Mach–Zehnder modulator, similar in design to that shown in Figure 1.6 and utilizes a traveling wave electrode configuration. The authors have measured an approximate –30 dB Error Vector Magnitude (EVM), which is a modulator performance metric that is much quicker to measure than the bit-error rate. They compare this performance to a lithium-niobate

Mach–Zehnder modulator, which has an EVM of –26 dB. Hence, the silicon modulator has an approximately 4 dB better EVM compared to commercially successful lithium-niobate modulators.

With the demonstration of a detector and modulator in silicon, the outlook for an integrated optical, electronic, and now wireless transceiver is promising. Such a device could have far-reaching applications where the need for mass-produced, high-data-rate RoF networks could be employed, such as picocell or femtocell networks that utilize a fiber-optic backbone.

While silicon photonics could enable further integration of other technologies with current electronic/optical integration efforts, it may also provide integrated, low-cost solutions to technologies not necessarily associated with the telecommunications industry. Much of the current focus in silicon photonics is in the telecom wavelength range. However, silicon is relatively low loss from 1.2 to 8 μm and from 24 to 100 μm [118], and therefore silicon photonic circuits can be used in mid- and far-infrared wavelength ranges. The mid-infrared (MIR) spectral region is particularly interesting, as the practical realization of optoelectronic devices operating in this wavelength range offers potential applications in a wide variety of areas, including optical sensing and environmental monitoring, free-space communications, biomedical engineering, thermal imaging, and IR countermeasures. Many pollutant and toxic gases and liquids exhibit bands of absorption lines in the MIR part of the spectrum. Consequently, the MIR is very attractive for the development of sensitive optical sensor instrumentation. In addition, there are two atmospheric transmission windows (3–5 and 8–14 μm) offering free-space optical communications. Of interest are also thermal imaging applications in both civil and military applications. MIR photonics also offers the potential for development of minimally invasive, effective, and safe diagnostic techniques. This spectral range is attractive for precise surgical procedures and medical ablation of tissue because of its high absorption in water and, hence, small penetration depth, especially for wavelengths around 3 μm, where the penetration depth can be as small as a few micrometers [119]. Group IV photonic devices can also find application in astronomy [120].

Furthermore, the free-carrier plasma dispersion effect is much stronger at wavelengths longer than 1.55 μm [121], and two photon absorption is significantly reduced at MIR wavelengths beyond the telecommunication band [122]. Additionally, more robust optical fibers are now available for this wavelength range. Finally, the larger dimensions of MIR devices mean that the dimensional tolerances will be more relaxed compared to those in the NIR, thus simplifying fabrication [18]. However, the advantages of this wavelength range have not been fully exploited owing to the limitations in current technology.

The fundamental challenge for group IV photonics in this wavelength range is the fact that the most popular material platform in the NIR range, that of silicon-on-insulator, cannot be used in most parts of the MIR, due to high material losses of SiO_2 in the 2.6–2.9 μm range and beyond 3.6 μm [123,124]. Therefore, other kinds of waveguides must be developed. Soref et al. proposed several waveguide structures suitable for mid- and long-IR spectral regions [125]: Si rib-membrane waveguides, Si on Si_3N_4 (SON), Si on sapphire (SOS), Ge-on-Si, Ge on SOI or GeSn-on-Si strip and slot waveguides, hollow waveguides with Bragg or anti-resonant cladding [126,127], and silicon on porous silicon waveguides [128,129]. Freestanding waveguides fabricated by a proton beam writing technique are also candidates for MIR silicon photonics [130]. As waveguide dimensions scale with the wavelength, in order to reduce the dimensions and facilitate CMOS compatibility of MIR devices, plasmonic waveguides and plasmonic waveguided components may prove to be a suitable solution [125]. It is expected that the plasmonic propagation loss decreases significantly as the wavelength of operation is increased into the MIR and far-IR [131].

Research in MIR group IV photonics is gathering pace, and there have been several recent results reported in the field: Raman amplification at the wavelength of 3.39 μm has been demonstrated [132], a spectrally selective (around 3.7 μm) mid-IR PbTe photodetector has been monolithically integrated on a silicon platform [133], and nonlinear effects at shorter MIR wavelengths have been investigated [134].

As in the NIR, one of the more challenging tasks in designing group IV photonic integrated circuits in the MIR will be the integration of optical sources. Monolithic integration could include SiGe/Si quantum cascade emitters integrated on a silicon chip and would represent a long-term solution, while the most expedient way may be hybrid integration of III-V sources [135,136] on group IV waveguide networks, which can include techniques developed in the NIR [137].

Since the field is still in its infancy, there are a number of challenges that need to be overcome in the future before group IV photonic integrated circuits are employed in a host of applications offered by the MIR spectral range. Nonetheless, significant progress has been made in a short time with respect to investigating the building blocks necessary to realize integrated circuits at MIR wavelengths. It is interesting to observe the parallels between the waveguides fabricated during the early 1990s for telecommunication device wavelengths and those for MIR applications herein. If MIR devices can make similar progress, then we should expect to see some rather remarkable advances in MIR optical components in the near future.

1.5 Conclusions

The field of silicon photonics has undergone a tremendous transformation since its inception in the mid-1980s. Almost from the beginning, integration to some degree was identified as important. Since then, there have been some impressive advances in nearly all aspects of the technology. One of silicon's biggest drawbacks, the lack of an efficient light source, is also being addressed. Commercial products with silicon photonic components are now on the market, and there is good reason to believe that future microelectronic devices will share space with optical counterparts. What is most impressive about these state-of-the-art silicon devices is not the fact that they can contend in some respects with the more conventional devices fabricated from III-V materials, but that the majority of the advances associated with these devices were achieved in the past 7 or 8 years. To attempt to put this into context with the microelectronics industry, the first commercial microprocessor sold by the Intel Corporation in 1970, the 4004, contained approximately 1000 transistors. Twenty years later, the i486 with approximately 1 million transistors was unveiled. In the same amount of time, silicon photonics has gone from micron-sized waveguides with tens of dBs of propagation loss and thermo-optic modulators running in the kHz regime to optical modulators with submicron waveguides operating at 40 Gbps and detectors with gain-bandwidth products of more than 300 GHz! While this is not a complete comparison of the two technologies, it nonetheless demonstrates how much technology can be advanced over a short period of time.

After a further 20 years of evolution, an Intel processor will soon have a chip with billions of transistors packed onto it. If one tries to imagine what the next 20 years will hold for silicon photonics, the opportunities are almost boundless. However, what has become clear is that silicon photonics no longer appears to be a technology based on "if," but rather on "when."

References

1. Texas Instruments, The Chip That Jack Built, http://www.ti.com/corp/docs/kilbyctr/jackbuilt.shtml, 2009.
2. Intel Corporation, Excerpts from a Conversation with Gordon Moore: Moore's Law, http://download.intel.com/museum/Moores_Law/VideoTranscripts/Excepts_A_Conversation_with_Gordon_Moore.pdf, 2005.
3. Moore, G. E., Cramming more components onto integrated circuits, *Proc. IEEE*, **86** (1), pp. 82–85, 1998.
4. Texas Instruments, First Integrated Circuit, http://www.ti.com/corp/graphics/press/image/on_line/co1034.jpg, 2009.
5. Texas Instruments, Jack Kilby Examines 300 mm Wafer, http://www.ti.com/corp/docs/kilbyctr/downloadphotos.shtml, 2009.
6. T. Ghani, Challenges and Innovations in Nano-CMOS Transistor Scaling, http://www.intel.com/technology/silicon/Neikei_Presentation_2009_Tahir_Ghani.ppt, 2009.
7. Intel Corporation, World's First 2 Billion Transistor Processor, http://www.intel.com/technology/architecture-silicon/2billion.htm?id=tech_mooreslaw+rhc_2b, 2009 .
8. International Technology Roadmap for Semiconductors, 2009 Edition, Executive Summary, http://www.itrs.net/Links/2009ITRS/2009Chapters_2009Tables/2009_ExecSum.pdf, 2009.
9. O. Semenov, A. Vassighi, and M. Sachdev, Impact of self-heating effect on long-term reliability and performance degradation in CMOS circuits, *IEEE Trans. Device and Materials Reliability*, **6**, pp. 17–27, 2006.
10. International Business Machines, Silicon Integrated Nanophotonics, http://domino.research.ibm.com./comm/research_projects.nsf/pages/photonics.index.html, 2009.
11. A. G. Rickman, G. T. Reed, and F. Namavar, Silicon on insulator optical rib waveguide loss and mode characteristics, *J. Lightwave Technol.*, **12**, pp. 1771–1776, 1994.
12. G. T. Reed, A. G. Rickman, B. L. Weiss et al., Optical characteristics of planar waveguides in SIMOX structures, *Proc. Mater. Res. Soc.*, **244**, pp. 387–393, 1992.

13. R. A. Soref, J. Schmidtchen, and K. Petermann, Large single-mode rib waveguides in GeSi-Si and Si-on-SiO$_2$, IEEE *J. Quant. Electron.*, **27**, pp. 1971–1974, 1991.

14. R. A. Soref and J. P. Lorenzo, All-silicon active and passive guided wave components for λ=1.3 and 1.6 μm, *IEEE J. Quant. Electron.*, **QE-22**, pp. 873–879, 1986.

15. R. A.Soref, D. L. McDaniel Jr., and B. R. Bennett, Guided-wave intensity modulators using amplitude-and-phase perturbations, *J. Lightwave Technol.*, **6**, pp. 437–444, 1988.

16. R. A. Soref and J. P. Lorenzo, Single-crystal silicon – a new material for 1.3 and 1.6 μm integrated-optical components, *Electron. Lett.*, **21**, pp. 953–954, 1985.

17. R. A. Soref, Silicon-based optoelectronics, *Proc. IEEE*, **81**, pp. 1687–1706, 1993.

18. G. T. Reed (Ed.), *Silicon Photonics: The State of the Art*, John Wiley & Sons, Chichester, U.K., 2008, ISBN-13 978–0470025796.

19. Y. Kang, H.–D. Liu, M. Morse, M. J. Paniccia, M. Zadka, S. Litski, G. Sarid, A. Pauchard, Y.–H. Kuo, H.–W. Chen, W. S. Zaoui, J. E. Bowers, A. Beling, D. C. McIntosh, X. Zheng, and J. C. Campbell, Monolithic germanium/silicon avalanche photodiodes with 340 GHz gain-bandwidth product, *Nat. Photonics*, **3**, pp. 59–63, 2008.

20. G. Abstreiter, Engineering the future of electronics, *Physics World, UK*, **5**, pp. 36–39, 1992.

21. B. L. Weiss, G. T. Reed, S. K. Toh, R. A. Soref, and F. Namavar, Optical waveguides in SIMOX structures, *IEEE Photon. Technol. Lett.*, **3**, pp. 19–21, 1991.

22. K. K. Lee, D. R. Lim, L. C. Kimmerling, J. Shin, and F. Cerrina, Fabrication of ultralow-loss Si/SiO$_2$ waveguides by roughness reduction, *Opt. Lett.*, **26**, pp. 1888–1890, 2001.

23. F. Xia, L. Sekaric, and Y. Vlasov, Ultracompact optical buffers on a silicon chip, *Nat. Photonics*, **1**, pp. 65–71, 2006.

24. S. P. Chan, C. E. Png, S. T. Lim, V. M. N. Passaro, and G. T. Reed, Single mode and polarisation independent SOI waveguides with small cross section, *J. Lightwave Technol.*, **23**, pp. 1573–1582, 2005.

25. M. M. Milosevic, P. S. Matavulj, B. D. Timotijevic, G. T. Reed, and G. Z. Mashanovich, Design rules for single-mode and polariza-tion-independent silicon-on-insulator rib waveguides using stress engineering, *J. Lightwave Technol.*, **26**, pp. 1840–1846, 2008.

26. M. Settle, M. Salib, A. Michaeli, and T. F. Krauss, Low loss silicon on insulator photonic crystal waveguides made by 193 nm optical lithography, *Opt. Express*, **14**, pp. 2440–2445, 2006.

27. A. Liu, R. Jones, L. Liao, D. Samara-Rubio, D. Rubin, O. Cohen, R. Nicolaescu, and M. Paniccia, A high-speed silicon optical modula-tor based on a metal-oxide-semiconductor capacitor, *Nature*, **427**, pp. 615–618, 2004.

28. R. A. Soref, D. L. McDaniel Jr., and B. R. Bennett, Guided-wave intensity modulators using amplitude-and-phase perturbations, *J. Lightwave Technol.,* **6,** pp. 437–444, March 1988.

29. G. V. Treyz, P. G. May, and J.-M. Halbout, Silicon optical modulators at 1.3 micrometer based on free-carrier absorption, *IEEE Electron Device Lett.,* **12,** pp. 276–278, 1991.

30. G. V. Treyz, P. G. May, and J. M. Halbout, Silicon Mach-Zehnder waveguide interferometers based on the plasma dispersion effect, *Appl. Phys. Lett.,* **59,** pp. 771–3, 1991.

31. C. K. Tang, G. T. Reed, A. J. Wilson, and A. G. Rickman, Simulation of a low loss optical modulator for fabrication in SIMOX material, *Silicon-Based Optoelectronic Materials Symposium, April 12–14, 1993,* San Francisco, CA, pp. 247–52, 1993.

32. C. K. Tang, G. T. Reed, A. J. Wilson, and A. G. Rickman, Low-loss, single-mode, optical phase modulator in SIMOX material, *J. Lightwave Technol.,* **12,** pp. 1394–1400, 1994.

33. A. Cutolo, M. Iodice, P. Spirito, and L. Zeni, Silicon electro-optic modulator based on a three terminal device integrated in a low-loss single-mode SOI waveguide, *J. Lightwave Technol.,* **15,** pp. 505–518, 1997.

34. P. D. Hewitt and G. T. Reed, Multi micron dimension optical p-i-n modulators in silicon-on-insulator, *Proc. of SPIE,* San Jose, CA, **3630,** pp. 237–243, 1999.

35. P. D. Hewitt and G. T. Reed, Improving the response of optical phase modulators in SOI by computer simulation, *J. Lightwave Technol.,* **18,** pp. 443–450, 2000.

36. P. D. Hewitt and G. T. Reed, Improved modulation performance of a silicon p-i-n device by trench isolation, *J. Lightwave Technol.,* **19,** pp. 387–390, 2001.

37. A. Irace, G. Breglio, and A. Cutolo, All-silicon optoelectronic modulator with 1 GHz switching capability, *Electron. Lett.,* **39,** pp. 232–233, 2003 .

38. C. E. Png, S. P. Chan, S. T. Lim, and G. T. Reed, Optical phase modulators for MHz and GHz modulation in silicon-on-insulator (SOI), *J. Lightwave Technol.,* **22,** pp. 1573–1582, 2004.

39. X. Qianfan, B. Schmidt, S. Pradhan, and M. Lipson, Micrometer-scale silicon electro-optic modulator, *Nature,* **435,** pp. 325–327, 2005.

40. S. Manipatruni, X. Qianfan, B. Schmidt, J. Shakya, and M. Lipson, High Speed Carrier Injection 18 Gb/s silicon micro-ring electro-optic modulator, *LEOS 2007. The 20th Annual Meeting of the IEEE,* pp. 537–538, 2007.

41. Q. Xu, S. Manipatruni, B. Schmidt, J. Shakya, and M. Lipson, 12.5 Gbit/s carrier-injection-based silicon micro-ring silicon modulators, *Opt. Express,* **15,** pp. 430–436, 2007.

42. W. M. J. Green, M. J. Rooks, L. Sekaric, and Y. A. Vlasov, Ultra-compact, low RF power, 10 Gb/s silicon Mach-Zehnder modulator, *Opt. Express,* **15,** pp. 17106–17113, 2007.
43. C. E. Png, Silicon-on-insulator phase modulators, Ph.D. thesis (University of Surrey, 2004).
44. F. Y. Gardes, G. T. Reed, N. G. Emerson, and C. E. Png, A submicron depletion-type photonic modulator in silicon on insulator, *Opt. Express,* **13,** pp. 8845–8854, 2005.
45. L. Liao et al., 40 Gbit/s silicon optical modulator for high speed applications, *Electron. Lett.,* **43,** pp. 1196–1197, 2007 .
46. L. Ling, L. Ansheng, J. Basak, H. Nguyen, M. Paniccia, Y. Chetrit, and D. Rubin, Silicon photonic modulator and integration for high-speed applications, *IEEE Electron Devices Meeting 2008,* pp. 1–4, 2008.
47. J. W. Park, J.-B. You, I. G. Kim, and G. Kim, High-modulation efficiency silicon Mach-Zehnder optical modulator based on carrier depletion in a PN diode. *Opt. Express,* **17,** pp. 15520–15524, 2009.
48. F. Y. Gardes, A. Brimont, P. Sanchis, G. Rasigade, D. Marris-Morini, L. O'Faolain, F. Dong, J. M. Fedeli, P. Dumon, L. Vivien, T. F. Krauss, G. T. Reed, and J. Martí, High-speed modulation of a compact silicon ring resonator based on a reverse-biased pn diode. *Opt. Express,* **17,** pp. 21986–21991, 2009.
49. D. M. Gill, M. Rasras, T. Kun-Yii, C. Young-Kai, A. E. White, S. S. Patel, D. Carothers, A. Pomerene, R. Kamocsai, C. Hill, and J. Beattie, Internal bandwidth equalization in a CMOS-compatible Si-Ring modulator, *IEEE Photon. Technol. Lett.,* **21,** pp. 200–202, 2009.
50. Lightwire, Inc., http://www.lightwire.com, 2009.
51. D. D'Andrea, CMOS Photonics Today & Tomorrow Enabling Technology, OFC Market Watch, Available at: http://www.ofcn-foec.org/conference_program/2009/images/09-DAndrea.pdf, 2009.
52. V. Laurent, J. Osmond, J.-M. Fédéli, D. Marris-Morini, P. Crozat, J.-F. Damelencourt, E. Cassan, Y. Lecunff, and S. Laval, 42 GHz p.i.n. germanium photodetector integrated in a silicon-on-insulator waveguide, *Opt. Express,* **17,** pp. 6252–6257, 2009.
53. J. Brouckaert, G. Roelkens, D. Van Thourhout, and R. Baets, Compact InAlAs-InGaAs metal-semiconductor-metal photodetectors integrated on silicon-on-insulator waveguides, *IEEE Photon. Technol. Lett.,* **19,** pp. 1484–1486, 2007.
54. P. Hyundai, A. W. Fang, R. Jones, O. Cohen, R. Omri, M. N. Sysak, M. J. Paniccia, and J. E. Bowers, A hybrid AlGaInAs-silicon evanescent waveguide photodetector, *Opt. Express,* **15,** pp. 6044–6052, 2007.
55. M. W. Geis, S. J. Spector, M. E. Grein, R. T. Schulein, J. U. Yoon, D. M. Lennon, S. Deneault, F. X. Kaertner, and T. M. Lyszczarz, CMOS-compatible all-Si high-speed waveguide photodiodes with high responsivity in near-infrared communication band, *IEEE Photon. Technol. Lett.,* **19,** pp. 152–154, 2007.

56. Electronic and Photonic Integrated Circuits (EPIC), http://www.darpa.mil/MTO/Programs/epic/index.html, 2009.

57. Ultraperformance Nanophotonic Intrachip Communications (UNIC), http://www.darpa.mil/MTO/Programs/unic/index.html, 2009.

58. M. Feldman, DARPA Places $44M Bet on Optical Interconnects, *HPC Wire*, March 28, 2008, available at: http://www.hpcwire.com/features/DARPA_Places_44M_Bet_On_Optical_Interconnects.html, 2008.

59. HELIOS, http://www.helios-project.eu/, 2008.

60. UK Silicon Photonics, http://www.uksiliconphotonics.co.uk, 2009.

61. D. J. Paul, Silicon photonics: A bright future?, *Electron. Lett.*, **45**, pp. 582–584, 2009.

62. Lightwire, Inc., Lightwire 10 Gbps Transceivers, http://www.lightwire.com/products.cfm, 2009.

63. Luxtera, http://www.luxtera.com/20091110193/luxtera-transforms-optical-industry-with-launch-of-first-single-chip-transceiver-for-motherboard-deployment.html, 2009.

64. Luxtera, Luxtera Introduces the industry's lowest power 40G AOC, http://www.luxtera.com/20091116194/luxtera-introduces-industrys-lowest-power-40g-aoc.html, 2009.

65. Kotura, Inc., http://www.kotura.com, 2009.

66. X. Zheng, P. Koka, H. Schwetman, J. Lexau, R. Ho, J. E. Cunningham, and A. V. Krishnamoorthy, Silicon photonic WDM point-to-point network for multi-chip processor interconnects, *5th IEEE International Conference on Group IV Photonics*, pp. 380–382, 2008.

67. A. W. Fang, H. Park, R. Jones, O. Cohen, M. J. Paniccia, and J. E. Bowers, A continuous-wave hybrid AlGaInAs-silicon evanescent laser, *IEEE Photon. Technol. Lett.*, **18**, pp. 1143–1145, 2006.

68. H. Rong, R. Jones, A. Liu, O. Cohen, D. Hak, A. Fang, and M. Paniccia, A continuous-wave Raman silicon laser, *Nature*, **433**, pp. 725–728, 2005.

69. J. Ahn, M. Fiorentino, R. G. Beausoleil, N. Binkert, A. Davis, D. Fattal, N. P. Jouppi, M. McLaren, C. M. Santori, R. S. Schreiber, S. M. Spillane, D. Vantrease, and Q. Xu, Devices and architectures for photonic chip-scale integration, *Appl. Phys. A*, **95**, pp. 989–997, 2009.

70. L. Pavesi, Optical gain in silicon and the quest for a silicon injection laser, Chapter 2 in *Optical Interconnects: The Silicon Approach*, L. Pavesi and G. Guillot (Eds.), Springer-Verlag, Berlin, Heidelberg, 2006.

71. R. Merritt, Potholes seen on the road to silicon photonics, *EE Times*, 28 January 2009, Available at: http://www.eetimes.com/news/latest/showArticle.jhtml?articleID=212903357.

72. X. Sun, J. Liu, L. C. Kimerling, and J. Michel, Toward a germanium laser for integrated silicon photonics, *IEEE J. Sel. Top. Quantum Electron.*, **16**, pp. 124–131, 2010.

73. The Silicon Laser Muri, http://mpc-web.mit.edu/index. php?option=com_content&view=article&id=74&Itemid=137, 2008.
74. J. Prat, C. Arellano, V. Polo, and C. Block, Optical network unit based on a bidirectional reflective semiconductor optical amplifier for fiber-to-the-home networks, *IEEE Photon. Technol. Lett.*, **17**, pp. 250–252, 2005.
75. R. Ho, K. W. Mai, and M. A. Horowitz, The future of wires, *Proc. IEEE*, **89**, pp. 490–504, 2001.
76. J. A. Davis, R. Venkatesan, A. Kaloyeros, M. Beylansky, S. J. Souri, K. Banerjee, K. C. Saraswat, A. Rahman, R. Reif, and J. D. Meindl, Interconnect limits on gigascale integration (GSI) in the 21st century, *Proc. IEEE*, **89**, pp. 305–324, 2001.
77. J. D. Meindl, Interconnect opportunities for gigascale integration, *Micro IEEE*, **23**, pp. 28–35, 2003.
78. D. A. B. Miller and H. M. Ozaktas, Limit to the bit-rate capacity of electrical interconnects from the aspect ratio of the system architecture, *J. Parallel Distrib. Comput.*, **41**, pp. 42–52, 1997.
79. K. C. Saraswat and F. Mohammadi, Effect of scaling of interconnections on the time delay of VLSI circuits, *IEEE J. Solid-State Circuits*, **17**, pp. 275–280, 1982.
80. M. Haurylau, C. Hui, Z. Jidong, C. Guoqing, N. A. Nelson, D. H. Albonesi, E. G. Friedman, and P. M. Fauchet, On-chip optical interconnect roadmap: Challenges and critical directions, *2nd IEEE International Conference on Group IV Photonics*, pp. 17–19, 2005.
81. J. G. Koomey, Estimating total power consumption by servers in the U.S. and the world, available at: http://blogs.business2.com/greenwombat/files/serverpowerusecomplete-v3.pdf, 2007.
82. A. L. Barroso, The price of performance, *Queue*, **3**, pp. 48–53, 2005.
83. G. Balamurugan, J. Kennedy, G. Banerjee, J. E. Jaussi, M. Mansuri, F. O'Mahony, B. Casper, and R. Mooney, A scalable 5-15 Gbps, 14-75 mW low-power I/O transceiver in 65 nm CMOS, *IEEE J. Solid-State Circuits*, **43**, pp. 1010–1019, 2008.
84. J. Poulton, R. Palmer, A. M. Fuller, T. Greer, J. Eyles, W. J. Dally, and M. Horowitz, A 14-mW 6.25-Gb/s Transceiver in 90-nm CMOS, *IEEE J. Solid-State Circuits*, **42**, pp. 2745–2757, 2007.
85. R. J. Drost, R. D. Hopkins, R. Ho, and I. E. Sutherland, Proximity communication, *IEEE J. Solid-State Circuits*, **39**, pp. 1529–1535, 2004.
86. W. R. Davis, J. Wilson, S. Mick, J. Xu, H. Hua, C. Mineo, A. M. Sule, M. Steer, and P. D. Franzon, Demystifying 3D ICs: The pros and cons of going vertical, *IEEE Design Test Comput.*, **22**, pp. 498–510, 2005.
87. A. Liu, L. Liao, D. Rubin, H. Nguyen, B. Ciftcioglu, Y. Chetrit, N. Izhaky, and M. Paniccia, High-speed optical modulation based on carrier depletion in a silicon waveguide, *Opt. Express*, **15**, pp. 660–668, 2007.

88. S. K. Selvaraja, D. Vermeulen, M. Schaekers, E. Sleeckx, W. Bogaerts, G. Roelkens, P. Dumon, D. V. Thourhout, and R. Baets, Highly efficient grating coupler between optical fiber and silicon photonic circuit, *CLEO 2009*, CTuC6, 2009.

89. W Bogaerts, D. Taillaert, P. Dumon, D. Van Thourhout, and R. Baets, A polarization-diversity wavelength duplexer circuit in silicon-on-insulator photonic wires, *Opt. Express*, **15**, pp. 1567–1578, 2007.

90. G. Roelkens, D. Vermeulen, F. Van Laere, S. Selvaraja, S. Scheerlinck, D. Taillaert, W. Bogaerts, P. Dumon, D. Van Thourhout, and R. Baets, Bridging the gap between nanophotonic waveguide circuits and single mode optical fibers using diffractive grating structures, *J. Nanosci. Nanotechnol.*, **10**, pp. 1551–1562, 2010.

91. M. R. Watts, D. C. Trotter, R. W. Young, and A. L. Lentine, Ultralow power silicon microdisk modulators and switches, *IEEE International Conference on Group IV Photonics*, pp. 4–6, 2008.

92. X. Zheng, J. Lexau, Y. Luo, H. Thacker, T. Pinguet, A. Mekis, G. Li, J. Shi, P. Amberg, N. Pinckney, K. Raj, R. Ho, J. E. Cunningham, and A. V. Krishnamoorthy, Ultra-low-energy all-CMOS modulator integrated with driver, *Opt. Express*, **18**, pp. 3059–3070, 2010.

93. A. Liu, L. Liao, D. Rubin, H. Nguyen, B. Ciftcioglu, Y. Chetrit, N. Izhaky, and M. Paniccia, High-speed optical modulation based on carrier depletion in a silicon waveguide, *Opt. Express*, **15**, pp. 660–668, 2007.

94. G. Roelkens, P. Dumon, W. Bogaerts, D. Van-Thourhout, and R. Baets, Efficient silicon-on-insulator fiber coupler fabricated using 248-nm-deep UV lithography, *IEEE Photon. Technol. Lett.*, **17**, pp. 2613–2615 2005.

95. A. Choudhury, TR Stanczyk, D. Richardson, A. Donval, R. Oron, and M. Oron, Method of improving light coupling efficiency between optical fibers and silicon waveguides, *IEEE Photon. Technol. Lett.*, **17**, pp. 1881–1883, 2005.

96. G. Z. Masanovic, G. T. Reed, W. Headley, B. Timotijevic, V. M. N. Passaro, R. Atta, G. Ensell, and A. G. R. Evans, A high efficiency input/output coupler for small silicon photonic devices, *Opt. Express*, **13**, pp. 7374–7379, 2005..

97. D. Taillaert, F. Van Laere, M. Ayre, W. Bogaerts, D. Van Thourhout, P. Bienstman, and R. Baets, Grating couplers for coupling between optical fibers and nanophotonic waveguides, *Jpn. J. Appl. Phys. Part 1*, **45**, pp. 6071–6077, 2006.

98. G. Roelkens, D. Vermeulen, D. Van Thourhout, R. Baets, S. Brision, P. Lyan, P. Gautier, and J.-M. Fedeli, High efficiency diffractive grating couplers for interfacing a single mode optical fiber with a nanophotonic silicon-on-insulator waveguide circuit, *Appl. Phys. Lett.*, **92**, p. 131101, 2008.

99. F. Van Laere, J. Schrauwen, D. Taillaert, P. Dumon, W. Bogaerts, D. Van Thourhout, and R. Baets, Compact grating couplers between optical fibers and silicon-on-insulator photonic wire waveguides with 69% coupling efficiency, *2006 Optical Fiber Communication Conference*, 2006.

100. D. Taillaert, H. Chong, P. I. Borel, L. H. Frandsen, R. M. De La Rue, and R. Baets, A compact two-dimensional grating coupler used as a polarization splitter, *IEEE Photon. Technol. Lett.*, **15**, pp. 1249–1251, 2003.

101. F. Van Laere, T. Stomeo, D. Taillaert, G. Roelkens, D. Van Thourhout, T. F. Krauss, and R. Baets, Efficient polarization diversity grating couplers in bonded InP-membrane, *IEEE Photon. Technol. Lett.*, **20**, pp. 318–320, 2008.

102. J. V. Galan, P. Sanchis, J. Marti, S. Marx, H. Schroder, B. Mukhopadhyay, T. Tekin, S. Selvaraja, W. Bogaerts, P. Dumon, and L. Zimmermann, CMOS compatible silicon etched V-grooves integrated with a SOI fiber coupling technique for enhancing fiber-to-chip alignment, *Proc. IEEE International Conference on Group IV Photonics*, pp. 148–150, 2009.

103. L. Zimmermann, H. Schroder, P. Dumon, W. Bogaerts, and T. Tekin, ePIXpack: Advanced smart packaging solutions for silicon photonics, *Proc. European Conference Integrated Optics*, pp. 33–36, 2008.

104. P. De Dobbelaere, Demonstration of First WDM CMOS Photonics Transceiver with Monolithically Integrated Photo-Detectors, *Proc. European Conference on Optical Communication*, pp. Tu.3.C.1, 2008.

105. Chiral Photonics, Inc., http://www.chiralphotonics.com/Web/default.html, 2008.

106. T. Baba, Slow light in photonic crystals, *Nat. Photonics*, **2**, pp. 465–473, 2008.

107. B. Corcoran, C. Monat, C. Grillet, D. J. Moss, B. J. Eggleton, T. P. White, L. O'Faolain, and T. F. Krauss, Green light emission in silicon through slow-light enhanced third-harmonic generation in photonic crystal waveguides, *Nat. Photonics*, **3**, pp. 206–210, 2009.

108. R. A. Soref, The opto-electronic integrated circuit, Chapter 1 in *Silicon Photonics: The State of the Art*, G. T. Reed (Ed.), John Wiley & Sons, Chichester, England, 2008.

109. M. L. Brongersma, J. A. Schuller, J. White, Y. C. Jun, S. I. Bozhevolnyi, T. Sondergaard, and R. Zia, Nanoplasmonics: Components, devices, and circuits, in *Plasmonic Nanoguides and Circuits*, S. I. Bozhevolnyi (Ed.), World Sci. Pub., 2009.

110. K. F. MacDonald, Z. L. Samson, M. I. Stockman, and N. I. Zheludev, Ultrafast active plasmonics, *Nat. Photonics*, **3**, pp. 55–58, 2008.

111. A. V. Krasavin and N. I. Zheludev, Active plasmonics: Controlling signals in Au/Ga waveguide using nanoscale structural transformations, *Appl. Phys. Lett.*, **84**, pp. 1416–1418, 2004.

112. G. Boeck, Design of RF-CMOS integrated circuits for wireless communications, *IEEE Int. Conf. on Microwaves, Communications, Antennas and Electronic System*, pp. 1–6, 2008.

113. H. Al-Raweshidy and S. Komaki (Eds.), *Radio Over Fiber Technologies for Mobile Communication Networks*, Artech House, London, 2002.

114. R. Jones, H. Rong, H.-F. Liu, and M. Paniccia, The Opto-electronic integrated circuit, Chapter 9 in *Silicon Photonics: The State of the Art*, G. T. Reed (Ed.), John Wiley & Sons, Chichester, England.

115. A. J. Seeds and K. J. Williams, Microwave photonics, *J. Lightwave Technol.*, **24**, pp. 4628–4641, 2006.

116. P. Lecoy and B. Delacressonniere, Design and realization of an optically controlled oscillator for radio over fiber at 5.2 GHz, *IEEE Int. Topical Meeting on Microwave Photonics*, 06EX1314, pp. 85–88, 2006.

117. F. Vacondio, M. Mirshafiei, J. Basak, A. Liu, L. Liao, M. Paniccia, and L. A. Rusch, A silicon modulator enabling RF over fiber for 802.11 OFDM signals, *IEEE J. Sel. Top. Quantum Electron.*, **16**, pp. 141–148, 2010.

118. G. J. Hawkins, Spectral characteristics of infrared optical materials and filters, Ph.D. thesis (University of Reading, 1998).

119. A. Krier (Ed.), *Mid-Infrared Semiconductor Optoelectronics*, Springer-Verlag, London, 2006.

120. L. Labadie and O. Wallner, Mid-infrared guided optics: A perspective for astronomical instruments, *Opt. Express*, **17**, pp. 1947–1962, 2009 .

121. R. A. Soref, S. J. Emelett, and W. R. Buchwald, Silicon waveguided components for the long-wave infrared region, *J. Opt. A*, **8**, pp. 840–848, 2006.

122. X. Liu, J. B. Driscoll, J. I. Dadap, R. M. Osgood, Y. A. Vlasov, and M. J. Green, Mid-infrared pulse dynamics in Si nanophotonic wires near the two-photon absorption edge, *CLEO 2009*, CFR5, 2009.

123. E. D. Palik, *Handbook of Optical Constants of Solids*, Vol. 1, Academic, London, 1985.

124. M. M. Milošević, P. S. Matavulj, P. Y. Yang, A. Bagolini, and G. Z. Mashanovich, Rib waveguides for mid-infrared silicon photonics, *J. Opt. Soc. Am. B*, **26**, pp. 1760–1766, 2009.

125. R. Soref, Towards silicon-based longwave integrated optoelectronics (LIO), *Proc. SPIE*, *6898*, 689809–1, 2009.

126. G. Z. Mashanovich, M. Milosevic, P. Matavulj, B. Timotijevic, S. Stankovic, P. Y. Yang, E. J. Teo, M. B. H. Breese, A. A. Bettiol, and G. T. Reed, Silicon photonic waveguides for different wavelength regions, *Semiconductor Sci. Technol.*, **23**, 064002, 2008.

127. R. M. Jenkins, B. J. Perrett, M. E. McNie, E. D. Finlayson, R. R. Davies, J. Banerji, and A. R. Davies, Hollow optical waveguide devices and systems, *SPIE Europe Security and Defence, Proc-SPIE*, **7113**, 71130E, 2008.

128. E. J. Teo, A. A. Bettiol, M. B. H. Breese, P. Y. Yang, G. Z. Mashanovich, W. R. Headley, G. T. Reed, and D. J. Blackwood, Three-dimensional fabrication of silicon waveguides with porous silicon cladding, *Opt. Express*, **16**, pp. 573–578, 2008.

129. E. J. Teo, A. A. Bettiol, P. Yang, M. B. H. Breese, B. Q. Xiong, G. Z. Mashanovich, W. R. Headley, and G. T. Reed, Fabrication of low loss silicon-on-oxidized porous silicon strip waveguides using focused proton beam irradiation, *Opt. Lett.*, **34**, pp. 659–661, 2009.

130. P. Y. Yang, G. Z. Mashanovich, I. Gomez-Morilla, W. R. Headley, G. T. Reed, E. J. Teo, D. J. Blackwood, M. B. H. Breese, and A. A. Bettiol, Free standing waveguides in silicon, *Appl. Phys. Lett.*, **90**, 241109, 2007.

131. R. Soref, R. E. Peale, and W. Buchwald, Longwave plasmonics on doped silicon and silicides, *Opt. Express*, **16**, pp. 6505–6514, 2009.

132. V. Raghunathan, D. Borlaug, R. R. Rice, and B. Jalali, Demonstration of a mid-infrared silicon Raman amplifier, *Opt. Express*, **15**, pp. 14355–14362, 2007.

133. J. Wang, J. Hu, X. Sun, P. Becla, A. M. Agarwal, and L. C. Kimerling, Spectral selective mid-infrared detector on a silicon platform, *Group IV Photonics*, **FB7**, pp. 235–237, 2009.

134. X. Liu, R. M. Osgood, Y. A. Vlasov, and W. M. J. Green, Broadband mid-infrared parametric amplification, .net off-chip gain, and cascaded four-wave mixing in silicon photonic wires, *IEEE Group IV Photonics Conference*, September 2009, San Francisco, USA, paper PD1.3, 2009 .

135. P. Q. Liu, A. J. Hoffman, M. D. Escarra, K. J. Franz, J. B. Khurgin, Y. Dikmelik, X. Wang, J.-Y. Fan, and C. F. Gmachl, Highly power-efficient quantum cascade lasers, *Nat. Photonics*, **4**, pp. 95–98, 2010.

136. Y. Bai, S. Slivken, S. Kuboya, S. R. Darvish, and M. Razeghi, Quantum cascade lasers that emit more light than heat, *Nat. Photonics*, **4**, pp. 99–102, 2010 .

137. D. Dai, A. Fang, and J. E. Bowers, Hybrid silicon lasers for optical interconnects, *New J. Phys.*, **11**, pp. 125016, 2009.

138. M. Notomi, T. Tanabe, A. Shinya, E. Kuramochi, H. Taniyama, S. Mitsugi,, and M. Morita, Nonlinear and adiabatic control of high Q photonic crystal nanocavities, *Opt. Express*, **15**, pp. 17458–17481, 2007.

139. S. Assefa, F. Xia, and Y. A. Vlasov, Reinventing germanium avalanche photodetector for nanophotonic on-chip optical interconnects, *Nature*, **464**, pp. 80–85, 2010.

140. V. R. Almeida, R. R. Panepucci, and M. Lipson, Nanotaper for compact mode conversion, *Opt. Lett.*, **28**, pp. 1302–1304, 2003.

141. S. J. McNab, N. Moll, and Y. A. Vlasov, Ultra-low loss photonic integrated circuit with membrane-type photonic crystal waveguides, *Opt. Express*, **11**, pp. 2927–2939, 2003.

142. K. Yamada, M. Notomi, I. Yokohama, T. Shoji, T. Tsuchizawa, T. Watanabe, J. Takahashi, E. Tamechika, and H. Morita, SOI-based photonic crystal line defect waveguides, *Proc. SPIE*, **4870**, pp. 324–338, 2002.
143. A. Sure, T. Dillon, J. Murakowski, C. Lin, D. Pustai, and D. W. Prather, Fabrication and characterization of three-dimensional silicon tapers, *Opt. Express*, **11**, pp. 3555–3561, 2003.
144. C. Gunn, CMOS photonics for high-speed interconnects, *IEEE Micro*, **26**, pp. 58–66, 2006.
145. D. Van Thourhout, G. Roelkens, R. Baets, W. Bogaerts, J. Brouckaert, P. Debackere, P. Dumon, S. Scheerlinck, J. Schrauwen, D. Taillaert, F. V. Laere, and J. V. Campenhout, Coupling mechanisms for a heterogeneous silicon nanowire platform, *Semiconductor Sci. Technol.*, **23**, 064004, 2008.

CHAPTER **2**

Silicon Plasmonic Waveguides

Richard Soref, Sang-Yeon Cho, Walter Buchwald,
Robert E. Peale, and Justin Cleary

Contents

2.1 Introduction

Silicon plasmonics has entered the plasmonics literature only recently. There are reports of metal-and-silicon structures for the near infrared, and a far-infrared composite has been simulated [1]. But there are wider possibilities. We consider the opportunities for silicon by examining a new group of plasmonic materials that can be manufactured in a silicon foundry, and we explore a wide-wavelength range of operation. Our work suggests that silicon plasmonics (or more accurately, Group IV plasmonics) is an important new technology for the "longwave" infrared that stretches all the way from 5 μm out to 500 μm. In this chapter, we examine a key component of on-chip "plasmonic network" integration: the plasmonic channel waveguide that can be bent, split, and coupled in a manner analogous to that of photonic waveguides. Generally, the waveguide is a composite structure of dielectric (D) and conducting (C) materials. Specifically, we have examined CD, CDC, and DCD strip structures. Generally, we find that the low loss fundamental mode is a composite of a long-range surface plasmon polariton (LRSPP) mode and a dielectric-confined mode (the latter using total internal reflection of the electromagnetic waves at the dielectric–air interfaces). Without adding propagation loss, we can make the lateral width (w) of the fundamental mode equal to the free-space wavelength (λ) simply by making the dielectric and conducting strips one wavelength wide in the CD, CDC, and DCD waveguides. As discussed later, there is a practical strategy of constraining the hybrid mode height (h) to be less than λ by choosing a sub-wavelength dielectric strip height, but some extra loss is thereby introduced. For all of the new conducting materials described in Section 2.2, the LRSPP will indeed propagate in the 3 to 5 μm mid-IR wavelength band. However, in that case the propagation losses will be very high, and the mode will travel only tens of microns. Longer wavelengths are a better choice. Our general findings are: (1) the high loss at ~3 μm decreases dramatically with increasing wavelength and becomes very low in the far-infrared Terahertz region and (2) a trade-off exists between mode size and mode attenuation. This is quantified in the following text. There are excellent possibilities for plasmonic and plasmonic–photonic "systems on a chip." Many different Group IV waveguided plasmonic components, both active and passive, are feasible and can be seamlessly interconnected to form a silicon (or germanium) plasmonic network-on-a-chip: a new kind of "integrated circuit." When integrated with on-chip electronics, the network becomes a plasmo-electronic integrated circuit: the PEIC. Ideally, the PEIC would include special electronic devices that can launch and detect SPPs without the "intervention" of photonics. Alternatively, photonic techniques can be woven into the plasmonic story. Looking ahead, we foresee an on-chip convergence of plasmonic and photonic structures in which the two technologies work together synergistically. As discussed later, this will lead to the plasmo-opto-electronic integrated circuit: the POEIC.

2.2 Plasmonic Materials for High-Volume Manufacture

"Factory experiments" on silicon photonics have been performed in Europe (IMEC and Ghent: the ePIXfab project) and the United States (BAE Systems and Freescale Semiconductor: the EPIC project). This work on application-specific OEICs shows that only slight modifications to standard Fab processes are required in order to produce commercially viable photonic waveguide components. We believe that the foundry viability discovered for photonics applies to Group IV plasmonics because the materials and processes are essentially the same in both technologies. In a series of papers [2–4], we have suggested that heavily doped silicon and the various metal silicides are excellent conductive materials (negative permittivity materials) for silicon plasmonics. Generally, we want all materials to be within Group IV and to be readily manufacturable in a Silicon CMOS fab. We believe that the foundry base will enable new cost-effective applications. Regarding the dielectrics, they will be undoped intrinsic crystalline Si, Ge, SiGe, SiGeSn, and layered heterostructures of elements and alloys, including quantum wells. The conductors will be metals, metal silicides, metal germanicides, heavily doped Si, heavily doped Ge, and doped alloys. The metals must be CMOS compatible, and those are usually Al, Cu, W, and Ti. Returning to the Group IV dielectrics, they could be noncrystalline (polycrystalline, dense nanocrystalline) and still retain most of their desired response in the CD, CDC, and DCD devices. Another practical aspect of these dielectric-and-conductor structures is the real possibility of creating integratable *active* plasmonic devices as discussed in the following text.

2.3 Plasmonic Structures for Normal-Incidence Free-Space Use

We do not have the space here to explore structures for shaping and filtering beams traveling unconfined in space—only in-plane devices are covered here—but at least we can list some of the principal device types. These normal-incidence 2D planar plasmonic structures are (1) subwavelength antennas, (2) subwavelength apertures, (3) meta-material arrays, and (4) Fano filters. The plasmonic nanoantenna concentrates the EM fields greatly and is used to enhance infrared emission or photodetection. The subwavelength aperture-in-conductor structure can be formed on the end facet of a semiconductor laser to provide shaping or focusing of the near-field or far-field radiative emission pattern [5]. Subwavelength-diameter conductive elements, such as split or bull's-eye rings, can be deployed in a periodic lattice array on the surface (or inside) of a uniform Si or Ge slab to create a metamaterial for infrared filtering. The beam traversing the array becomes "shaped" by making the array spatially nonuniform. Plasmonic cloaking is feasible. We define the plasmonic Fano filter as a "plasmonic crystal" with in-plane waveguiding. This crystal is formed using a 2D lattice array of very thin conductive disks whose diameter is slightly subwavelength. The slab supports various guided modes. Infrared radiation normally incident upon the slab excites several of those resonant modes. As a result, this slab has a unique free-space reflectance and transmission spectrum, both of which can be shaped by the Fano designer. Reflectance of 99.9% over a narrow band can be achieved.

2.4 Low Loss Group IV Plasmonic Waveguide Structures

2.4.1 Plasmonic Definitions

The surface plasmon polariton (SPP) is a hybrid of an optical wave and the collective density oscillations of conduction electrons at a dielectric–conductor interface. The dielectric here is an intrinsic semiconductor. The bound SPP is a long-range TM-like mode that propagates along an elongated conductor or channel. Examining the EM mode field strengths in the direction perpendicular to the conductor surface, we see that those fields are maximum at the interface and fall off exponentially into the dielectric. Later in this chapter, we propose waveguiding of surface phonon polaritons (SPnPs), which are hybrids of an infrared wave and the collective density oscillation of ions in a compound semiconductor or insulator operating in the LWIR reststrahl wavelength region.

2.4.2 Launching and Detecting LRSPPs

An important method is to use an electron beam to create SPPs that propagate toward appropriate out-coupling structures where plasmons are converted to free photons that may be analyzed spectroscopically [6–8]. This cathodoluminescence is valuable for studying plasmonic properties, but is poorly suited for the chip-IC context. Launching and detecting SPPs via photonics or electronic devices is more practical for the IC. Light can be in-coupled to the waveguide by a grating [9] or other means to generate the SPP. The coupling technique must compensate the Δk momentum mismatch between the light and the SPP. When the polariton mode reaches its waveguided destination, the SPP regenerates photons that are radiated and collected for detection.

Evanescent-wave or end-fire coupling between photonic dielectric waveguides and plasmonic waveguides is also feasible. Alternatively, photonic launching can be avoided. There is evidence in the literature that a transistor-like electronic device such as the CMOS-driven tunnel junction [10] integrated on the plasmonic waveguide input end can excite the SPP, while a related electronic device senses the SPP arriving at the waveguide's output end. It is known that a <100 nm silicon MOSFET can, at room temperature, detect radiation whose frequency is around 1 THz [11]. This suggests that THz SPPs could be detected by a similar transistor.

2.4.3 Waveguide Geometries

The plasmonic waveguide is a striplike composite structure of conductors and dielectrics that is elongated in the propagation direction. Similar to optical channel waveguides, plasmonic waveguides can be fashioned into bends, splitters, combiners, couplers, rings, intersections, etc.—an important benefit of plasmonics. Another virtue of the plasmonic channel waveguide is its relatively small cross-sectional dimensions, which can be "subwavelength." To describe the waveguide, let us define the Z axis as the direction of SPP mode propagation along the interface, while the lateral axes Y and X are used to define the height and width of the conductors and the SPP mode. The subwavelength mode dimension is usually along Y. The nonsymmetric CD waveguide geometry features a thick conductor. The symmetric guides are the conductor-dielectric-conductor (CDC) structure featuring thick conductors and the dielectric-conductor-dielectric structure (DCD) featuring an ultrathin C-ribbon. A detailed behavior of these structures is described later.

2.4.4 The Complex Permittivity of Conductors

Our research [12] shows that metal-silicides and heavily doped silicon (p-type or n-type with a concentration above 10^{20} cm^{-3}) combine well with intrinsic silicon for plasmonic structures. Regarding waveguide materials, our focus here is the silicon base; therefore, the waveguide structures investigated are silicon/silicide/silicon, silicon/p-Si/silicon, silicide/Si/silicide, and so forth. More generally, we are proposing Group IV CDCs and DCDs. This means that the D layers utilize the elements, alloys, and heterostructures in Group IV such as Si, Ge, SiGe, SiGeC, SiGeSn, and Ge/SiGe multi-quantum-wells. When Si is the dielectric and when metal is desired for C, the metal should be one of four compatible with the CMOS foundry processing (Cu, Al, W, and Ti) rather than Au or Ag. Silicon combined with silicide or doped-Si conductors is quite manufacturable. Taking Ge as the dielectric, C can be a metal or p-Ge or n-Ge or any germanicide. Exploration of silicides and doped silicon as SPP hosts has been made theoretically [2,12], and recently Cleary et al. [13] measured the complex permittivity of four important silicides over a wide range of photon energy. Figures 2.1a and 2.1b present this experimental Re[ε] and Im[ε] data with a comparison to Ag data. We then plotted Re[ε] and Im[ε] on the same graph in Figure 2.2 for the representative

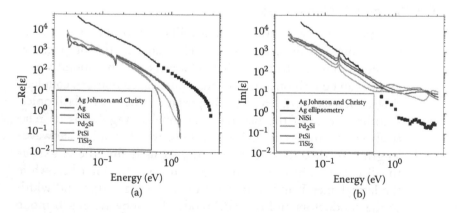

FIGURE 2.1 Comparison of (a) real and (b) imaginary parts of permittivity for four silicides with those of Ag.

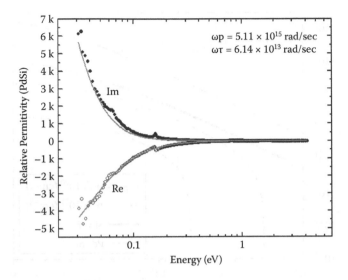

FIGURE 2.2 Experimentally determined permittivity for Pd$_2$Si. The solid line is a curve fit to Drude theory used to extract the bulk plasmon frequency and relaxation frequency.

silicide Pd$_2$Si. The thin solid curves in Figure 2.2 show our fitting of the Drude model to the data points, from which the Drude parameters ω_p and ω_τ are extracted.

Let us now compare the complex permittivity of doped silicon with that of other conductors. In Figures 2.3 and 2.4, taking n-type Si as representative, we have plotted Re[ε] and Im[ε] versus λ for doped Si, along with the curves for Pd$_2$Si and Ag. Impurity concentrations of 10^{20} and 10^{21} cm^{-3} were examined. We see that Re[ε] is negative as desired only for $\lambda > 2$ μm (when N = 10^{21} cm^{-3}) and for $\lambda > 6.5$ μm (when N = 10^{20} cm^{-3}). Table 2.1 presents our empirically determined parameters ω_p and ω_τ for three approaches, revealing a ω_p hierarchy of metal, silicide, and doped-Si, plus a similarity in the values of ω_τ. The experiments, modeling, and literature results utilized in constructing Table 2.1 are discussed in Cleary et al. [13].

2.5 Waveguide Modeling and Simulation

2.5.1 The Unsymmetric CD Structure

The simplest waveguide is the unsymmetric CD structure shown in Figure 2.5 (left). The strip conductor is thick in the Y direction, while the dielectric is several wavelengths thick in Y. We use the

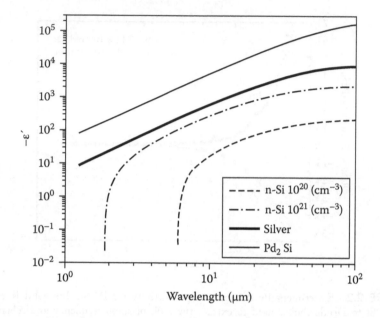

FIGURE 2.3 Comparison of $-\mathrm{Re}[\varepsilon]$ for doped-Si, silicide, and metal conductors.

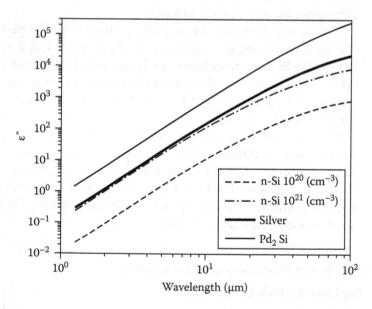

FIGURE 2.4 Comparison of $\mathrm{Im}[\varepsilon]$ for doped-Si, silicide, and metal conductors.

TABLE 2.1   Drude Parameters: Bulk Plasmon Frequency (ω_p) and Relaxation Frequency (ω_τ) for Three Kinds of Group IV Compatible Conductors

Plasmonic conductor	ω_p ($\times 10^{15}$ rad/s)	ω_τ ($\times 10^{14}$ rad/s)
Silver	13.7	0.27
Pd$_2$Si	4.71	0.46
TiSi$_2$	6.38	0.22
NiSi	5.77	0.53
PtSi$_2$	6.71	1.31
n-type Si 10^{20} cm^{-3}	0.32	0.68
n-type Si 10^{21} cm^{-3}	1.00	0.70
p-type Si 10^{20} cm^{-3}	0.27	0.96
p-type Si 10^{21} cm^{-3}	0.86	0.99

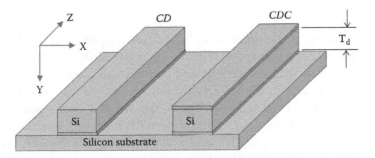

FIGURE 2.5   Perspective view of two plasmonic channel waveguides for the example of a silicon dielectric.

Drude model of this composite. The complex wavevector of the surface plasmon is given by

$$k_{sp}(\omega) = \sqrt{\frac{\varepsilon_d(\omega)\varepsilon_c(\omega)}{\varepsilon_d(\omega)+\varepsilon_c(\omega)}} \qquad (2.1)$$

According to Drude theory, the complex permittivity of the conductor is represented as

$$\varepsilon_c(\omega) = \varepsilon_\infty\left[1 - \frac{\omega_p^{\,2}}{\omega_\tau^{\,2}(1+\omega^2/\omega_\tau^{\,2})} + i\frac{\omega_p^{\,2}}{\omega\omega_\tau(1+\omega^2/\omega_\tau^{\,2})}\right] \qquad (2.2)$$

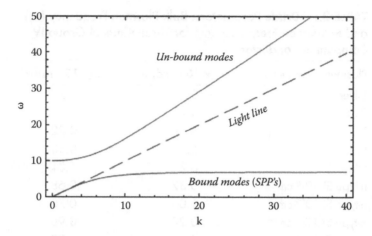

FIGURE 2.6 Plasmon dispersion relations for a CD waveguide.

where ε_∞ is the permittivity at infinite frequency. If we then plot the plasmon dispersion relation, ω-*versus*-*Re[k]*, we find unbound modes above the light line and bound modes below the light line as indicated in Figure 2.6. Bound modes are the subject of this chapter. For the CD waveguide, the 1/e SPP power propagation loss L_z in dB/cm is 8.68 Im[k(ω)] with k in cm⁻¹, and the 1/e lateral penetration depth along *Y* of this mode into the conductor or dielectric (the mode height *h* along *Y*) is

$$h_{d,c}(\omega) = \frac{1}{Re\left[\beta_{d,c}(\omega)\right]} \tag{2.3}$$

where

$$\beta_{d,c}(\omega) = \sqrt{k_{sp}(\omega)^2 - \left(\frac{\omega^2}{c^2}\right)\varepsilon_{d,c}(\omega)} = \left(\frac{\omega}{c}\right)\sqrt{\frac{-\varepsilon_{d,c}(\omega)^2}{\varepsilon_d(\omega)+\varepsilon_c(\omega)}} \tag{2.4}$$

Utilizing these relations and the wavelength-dependent permittivity data presented in Section 2.4, we predict the propagation loss shown in Figure 2.7 for the three CD waveguide types. Also, the predicted fundamental-mode height for those four waveguides is given in Figure 2.8 (semi-infinite Si is assumed in this calculation). Figure 2.8 also shows the depth of mode penetration into the conductor. Note that Figure 2.7 does not

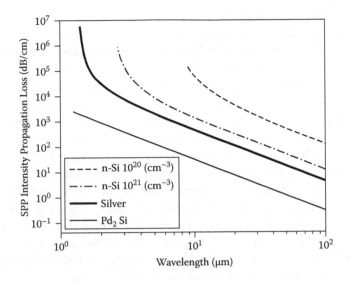

FIGURE 2.7 Power propagation loss of SPP mode in dB/cm as a function of free-space wavelength for CMOS-compatible silicide/Si and n-Si/Si CD channel waveguides. Propagation loss for Ag/Si channel waveguide is shown for comparison purposes.

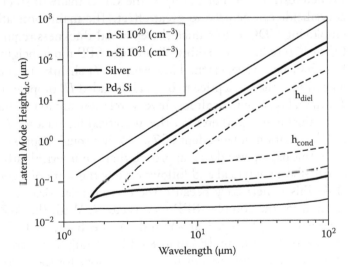

FIGURE 2.8 SPP mode height in dielectric and in conductor for the CMOS-compatible Pd_2Si/Si and n-Si/Si channel waveguides. Mode height for the Ag/Si channel waveguide is shown for comparison purposes.

include the phonon-related absorption of intrinsic crystal Si in the 8 to 20 μm range, loss that is discussed later.

If we say that the "maximum permissible" attenuation in Figure 2.7 is about 100 dB/cm, then we find that the unsymmetric metal/Si and silicide/Si CD waveguides are quite acceptable at wavelengths longer than 5 and 20 μm, respectively, whereas the doped-Si/Si CD waveguides work well only at long wavelengths (longer than 30 μm for 10^{20} cm^{-3}). The strong increase of mode size with λ is revealed in Figure 2.8. Looking at Figures 2.7 and 2.8, we see an inevitable compromise between falling loss and rising mode height. Generally, the rapidly growing mode height is a deficiency of the CD channel. Thus, Figure 2.8 motivates exploration of symmetric structures. In the foregoing figures, the SPP wavelength λ_{spp} is always less than the photon wavelength λ because λ_{spp} is given by $2\lambda/k_{spp}$ for a mode below the light line. In Figure 2.7, inclusion of the "dielectric loss" would boost the curves only slightly over the 8 to 20 μm region.

2.5.2 The Symmetric CDC Channel Waveguide

We turn now to the CDC waveguide of Figure 2.5 (right), sometimes called the "double plasmonic," which has had great success in the III-V quantum cascade laser and is quite applicable to silicon plasmonics, where the dielectric can be Si or Ge or a group IV heterostructure. For example, the CDC made of silicide/silicon/silicide can be used to great effect. The main motivation for employing CDC is to reduce the epitaxial thickness requirement (T_d in Figure 2.5) by making T_d and the SPP mode height much less than one wavelength. However, as T_d shrinks to subwavelength size, there is a price to be paid in added *IR* propagation loss L_z. In reality, L_z has an almost inverse relationship to the dielectric layer thickness T_d. Therefore, the analytical formula for L_z given in [1], a relation independent of T_d, is only a rough approximation.

In this paper, we take an approach that is numerical rather than closed-form analytical and follows closely the work of Economou [14]. This is necessary because there is no analytical dispersion relation for the complex SPP wavevector k. With the X-Z plane conductors C deployed at +Y and −Y, we have, as before, the propagation loss L_z in dB/cm as 8.68 Im [$k(\omega)$]; however, here we must solve for k numerically using expressions for the exponential decay coefficients β_c and β_d in the *Y* direction perpendicular to the interfaces, the expressions containing the complex permittivities of conductor and dielectric:

$$\pm exp(-\beta_d T_d) = \frac{(1-R)}{(1+R)} \qquad (2.5)$$

with

$$R = -(\beta_d / \varepsilon_d) / (\beta_c / \varepsilon_c) \qquad (2.6)$$

where T_d is the thickness of the dielectric between the conducting strips and β_d and β_c are defined earlier. In practice, the CDC ribbon width along X is going to be only one (or a few) wavelengths. Let us now consider a specific example in which CDC = NiSi/Si/NiSi, with T_d fixed at 25 μm. First, what is the shortest wavelength of operation? We can determine this by defining λ_{min} as the wavelength for which the $1/e$ intensity propagation length is $2 \lambda_{min}$. Using that relation, this criterion gives λ_{min} = 2.25 μm, where L_z = 9640 dB/cm. Plotting L_z as a function of wavelength, we find that L_z decreases with increasing λ. Then L_z reaches a low plateau of 197 dB/cm at λ = 39 μm and stays constant as λ approaches 100 μm. For this particular CDC, we fixed the wavelength at 80 μm (3.75 THz) and then varied the thickness of the Si strip from 20 to 100 μm to determine how T_d affects the mode loss. These L_z results are presented in Figure 2.9.

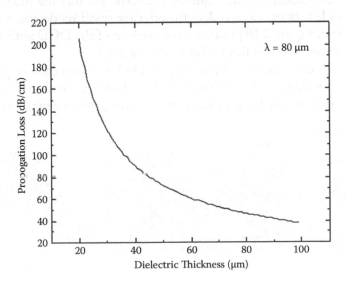

FIGURE 2.9 Propagation loss in dB/cm for a NiSi/Si/NiSi CDC waveguide as a function of central Si thickness at a free space wavelength of 80 μm.

Since the mode height h is about the same as T_d, the trade-off of L_z versus h is seen in the graph, and the choice $h \sim \lambda/2$ seems to be a sweet spot.

2.5.3 The New Buried-Ribbon DCD Channel Waveguide

During our research on DCDs, we discovered a low-loss waveguide whose operation can be explained by considering composite modes comprising plasmonic and dielectric portions. There are reports in the literature of long-range surface plasmon polaritons (LRSPPs) that cling to the surfaces of a thin metal strip. This long-range behavior applies also to thin silicide, germanicide, and doped silicon. We call the thin conductor strip a ribbon because it can be curved, split, combined, coupled, and resonated in the manner of a dielectric strip waveguide. The ribbon thickness is much less than a wavelength—so thin that this conductor does not induce much mode attenuation even though the SPP surrounds and penetrates the ribbon. This is the long-range $ss_b°$ mode reported by Berini [15]. If a Si, Ge, or alloy slab is bounded above and below by air, that slab is a conventional photonic waveguide with vertical confinement of light. Now, if we embed the conductive ribbon at the midlevel of this slab, then infrared fields of the LRSPP launched in the ribbon will fringe out into the slab, and the propagating mode will be a composite of SPP and vertical dielectric. The bounded dielectric shrinks the mode height to less than one wavelength, offering small mode size with low loss. Figure 2.10 presents a comparison of the DCD with the two waveguides studied earlier in this chapter.

How will this ribbon waveguide be constructed in practice? One choice is to "undercut" SOI or GeOI (to remove the buried oxide locally beneath the thin top layer), producing a suspended

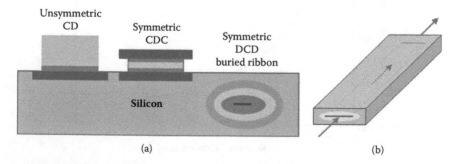

(a)   (b)

FIGURE 2.10 (a) Cross-section end view of the plasmonic channel waveguides investigated here; (b) perspective view of the buried ribbon waveguide.

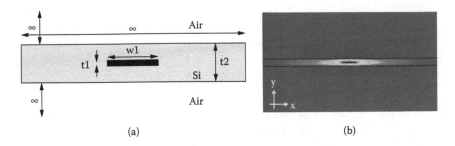

FIGURE 2.11 (a) Cross-sectional view of DCD Si/silicide/Si plasmonic ribbon wave-guides and (b) calculated transverse field profile (H_x) of the composite LRSPP mode.

Si or Ge membrane as in Figure 2.11a. Another choice is to operate at THz in SOI or GeOI. Since SiO_2 has low-to-moderate loss in the far IR, the group IV slab could be the top film in actual SOI or GeOI for the 1 to 3 THz range. The eigenmodes of a silicide ribbon in Si are calculated [16] using a commercial finite-element-method tool, *COMSOL Multiphysics*™. Among existing eigenmodes, the composite LRSPP mode is characterized by its symmetric transverse field profile and by the TM–like the orientation of the electric field vector. Figure 2.11b shows the transverse field profile of the composite LRSPP mode in a silicide ribbon waveguide. Detailed results are given in [16]. The Table 2.1 Drude-model Pd_2Si is used, and Palik [17] gives the optical properties of Si.

The spectral characteristics of the LRSPP mode are calculated here for mid- and far-infrared applications (λ of 5 to 100 μm). The plasmonic portion of the overall propagation loss is due to the material absorption by the silicide ribbon because scattering loss from imperfect surfaces is ignored. In order to find the main contributor to the propagation loss, the losses of the silicide ribbon waveguide are calculated with and without the actual phonon-related absorption [17] of intrinsic Si.

In an Si slab of thickness $\lambda/2$, we calculated the spectral characteristics of the composite LRSPP mode for a ribbon of width λ and thickness $\lambda/1000$. As shown in Figure 2.12a, we find that the height of the composite mode is reduced to 0.3 λ due to slab confinement, an important result. Two loss curves are plotted in Figure 2.12b. The lower curve shows only plasmonic loss. The upper curve in Figure 2.12b includes Si dielectric loss [17] and plasmonic loss. Here, the overall propagation loss at $\lambda = 100$ μm is only 2.18 dB/cm. The reader is referred to our Optics Letters paper [16] for a discussion of effective refractive index, ribbon

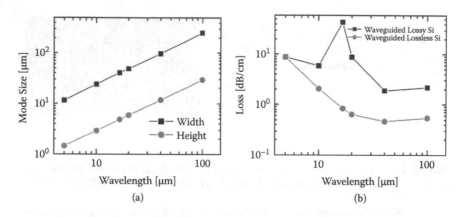

FIGURE 2.12 (a) Mode size and (b) propagation loss of the LRSP in silicide ribbon in $\lambda/2$ Si slab for $t_1 = \lambda/1000$ and $w_1 = \lambda$.

thickness effects, and slab thickness effects. To summarize: we have predicted low propagation losses and a composite-mode cross section of $0.3 \lambda \times 1.0 \lambda$ over the 5 to 100 µm wavelength range in a Group IV "membrane" containing a thin silicide ribbon. On-chip ribbon networks are expected to be useful in many longwave and THz applications.

2.5.5 Waveguiding in 2D Plasmonic Crystals

It is useful to construct polaritonic-crystal (PolC) lattices whose waveguiding functions are analogous to those of a photonic-crystal (PhtC). The modes found in PolC are either plasmon-polariton or phonon-polariton (the former is known as a *plasmonic crystal*). Looking at the 2D PolC, this lattice can be contained in a silicon slab that includes a periodic 2D array of silicide micro-disks of $\sim \lambda/2$ diameter buried at midlevel, thereby creating the plasmonic bandgap crystal. If we consider several rows of conducting disks in the crystal, these side-by-side chains of nano particles form an inline plasmonic waveguide because SPPs are localized at each nano particle and couple from one nano structure to the next. Line-defect rows that sandwich the chains can confine the mode laterally, thereby giving channel waveguiding. A single line defect can form a "slot" within a multirow chain waveguide.

In some applications, it will be useful to have signal communications between a photonic waveguide network and a polaritonic waveguide network; essentially, a waveguide-to-waveguide signal transfer through a dielectric. For photon guides, there is a dielectric core. This core "D" can be shared with the polaritonic waveguide, acting as a portion of the CDC or DCD plasmonic structure. Therefore, simple end-fire coupling is one strategy for signal coupling between the two types of waveguides. Another good approach is side-by-side evanescent-wave coupling obtained by placing the photonic waveguide in local proximity to the polaritonic waveguide.

2.5.6 Plasmonic Waveguide Devices (Including "Actives")

Infrared modulators and switches can be made from initially "passive" CD, CDC, and DCD structures by injecting electrons and holes into the semiconductor near the CD interfaces. This can be done optically or electrically using (respectively) above-bandgap light focused into the desired region or a lateral PIN diode structure integrated into the intrinsic semiconductor at the localized region. Each serves to perturb the complex index $\Delta n + i\Delta k$ of Si or Ge. Since the Δk portion tends to dominate at LWIR, the induced modulation is more likely to be electroabsorption than electrorefraction. (By contrast, the thermo-optic effect is primarily electrorefractive.)

Many shapes and functionalities are feasible with SPP waveguided devices. A few that come to mind are: the 1×1 and 2×2 plasmonic ring resonator structures with rings coupled to bus waveguides via waveguide apertures [18], the 1×2 and 2×2 multimode interference plasmonic devices [19], the slotted plasmonic waveguide devices useful for enhanced nonlinear effects and sensing, and the plasmonic biosensor devices such as the Mach–Zehnder interferometer [20]. All of these are subjects for future research.

2.5.7 Subwavelength Cross Section of the Plasmonic Mode

As discussed earlier, the plasmonic mode propagates along Z and is laterally confined to a small area in the X-Y plane. An oft-cited virtue of plasmonic waveguides is their ability to confine EM waves to an area much smaller than the diffraction-limited focal spot of conventional optics. There are actually several ways to obtain subwavelength mode dimensions in both X and Y. One way is to make a subwavelength slot in the conductor of CD or DCD for slot-concentrated fields. Another method is to reduce the X-width of the conductive ribbon, as in Figure 2.10c, to well below one wavelength, such as $0.1\ \lambda$. In that case, the ribbon becomes a "micro wire," and the surrounding dielectric can then be shrunk in its cross section. For example, the surround could be reduced to a dielectric cylinder whose diameter is 0.3 λ (with a waveguide loss penalty for such shrinkage). In this wire/cylinder case, we have created a subwavelength plasmonic waveguide with an approximate λ/4 × λ/4 mode area, and the overall structure is quite analogous to the coaxial cable waveguide that is used at radio frequencies.

A third approach is the double plasmonic structure of Figure 2.5b. We see in Figure 2.9 that a 0.25 λ mode height is readily attained in CDC with a loss penalty. For further mode confinement, it is quite feasible to shrink the width of both conductors and the dielectric to about 0.3 λ, but that mode area reduction comes with additional loss. The mode size discussed here is actually used in the quantum cascade laser and would be viable for Ge/SiGeSn THz QCLs.

A fourth possibility is to create a group IV plasmonic-waveguide laser device with an extremely confined hybrid mode that enhances spontaneous and stimulated emission via EM field concentration. This would be done using the wire/dielectric/conductor structure proposed by Oulton et al. [21]. If we reason by analogy to the CdS/MgF$_2$ structure of Oulton, then our plasmonic laser would consist of three parts: (1) a gain wire made of direct-bandgap material such as GeSn alloy, (2) an ultrathin lower-index Si or Ge layer upon which the wire rests, and (3) a conductive layer such as Pd$_2$Si that contacts the lower surface of the ultrathin Si. A silicide-based composite plasmonic-and-dielectric mode forms in the ultrathin Si. The ultrathin layer prevents the wire from touching the conductor. The wire is the infrared emission medium, and the idea here is to make the emitter part of the EM mode the "confiner." The wire helps trap the mode and the fields emitted from the wire fringe out into the mode volume This guided emission has a cross section of about λ/20 × λ/40, more than 100 times smaller than the diffraction limit. In principle, laser action is initiated by injecting electrons and holes into the GeSn wire. Another idea is that induced gain could compensate passive waveguide loss at injection less than the lasing threshold.

A fifth type of subwavelength plasmonic waveguide has been discussed by Benjamin Williams in his Ph.D. thesis on THz quantum cascade lasers [22]. He uses the "single plasmonic" CD waveguide of our Figure 2.5(left) except that he inserts an ultrathin electrical contact layer within the dielectric strip D (his Figures 4.6c and 4.10). By positioning that layer close to the top conductor, the mode height is reduced to around λ/10. Another improvement, as we have suggested here, is to reduce the lateral extent of the three strips (C, D, contact) to about $\lambda/3$, yielding a $\lambda/10 \times \lambda/3$ mode with acceptable loss at THz frequencies.

2.5.8 Surface Phonon Polariton (SPnP) Waveguides

Transverse optical (TO) phonons play a key role in the absorption-and-dispersion spectra of *semiconductors* and *insulators* in the LWIR reststrahl region. If we examine ε versus λ, we see one or two large Im[ε] absorption peaks and the Kramers-Kronig-associated Re[ε] "derivative shape." The Re[ε] spectrum includes a narrow region in which the permittivity dips down and becomes negative—a metal-like response linked to the TO phonon. We have already discussed this negative-ε property in [23], where the large contrast in real index between a reststrahl material and a transparent semiconductor slab created a "phonon-induced" 2D photonic crystal having line-defect waveguides.

We believe that the negative-ε property can be exploited to make LWIR strip-shaped channel waveguides employing a surface phonon polariton (SPnP) bound to the interface surface between the "phononic material" and a "clear" dielectric [24]. Denoting the phononic material as Pn, the principal waveguide structures are DPnD and PnDPn by analogy to plasmonics. To be definite, let us examine the DPnD ribbon waveguide, which uses a very thin Pn film. With our Group IV emphasis, D will be Si or Ge or an alloy or heterostructure. Since Si has its own phonon-related absorption features in the vicinity of 520 wavenumbers (Ge exhibits absorption around 300 wavenumbers), the choice of Si or Ge must harmonize with the transverse vibrational frequency ν_{to} of Pn; that is, the Si or Ge must be transparent and have low dispersion at ν_{to}.

The principal Pn insulators of interest here are the high-K dielectrics HfO_2 and ZrO_2 from the silicon industry, as well as SiO_2, Si_3N_4, GeO, and Al_2O_3. The main Pn semiconductors are GaP and GaAs, which lattice-match (respectively) Si and Ge. In addition, the nonmatched materials SiC, GaN, AlN, InP, and GaSb might have value.

The operation wavelength of SPnP-mode materials is ~25 μm for GaP, ~34 μm for GaAs, ~10.8 μm for 6H SiC, ~11 μm for 3C SiC, ~20.4 μm for glassy SiO_2 (second dip in Re[ε]), ~30 μm for InP, and ~12 μm for GaN. For the cases of HfO_2-on-Si and ZrO_2-on-Si, the Re[ε] has several longwave dips because the polariton utilizes TO and LO modes combined in a soft surface-optic phonon mode [25].

There are two caveats for these SPnP waveguides: (1) the spectral bandwidth of operation is only a few percent of the center wavelength, and (2) there is an IR absorption loss associated with TO that is about 200 dB/cm according to [23]. Any LWIR scattering loss associated with interface defects will be much less than the TO loss.

2.6 Chip-Scale Plasmo-Opto-Electronic Integration

How will chip-scale integration evolve in the future? Our answer is that the chip of the future will contain a "convergence" of microphotonics, microplasmonics, and microelectronics. In this section, we propose ways in which the integrated photonic-and-polaritonic chips can be built. The chip will be a vertically stacked 3D structure—a semiconductor parallelepiped that we call a *cube*. Various layers within the cube will provide bidirectional photonic-polaritonic interactions if desired while other photonic and polaritonic layers can act independently if that is wanted. The cube's materials of construction will be largely the Group IV elements, alloys, and heterostructures.

2.6.1 System in a Cube

Three technologies are interleaved in the "system-in-a-cube." We are suggesting a multispectral chip whose height is something like 50 wavelengths at the longest wavelength of operation. This cube will give higher levels of functionality and performance than those attained with any single-technology chip. The cube is really a new kind of integrated circuit that we call a three-dimensional plasmo-opto-electronic integrated circuit (3D-POEIC).

2.6.2 3D Chip Construction

The cube is an application-specific IC using layer-by-layer construction rather than a complex 3D lattice etching. There are four types of layers, as follows: (a) *photonic layer*: in-plane dielectric strip or ridge waveguides plus active and passive components that are wavelength-scale or subwavelength in size. No dielectric lattice is present; (b) *photonic-crystal-lattice layer*: contains 2D PhtC waveguides and components, both active and passive; (c) *polaritonic layer*: such as plasmonic ribbon channel waveguides, plus active and passive plasmonic components. No polaritonic lattice is present; (d) *polaritonic-crystal-lattice layer*: contains 2D PolC waveguides and components, both active and passive. Two or three technologies blended into one layer are feasible too.

2.6.3 The Cube's Layout

Figure 2.13 shows schematically the construction—one possible stacking among many. The electronics layer can be placed in several locations; for example, on the bottom as shown in Figure 2.14, where the electronic control, drive, bias, and amplification signals are routed electrically to their appropriate photonic-polaritonic destinations by means of the through-layers conductive vias illustrated schematically in Figure 2.14. The on-chip transistors (probably CMOS) provide on-chip computing and signal processing to both plasmonics and photonics, along with drive and control. The cube has five faces exposed to free space for normal or near-normal

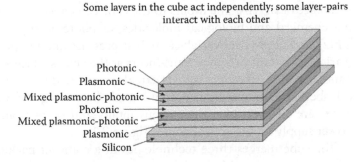

FIGURE 2.13 Layered construction of 3D chip "cube" (Plasmo Opto Electronic Integrated Circuit).

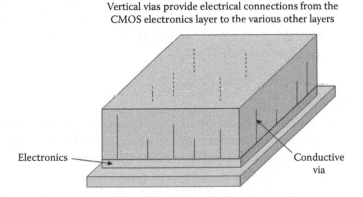

FIGURE 2.14 Through-layers-via technique for contacting and routing multiple electrical signals from electronics in POEIC to the various individual layers.

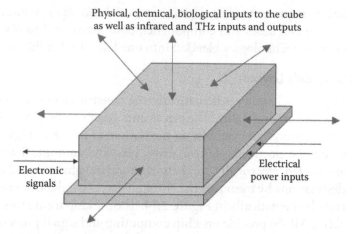

Physical, chemical, biological inputs to the cube
as well as infrared and THz inputs and outputs

Electronic
signals

Electrical
power inputs

FIGURE 2.15 POEIC cube "interacts with the world" in many ways via its five exposed surfaces, as illustrated schematically by the arrows.

incidence of signals. The inputs to the cube can consist of impinging chemical and biological molecules, of photonic signals (IR, THz), and of physical variables such as pressure and temperature (variables that may require transducers at the cube surface such as nano-MEMS). Photonic signals can also come and go from optical fibers. Electronic input/output signals (including RF, microwave) are also feasible. The various inputs-and-outputs and the power supply inputs are shown in Figure 2.15.

The cube merges three technologies in a compact package. It offers, among other things, a way to integrate multiple-wavelength input and output couplers into the exposed surfaces of the cube. The cube is self-contained unless in-cube lasers are too hot or power-hungry. Then the required laser diodes must be located off-cube in a separate "photonic power supply." Among the numerous possible cube applications, we would cite the lab-on-a-chip sensor and the multi-core CPU chip.

2.7 Conclusions

Very few silicon-based plasmonic waveguides have actually been built and tested; therefore, this plasmonics field is "wide open" for experimentalists. The main point of this chapter is that CMOS-manufacturable silicide/Group IV plasmonics are uniquely suited for applications at any infrared wavelength from 1.55 μm out to 500 μm, a huge unexploited range. This chapter presents a plan of action for researchers because the modeling and simulation in this chapter suggest that there are practical candidates for low-loss plasmonic waveguiding in Group IV materials such as Si, Ge, SiGe, SiGeSn, and their heterostructures—structures that could, and should, be realized.

The motivation for experimental development is to provide an on-chip network of waveguides and active components for plasmo-electronic integrated circuits, or more generally for plasmo-opto-electronic ICs. Specifically, we show how the three longwave-infrared technologies of microplasmonics, microphotonics, and microelectronics can converge in a 3D layered Group IV chip that we call a *cube*. This POEIC merges three highly capable techniques in a tiny package that interacts well with the external world. The cube's inputs and outputs can be physical, chemical, biological, RF, microwave, photonic, and electronic. As mentioned, an extremely wide wavelength range is feasible.

We have examined the CD, CDC, and DCD channel waveguide structures using CMOS silicide as a conductor example along with silicon dielectric strips of width $\leq \lambda$ and height T_d. The finding for all three is that attenuation of the fundamental propagating TM mode decreases beneficially with increasing wavelength, while the mode height h increases with wavelength. This implies that the mode losses that have plagued "traditional plasmonics" in the near infrared are not a problem for our plasmonics in the LWIR. In CD, $h \sim \lambda$, whereas in CDC and DCD, T_d governs h and allows subwavelength h. However, in CDC, there is the penalty of added loss for constraining $h < \lambda$ as quantified here. The DCD ribbon waveguide (built in a slab) is unique because we can get $h \sim 0.3\,\lambda$ without significant loss; L_z is dominated by the phonon-related absorption in silicon, not by plasmonic loss. The caveat here is that a suspended membrane of Si is required for DCD rather than a traditional thick silicon substrate—although unsuspended DCD built into the top layer of conventional SOI will work well in the 1 to 3 THz range because of the loss of the glassy SiO$_2$ "lower cladding" ranges from 11 to 87 dB/cm, which will not induce excessive loss in the core.

Acknowledgment

The authors wish to thank the Air Force Office of Scientific Research, Dr. Gernot Pomrenke, program manager, for continuing support of this research.

References

1. Ikonic, Z., Kelsall, R. W., and Harrison, P., Waveguide design for mid- and far-infrared p-Si/SiGe quantum cascade lasers, *Semiconductor Sci. Technol.*, **19**, 76–81, 2004.

2. Soref, R., Peale, R. E., Buchwald, W., and Cleary, J., Silicon plasmonic waveguides for the infrared and terahertz regions, paper MTuD7 presented at the *OSA Topical Conf. on Plasmonics and Metamaterials*, Rochester, New York, October 21, 2008.

3. Soref, R., Buchwald, W., and Peale, R. E., Silicon plasmonic waveguides for infrared and terahertz applications, invited paper, SPIE Photonics North, Session on Photonics Design and Simulation, Quebec City, Canada, May 25, 2009.

4. Soref, R., Towards silicon-based longwave integrated optoelectronics (LIO), invited paper 6898-5, SPIE Photonics West, Silicon Photonics III Conference, San Jose, CA, January 21, 2008, *SPIE Proceedings*, vol. 6898.

5. Moon, J. S., Son, K. A., Chattopadhay, G., and Ting, D. Z., Development of optical antenna-coupled Si CMOS-based detector at 30 THz, SPIE Defense, Security and Sensing 2009, paper 7311-11 Orlando, FL, April14, 2009.

6. Peale, R. E., Lopatiuk, O., Cleary, J., Santos, S., Henderson, J., Clark, D., Chernyak, L., Winningham, T. A., Del Barco, E., Heinrich, H., and Buchwald, W. R., Propagation of high-frequency surface plasmons on gold, *J. Opt. Soc. Am. B*, **25** (10), 1708–1713, 2008.

7. Bashevoy, M. V., Jonsson, F., Krasavin, A. V., Zheludev, N. I., Chen, Y., and Stockman, M. I., Generation of traveling surface plasmon waves by free-electron impact, *Nano Lett.*, **6**, 1113–1115, 2006.

8. van Wijngaarden, J. T., Verhagen, E., Polman, A., Ross, C. E., Lezec, H. J., and Atwater, H. A., Direct imaging of propagation and damping of near-resonance surface plasmon polaritons using cathodoluminescence spectroscopy, *Appl. Phys. Lett.*, **88**, 221111-1–221111-3, 2006.

9. Peale, R. E., Cleary, J. W., Shelton, D., Boreman, G., Soref, R., and Buchwald, W., Silicides for infrared surface plasmon resonance biosensors, *Proc. Materials Research Society*, vol. 1133 Symposium AA10-03 Fall MRS meeting, Boston, MA, December 1, 2008.

10. Brongersma, M. L., An illustration of tunnel-junction excitation of SPPs is presented at his website: http//brongersma.stanford.edu/plasmonmuri/BrongersmaGroup.

11. Tauk, R. et al., Plasma wave detection of terahertz radiation by silicon field effect transistors: Responsivity and noise equivalent power, *Appl. Phys. Lett.*, **89**, 253511, 2006.

12. Soref, R., Peale, R. E., and Buchwald, W. R., Longwave plasmonics on doped silicon and silicides, *Opt. Express*, **16**, 6507–6512, April 28, 2008.

13. Cleary, J. W., Peale, R. E., Shelton, D., Boreman, G., and Buchwald, W. R., IR permittivities for silicides and doped silicon, *J. Opt. Soc. Am. B*, **27** (4), 730–734, 2010.

14. Economou, E. N., Surface plasmons in thin films, *Phys. Rev.*, **182** (2), 539–554, 1969.

15. Berini, P., Plasmon-polariton waves guided by thin lossy metal films of finite width: Bound modes of symmetric structures, *Phys. Rev. B*, **61**, 10484, April 15, 2000-I.

16. Cho, S. -Y. and Soref, R., Low-loss silicide/silicon plasmonic ribbon waveguides for mid- and far-infrared applications, *Opt. Lett.*, **34**(12), 1759–1761, 2009.

17. Palik, E. D., *Handbook of Optical Constants of Solids*, vol. 1, Academic Press, Orlando, FL, 1985.

18. Han, Z., Van, V., Herman, W. N., and Ho, P. T., Aperture-coupled MIM plasmonic ring resonators with sub-diffraction modal volumes, *Opt. Express*, **17**, 12678–12684, July 20, 2009.

19. Yuan, G., Wang, P., and Ming, H., Multimode interference splitter based on dielectric-loaded surface plasmon polariton waveguides, *Opt. Express*, 17, 12594–12600, July 20, 2009.

20. Hoa, X. D., Kirk, A. G., and Tabrizian, M., Towards integrated and sensitive surface plasmon resonance biosensors: A review of recent progress, *Biosens. Bioelectron.*, Elsevier B.V., **23**, 151–60, 2007.

21. Oulton, R. F., Sorger, V. J., Zentgraf, T., Bartal, G., and Zhang, Z., Towards sub-wavelength plasmonic laser devices, paper JTuA3, *OSA Topical Conf. on Integrated Photonics and Nanophotonics Research and Applications*, Honolulu, HI, July 12–17, 2009.

22. Williams, B. S., Terahertz quantum cascade lasers, Ph.D. thesis submitted to the Massachusetts Institute of Technology (Dept. EECS), August 2003.

23. Soref, R., Qiang, Z., and Zhou, W., Far infrared photonic crystals operating in the Reststrahl region, *Opt. Express*, **15**, 10637–10648, August 20, 2007.

24. Schuller, J. A., Zia, R., Taubner, T., and Brongersma, M. L., Dielectric metamaterials based on electric and magnetic resonances of silicon carbide particles, *Phys. Rev. Lett.*, **90**, 107401, September 7, 2007.

25. Barker, J. R., Watling, J. R., and Ferrari, G., SO phonon scattering rates at the Si-HfO$_2$ interface in Si MOSFETs, *J. Phys.*, conference series 38, 184–187, npms-7/sImd-5, Maui, 2005.

C H A P T E R 3

Stress and Piezoelectric Tuning of Silicon's Optical Properties

Kevin K. Tsia, Sasan Fathpour, and Bahram Jalali

Contents

3.1 Introduction

It is well known that mechanical stress (or strain) can alter material properties, for example, mechanical, electrical, and optical properties. Such phenomena have been harnessed in a wide range of areas. For example, applying strain in the channel of a complementary metal-oxide-semiconductor (CMOS) silicon transistor has been widely adopted to enhance integrated circuit performance because of the strain-dependent carrier mobility [1]. Moreover, oscillations of stress, that is, acoustic waves, have

77

long been exploited to create versatile devices including radio-frequency (RF) electronic filters used in mobile cell phones and sensors for automotive, industrial, and medical applications [2].

In optics, stress modifies the material's optical properties through the photoelastic effect: a process in which the refractive index varies with the applied stress. More interestingly, acoustic waves, which result in refractive index gratings in the material, lead to an important class of photonic devices called acousto-optic devices, which have been used as optical modulators, beam deflectors, frequency shifters, and optical couplers [3].

When a material experiences an anisotropic stress, it can exhibit stress-induced birefringence, which can be employed to engineer the birefringence of the photonic devices. This is particularly relevant to the polarization-sensitive devices. In the context of silicon photonics, the stress within a silicon waveguide core induced by a silicon dioxide cladding layer has been previously used to eliminate the modal birefringence [4]. However, the stress induced by such strained dielectric cladding layers is static; in other words, it is not dynamically tunable. Direct application of mechanical stress via a load cell has been reported for dynamic tuning of the modal birefringence [5]. While this approach was useful in demonstrating the concept, the use of an external load cell is not a practical approach from the point of view of packaging size and cost.

Integration of a piezoelectric transducer into a silicon photonics platform solves this problem. We recently demonstrated a CMOS-compatible approach to achieve chip-scale dynamical tuning of silicon's optical properties based on piezoelectricity. In particular, we demonstrated electrical tuning of birefringence in silicon-on-insulator (SOI) waveguides by applying the stress induced by a thin-film piezoelectric transducer integrated on top of the waveguide [6,7]. We also showed that the birefringence induced by such a piezoelectric transducer can be used to actively control the phase matching of nonlinear optical processes—a major benchmark for determining the efficiency of the processes. Utilizing piezoelectricity to dynamically fine-tune both the linear and nonlinear optical interactions within a waveguide is of immense value in silicon photonics as it would make it possible to compensate for undesirable effects arising from the fabrication-induced uncertainties in waveguide dimensions, for example, birefringence and phase mismatch. More importantly, because of the negligible leakage current in typical on-chip piezoelectric transducers, the static power consumption of the present piezo-electric tuning technique is expected to be small, compared to other effects that are also capable of modifying the optical properties of silicon, such as the thermo-optics effect and the free-carrier effect. As a result, a particularly intriguing capability offered by such a technology is power-efficient adaptive control of the optical response using electronic intelligence.

In general, stress can modify not only the linear optical response of a material, but also the nonlinear counterparts. Stress engineering of nonlinear optical properties has especially great implications for centrosymmetric materials, such as bulk silicon. Due to its centrosymmetry, silicon crystal lacks second-order optical nonlinearity $\chi^{(2)}$, the foundation of nonlinear optics. Hence, its lowest-order nonlinearity originates from the third-order susceptibility, $\chi^{(3)}$, which gives rise to Raman and Kerr effects [8,9]. Nevertheless, silicon's crystal symmetry can be broken to create $\chi^{(2)}$ by applying mechanical stress. For instance, it has been demonstrated that silicon nitride can be used as a straining cladding layer to break the symmetry of the silicon crystal and thus create a linear electro-optic effect in silicon [10].

An appealing property of parametric $\chi^{(2)}$ processes is the possibility of achieving quasi-phase matching (QPM) by periodic poling, a technique that enhances the efficiency of nonlinear interactions. But conventional poling processes, such as those used for lithium niobate (LiNbO$_3$) and nonlinear polymers, do not apply to silicon because it lacks $\chi^{(2)}$ in its native form. We recently proposed an approach for realizing, what is in effect, *periodically poled silicon* (PePSi)—a new technology for efficient second-order nonlinear processes [11,12]. We achieve this by creating alternating stress fields along a silicon waveguide using a periodic arrangement of stress-inducing thin films. Such a structure, with an appropriate design, creates appreciable $\chi^{(2)}$ and, simultaneously, achieves QPM. It should also be noted that PePSi can also be combined with a piezoelectric stressed layer, which offers the capability to dynamically control $\chi^{(2)}$ in silicon using electronics. The technology presented here could open a new paradigm for exploring efficient $\chi^{(2)}$ processes in silicon photonics.

This chapter presents the two aforementioned stress-engineering technologies in silicon photonics: piezoelectric-tuning technique and PePSi (Figure 3.1). In the following sections, we first present the concept and implementation of integrating piezoelectric transducers with silicon optical waveguides, and

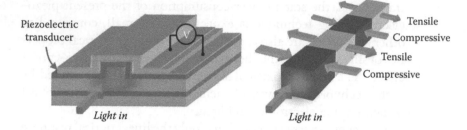

FIGURE 3.1 Schematics of (a) the piezoelectric-tuning technique in silicon waveguide and (b) periodically poled silicon (PePSi).

the dynamical birefringence tuning and active phase matching in nonlinear parametric processes in such silicon waveguides. We will then describe the concepts and designs of a PePSi waveguide and its potential applications.

3.2 Piezoelectric-Transducer-Integrated Silicon Waveguides

Piezoelectricity describes the phenomenon of electric charge being induced on the surface of a material subjected to a mechanical load. Conversely, when a voltage difference is introduced across the material, it deforms accordingly. This effect has been used to create devices such as ultrasonic transducers in ultrasound imaging, actuators used in atomic force microscopes, surface acoustic wave or bulk acoustic wave devices used in electronic filters, or oscillators for wireless mobile phones [10]. Piezoelectricity has also been incorporated in photonic devices for dynamical resonance tuning in the deformable photonic crystal microcavities [13], optical micro-electro-mechanical systems (MEMS) actuation [14], and phase modulation in silica (glass) waveguides [15]. Piezoelectricity is applied in this work to induce birefringence in standard SOI waveguides and make use of such stress-induced birefringence to dynamically fine-tune the phase mismatch of nonlinear optical processes.

3.2.1 PZT Transducer Design

Among the important piezoelectric materials, lead zirconate titanate (PZT) is chosen to integrate with silicon optical waveguides. This is because PZT is well known for its strong piezoelectric effect [9,10], and it has been widely used in MEMS technologies such as micro-actuators, sensors, and ultrasonic transducers [14]. It has also been demonstrated to be one of the promising materials for making high-density nonvolatile ferroelectric random-access memories (FRAM) integrated in CMOS logic circuitry [16]. Such CMOS compatibility is one of the motivations for employing PZT in the present work.

The device structure consists of an oxide-clad silicon waveguide and a PZT capacitor formed on top of it. The PZT capacitor consists of a thin-film PZT layer sandwiched between a top and a bottom platinum (Pt) electrode. Pt is one of the common electrode materials for fabricating PZT capacitors because of its relatively high electrical conductivity, good stability at high temperature, and oxidizing ambient, which is inevitable in the subsequent annealing processes of the PZT and Pt films [17]. An oxide overcladding is required to minimize the optical absorption loss resulting from the bottom electrode. Based on the full-vectorial beam propagation numerical simulation (by BeamProp), it was found that the metal electrode causes insignificant optical loss, as low as 0.01dB/cm.

The stress distribution in the waveguide is simulated by using a finite-element analysis package (ANSYS) that is capable of incorporating structural, material, electromagnetic, and piezoelectric analyses. We have designed an oxide-clad (500 nm thick) single-mode waveguide with a width of 2 μm, a rib height of 2 μm, and a slab height of 1.1 μm. The top and bottom Pt electrodes have thicknesses of 100 nm and the PZT layer is 500 nm thick. Figures 3.2a and 3.2b show the piezoelectric-induced stress distributions in the x-direction ($\sigma_{\text{piezo},X}$) and the y-direction ($\sigma_{\text{piezo},Y}$), respectively, within the structure when a voltage of 12 V is applied across the PZT capacitor. It should be emphasized that the shown stresses are only the tunable component arising from the applied field to the PZT layer. In reality, there exists an additional stress induced by the strained cladding layers, which could have higher values [4,10]. However, this residual stress has a constant value (as discussed in a later section); thus, it is not relevant to the scope of this work, that is, dynamical stress tuning. The average values of $\sigma_{\text{piezo},X}$ and $\sigma_{\text{piezo},Y}$ within the waveguide core area are calculated to be –1 MPa and +12 MPa, respectively. The anisotropic stress induced by the piezoelectric effect in PZT breaks the centrosymmetry of the silicon crystal and leads to birefringence in silicon, as governed by the photoelastic effect.

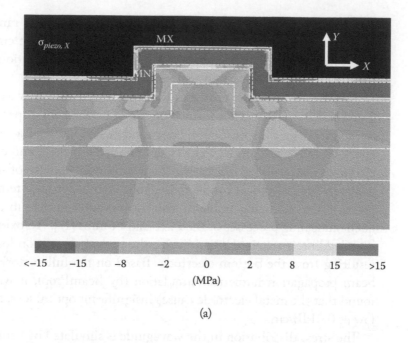

(a)

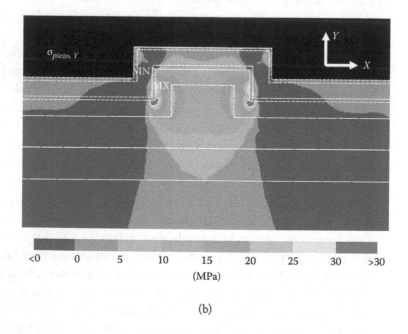

(b)

FIGURE 3.2 Stress distribution in the (a) horizontal direction ($\sigma_{piezo,X}$) and (b) vertical direction ($\sigma_{piezo,Y}$) in the waveguide structure, when a voltage of 12 V is applied across the PZT capacitor. (From K. K. Tsia et al., *Appl. Phys. Lett.* 061109, 2008.)

3.2.2 Fabrication and Characterization of PZT-Integrated SOI Waveguides

We have developed a CMOS-compatible process for fabricating SOI waveguides integrated with PZT capacitors. The silicon waveguide is patterned by standard photolithography and a plasma dry-etch process. An oxide cladding layer (500 nm) is then deposited by plasma-enhanced chemical vapor deposition (PECVD). On top of the oxide cladding, a 100 nm/10 nm platinum/titanium (Pt/Ti) bilayer (the bottom electrode) is deposited by electron beam evaporation with a subsequent rapid thermal annealing (RTA) at 700°C for 30 s in oxygen ambient. A thin Ti buffer layer is also introduced here because it improves the adhesion between the oxide and the Pt film and prevents the formation of Pt silicide [17].

We adopted the RF magnetron sputtering technique to deposit a thin PZT film [16]. PZT sputtering is performed using a ceramic PZT (Zr/Ti = 52/48) target at room temperature [18]. PZT with such a compositional Zr/Ti ratio is well known to exhibit the strongest piezoelectric effect [18]. The 500-nm-thick sputtered PZT film is then crystallized by RTA at 650°C for 60 s in oxygen ambient to achieve the peroskite structure—a prerequisite of exhibiting piezoelectric effect. The sputtered PZT film thickness deviates less than 50 nm along a 2-cm-long waveguide as measured from scanning electron microscope (SEM) images. The right composition of the deposited PZT ($Pb(Zr_{0.52},Ti_{0.48})O_3$) was confirmed with the Rutherford backscattering technique. The x-ray diffraction (XRD) pattern of the annealed PZT film was also studied. Typical peaks in the (111), (100), (110), and (200) directions are observed with a preferential orientation along the (111) direction. A 100 nm/10 nm Pt/Ti top electrode layer is finally deposited on top of the PZT layer by electron-beam evaporation. The PZT capacitor was then patterned by Cl_2-based plasma dry etching [19]. Figure 3.3 shows the images of the fabricated device structure at a cleaved facet.

We characterized the PZT film quality by measuring its ferroelectric, piezoelectric properties, and leakage current. A ferroelectric hysteresis loop of the PZT film is measured by a Sawyer-Tower circuit [20] at 1 kHz. The remnant polarization is 20 μC/cm², and the coercive field is ~55 kV/cm (Figure 3.4a). On the other hand, the PZT film has a piezoelectric coefficient d_{33} ~ 130 pC/N measured by the normal load technique (Figure 3.4b)

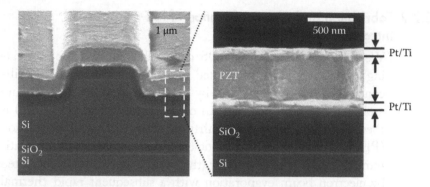

FIGURE 3.3 (a) Scanning electron microscope image of the SOI waveguide with the PZT capacitor on top of it. (b) The enlarged cross section of the PZT capacitor. The waveguide has a width of 1.5 μm, rib height of 2 μm, and slab height of 1.1 μm. The oxide cladding is 500 nm thick. The PZT thickness is 500 nm. Both top and bottom Pt/Ti electrodes are 100 nm/10 nm thick. (From K. K. Tsia et al., *Opt. Express* 9838–9843, 2008.)

[21]. The leakage current density of the PZT film is kept as low as $\sim$$10^{-7}$–$10^{-5}$ A/cm^2 before its breakdown voltage of $\sim$10 V (Figure 3.4c). The leakage current can be improved through further optimization of the PZT annealing process conditions [17]. All the above-measured values are comparable with previously reported values of sputtered PZT thin films [14,16].

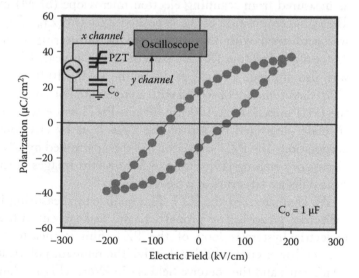

FIGURE 3.4 (a) Ferroelectric hysteresis loop of the PZT capacitor measured by a Sawyer-Tower circuit at 1 kHz. The inset shows the measurement schematic. (*continued*)

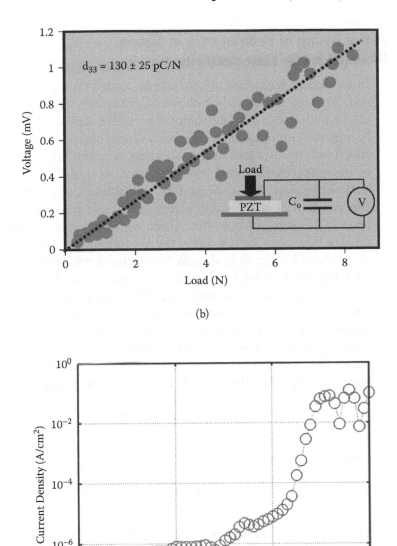

FIGURE 3.4 (CONTINUED) (b) Measured piezoelectric effect of the PZT capacitor by direct normal load technique. (c) Measured leakage current of the PZT capacitor as a function of applied voltage.

3.3 Active Tuning of Birefringence in Silicon Waveguides by Piezoelectricity

In an optical waveguide, birefringence leads to a relative phase shift between orthogonally polarized light beams, a property that can be undesirable for polarization-insensitive applications or can be exploited for phase matching in nonlinear optical interactions, which will be discussed in the following sections. Birefringence, or more specifically modal birefringence, in a PZT-integrated SOI waveguide, Δn, is defined as the difference between the effective indices of transverse electric (TE) and transverse magnetic (TM) waveguide modes. Three components contribute to Δn. The first component is due to the asymmetric geometry of the rib waveguide. The second term is associated with the material birefringence induced by the residual stress of the cladding layers (oxide, electrodes, and unbiased PZT) that cover the rib. The birefringence caused by these two components is not tunable and can be expressed as a constant Δn_o. The third component contributing to Δn is the tunable material birefringence, Δn_{piezo}, due to the tunable stress induced by an applied electric filed to the PZT layer. Consequently, the modal birefringence can be expressed as

$$\Delta n = n_{TE} - n_{TM} = \Delta n_o + \Delta n_{piezo} \qquad (3.1)$$

where

$$\Delta n_{piezo} = (C_2 - C_1)(\sigma_{piezo,X} - \sigma_{piezo,Y}) \qquad (3.2)$$

where C_1 and C_2 are the stress-optic coefficients related to Young's modulus, Poisson's ratio, and the photoelastic tensor elements of the material [4]. Based on the finite-element numerical simulation results (Figure 3.2), the stress-induced birefringence caused by the piezoelectric effect is estimated to be $\Delta n_{\text{piezo}} \approx 3 \times 10^{-4}$, at a DC bias voltage of 12 V. As discussed later, this value is large enough to be useful in applications requiring dynamic tuning of birefringence.

We use the Fabry–Pérot (FP) resonance technique to measure the waveguide loss and, more importantly, the birefringence induced in the waveguide by piezoelectricity. The input light from an external cavity tunable laser is coupled into and out of the 2-cm-long PZT-integrated SOI waveguide by two identical objective lenses (numerical aperture (NA) = 0.4). The polarization state is controlled by a polarizer positioned before the input of the waveguide. The silicon waveguide integrated with the PZT capacitor exhibits a low linear propagation loss of 0.6 dB/cm.

The group indices $n_{g,TE}$ and $n_{g,TM}$ (for TE and TM modes, respectively) were measured from the free spectral range of FP resonances caused by reflection from the waveguide facets ($n_{g,TE} \approx n_{g,TM} = 3.63 \pm 0.01$). Taking the waveguide dispersion into account ($dn/d\lambda \sim 0.1\ \mu m^{-1}$ from BeamProp simulations), the effective refractive indices for TE and TM modes were extracted to be $n_{o,TE} \approx n_{o,TM} = 3.46 \pm 0.01$. We note that these values are consistent with the simulation results.

The wavelength shift for TE ($\Delta\lambda_{TE}$) and TM ($\Delta\lambda_{TM}$) modes in the FP spectra were measured when a DC bias was applied to the PZT capacitor. By using the relations $\Delta\lambda_{TE}/\Delta\lambda_{TE} = \Delta n_{TE}/n_{o,TE}$ and $\Delta\lambda_{TM}/\lambda_{TM} = \Delta n_{TM}/n_{o,TM}$, we can extract the stress-induced birefringence due to the piezoelectric effect, that is, $\Delta n_{piezo} = \Delta n_{TE} - \Delta n_{TM}$. Figure 3.5 illustrates this measured birefringence as a function of the DC voltage, showing a linear behavior. Varying the PZT voltage from –10 to 5 V, the overall birefringence tuning range was measured to be $\Delta n_{piezo} = 1 \times 10^{-5}$. This is lower than the simulated value of 4×10^{-4} from ANSYS simulations. The discrepancy can be attributed to the fact that the PZT piezoelectric coefficients used in the simulation are taken from the bulk values, which are different from the thin-film values [14]. Also, the

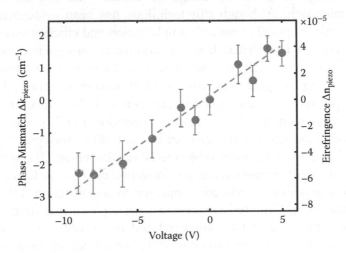

FIGURE 3.5 Phase mismatch (left axis) and the corresponding birefringence (right axis) of the SOI waveguide due to the piezoelectric effect from PZT as a function of DC voltages. (From K. K. Tsia et al., *Opt. Express* 9838–9843, 2008.)

nonconformality of PZT and Pt/Ti films over the rib waveguide may have caused further complications. The leakage current of the PZT capacitor in the present devices prevents applying voltages of above 10 V. Further optimization of the structural design resulting in larger difference between $\sigma_{piezo,X}$ and $\sigma_{piezo,Y}$ can enhance the birefringence tuning range.

3.4 Active Control of Parametric Processes in Silicon Waveguides by Piezoelectricity

Strong third-order optical nonlinearity in silicon has led to a wide range of functionalities that were believed to be absent in silicon [8,9]. Among them are nonlinear parametric processes, such as wavelength conversion, parametric amplification, and oscillation. Such processes are based on either four-wave mixing (FWM) using the Kerr effect, or coherent anti-Stokes Raman scattering (CARS). The latter is a form of FWM mediated by the Raman effect. For successful operation, these processes require dispersion engineering because dispersion-induced phase mismatch lowers the efficiency of power transfer between the interacting waves.

A number of techniques for achieving phase matching in silicon waveguides have been demonstrated. One approach relies on engineering the group-velocity dispersion (GVD) by proper waveguide geometry design to balance the nonlinear phase mismatch. Although this technique has been demonstrated to achieve optical parametric amplification and efficient wavelength conversion, it demands precise control of waveguide dimensions (<50 nm) [22,23] in order to achieve the prerequisite dispersion, placing stringent requirements on fabrication tolerances. Another approach is to use modal birefringence of the interacting waves to compensate the chromatic dispersion [24,25]. Controlling birefringence can be achieved by controlling waveguide dimensions, which is, again, subject to fabrication errors. The dynamical tuning of stress-induced birefringence by a piezoelectric transducer is thus an attractive approach to actively control the phase matching. This technology can ensure phase matching even in the presence of fabrication-induced uncertainties in waveguide dimensions. It can also be used within a feedback loop to maintain phase matching in the face of environment-induced variations. We note that varying the stress in the waveguide alters the dispersion for cross-polarized beams. Hence, this technology can be useful for GVD engineering, as well. Here, we present such an active phase-matching technique in the context of wavelength conversion based on CARS.

Efficient CARS wavelength conversion in silicon waveguides requires phase matching among Stokes, anti-Stokes, and pump waves, which span ~200 nm in wavelength at near-infrared wavelengths [25]. The phase matching condition for the CARS process is

$$k_A^j = 2k_P^i - k_S^j \qquad (3.3)$$

where k_P, k_S, and k_A are the propagation constant at pump, Stokes, and anti-Stokes wavelengths, respectively. The superscripts i and j refer to TE or TM polarizations. Deviations from phase matching can be expressed as $\Delta k = 2\Delta k_B + \Delta k_{WG} + \Delta k_{MAT}$, where Δk_B is the phase mismatch due to modal birefringence at pump wavelength, Δk_{WG} is due to the chromatic dispersion associated with wavelength dependence of modal area, and Δk_{MAT} is due to material dispersion. The waveguide considered here operates in the normal dispersion regime. While the phase mismatch due to waveguide and material dispersion, that is, $\Delta k_{WG} + \Delta k_{MAT}$, is on the order of –100 cm^{-1} [24,25], birefringence phase mismatch, Δk_B, can be used to compensate Δk_{WG} and Δk_{MAT} and, consequently, the phase matching condition can be attained. It should be noted that we neglect the nonlinear phase mismatch in Equation 3.3 as it has a negligible effect on the overall phase-matching condition for the present waveguide design [25].

Two main components contribute to Δk_B. The first component is the static birefringence, Δn_o, due to the asymmetric geometry of the rib waveguide and the material birefringence induced by the residual stress of the cladding layers that cover the rib. The second component is the tunable birefringence due to the stress induced by the PZT, Δn_{piezo} (see Equation 3.2). As such, Δk_B can be expressed as

$$\Delta k_B = \frac{2\pi}{\lambda_p}(n_{TE} - n_{TM}) = \frac{2\pi}{\lambda_p}(\Delta n_0 + \Delta n_{piezo}) = \Delta k_{B0} + \Delta k_{piezo} \qquad (3.4)$$

where $\Delta k_{Bo} = 2\pi\Delta n_o/\lambda_p$ and $\Delta k_{piezo} = 2\pi\Delta n_{piezo}/\lambda_p$ are phase mismatches corresponding to the static birefringence and the tunable birefringence, respectively, and λ_p is the pump wavelength.

It should be emphasized that the role of Δk_{piezo} is to provide an additional degree of freedom to fine-tune the phase mismatch arising from fabrication-induced errors, such as uncertainties in waveguide dimension and the stress of the cladding layers. Based on the previously described FP measurement (with measured $\Delta n_{piezo} \approx 1 \times 10^{-4}$ (see Figure 3.5)), the overall piezoelectric-induced phase-mismatch tuning range is measured to be 4 cm⁻¹ (Figure 3.5). The sensitivity of birefringence on waveguide dimensions was quantified in [24]. Using the results of that work, our measured birefringence tuning value of $\Delta n_{piezo} \approx 1 \times 10^{-4}$ is sufficient for compensating the phase mismatch caused by up to ~50 nm in fabrication-induced errors in the waveguide dimension.

In the CARS wavelength conversion experiment, the pump beam is from an external cavity diode laser, amplified by an erbium-doped fiber amplifier at 1538 nm (TE polarized). The Stokes signal is from a distributed feedback (DFB) fiber laser at 1673 nm (TM polarized). The pump and Stokes are combined by a wavelength division multiplexer (WDM) and coupled into and out of the waveguide by two identical objective lenses (NA = 0.40). The polarization states of the pump and Stokes signals are controlled by polarization controllers before the WDM. The output light from the waveguide is sent to an optical spectrum analyzer (OSA). The polarization state of the converted anti-Stokes signal at 1424 nm (TM-polarized) is verified by an analyzer before the OSA.

Figures 3.6a and 3.6b show the measured anti-Stokes spectra at different bias voltages applied across the PZT capacitor, under two different input polarization conditions. The coupled input pump power is ~1.2 W. Conversion efficiency is defined as the ratio of the Stokes signal to the anti-Stokes signal at the waveguide output [22,23]. We note that both Stokes and anti-Stokes signals suffer from free-carrier loss caused by the pump. By comparing the Stokes power at the output for on (1.2 W) and off pumps, we measure an on–off loss of ~1 dB for the Stokes signal.

By varying the DC bias applied to the PZT capacitor from 0 to 5 V, we attain an improvement of ~6 dB in conversion efficiency (Figure 3.6a). An enhancement in efficiency only occurs when the pump and the Stokes are cross-polarized (Figure 3.6a) as compared with the copolarized case (Figure 3.6b). This verifies that the effect originates from the birefringence induced by the PZT capacitor. The wide 3 dB bandwidth of 0.2–0.3 nm of the con-

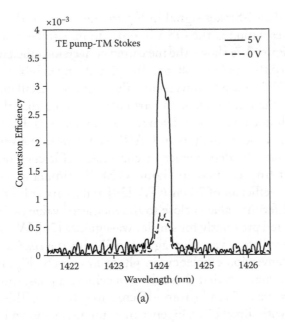

(a)

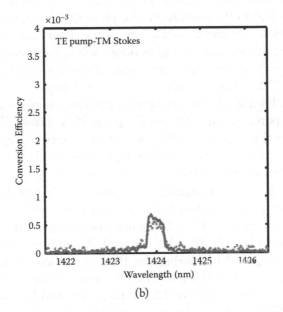

(b)

FIGURE 3.6 Anti-Stokes spectra (TM polarized) of wavelength conversion based on Coherent anti-Stokes Raman scattering at voltage biases of 0 (red) and 5 V (blue) applied to the PZT capacitor. Two input conditions are studied: (a) TE-polarized pump and TM-polarized Stokes; (b) TE-polarized pump and TE-polarized Stokes. The coupled input pump power is ~1.2 W. (From K. K. Tsia et al., *Opt. Express* 9838–9843, 2008.)

verted anti-Stokes signal in Figure 3.6 is due to the intentionally low resolution at the OSA in order to maximize its sensitivity.

Figure 3.7a shows the measured conversion efficiency as a function of the applied voltage. The plot clearly exhibits the hallmark of FWM-based conversion efficiency: an oscillatory variation with phase mismatch. In particular, we note that the oscillation depth in conversion efficiency is as high as ~20 dB. Based on the coupled-mode analysis of CARS wavelength conversion [5], we calculate the dependence of conversion efficiency on phase mismatch (the red curve in Figure 3.7b). The simulation uses a Raman gain coefficient of 20 cm/GW [24] and a carrier lifetime of 20 ns. The lifetime value is close to the measured value of 15 ns we have reported previously for similar waveguides [26]. We observed that the oscillatory variation agrees well with the $sinc^2$ dependence of the conversion efficiency on phase mismatch (Figure 3.7b). The figure suggests that these devices achieve a phase-mismatch tuning range of 5 cm^{-1} (from –15 cm^{-1} to –10 cm^{-1}). This value agrees well with direct birefringence measurements shown in Figure 3.4. The tunable range reported here is limited by the bias range of 15 V applied to the transducer, the piezoelectric coefficient of PZT, as well as the stress compliance of the silicon core in the present waveguide structure. Higher biases and optimizing the structural design of the waveguide by a finite-element analysis package such as ANSYS [6] can result in higher tunability.

In Figure 3.8, the conversion efficiency is plotted versus coupled pump power for two different PZT biases. The plot also shows the conversion efficiency of an air-clad waveguide, that is, a bare SOI waveguide before depositing the cladding layers. The air-clad waveguide has the lowest efficiency, as it is farthest away from the phase-matching condition (gray triangles). The residual stresses of the cladding layers (i.e., oxide, electrodes, and unbiased PZT) introduce a birefringence component that improves the efficiency by 7–8 dB (blue squares). An additional tunable stress can be applied by biasing the PZT, by which a further 5–6 dB enhancement of efficiency can be achieved (red circles). Hence, a total improvement of 12–14 dB is obtained by the addition of the PZT capacitor. The calculations based on the aforementioned model also match with experimental results at 0 and 5 V (blue and red curves), which correspond to phase mismatches of –12 cm^{-1} and –10.5 cm^{-1}, respectively.

It is evident from Figure 3.7b that the dimensions of the present waveguide render it far from the phase-matching condition

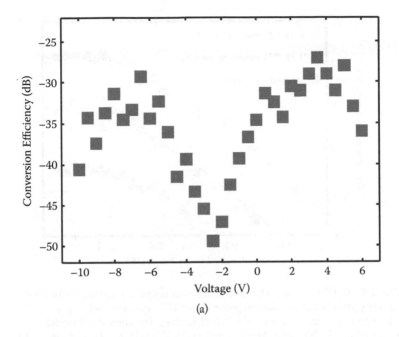

(a)

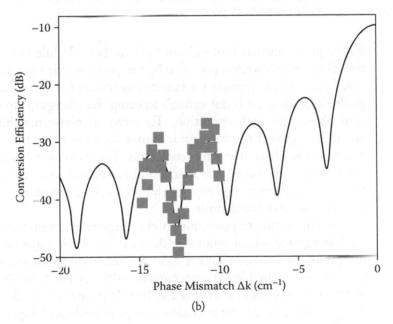

(b)

FIGURE 3.7 (a) Measured CARS wavelength conversion efficiency versus PZT voltages at coupled pump power of ~1.2 W. (b) Calculated dependence of CARS conversion efficiency on phase mismatch (red curve). The measured data (blue square dots) are fit with the model. (From K. K. Tsia et al., *Opt. Express* 9838–9843, 2008.)

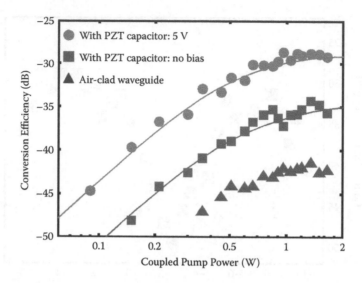

FIGURE 3.8 CARS wavelength conversion versus coupled pump power for air-clad SOI waveguide (grey triangles), the waveguide with PZT capacitor cladding without voltage bias (blue squares), and at a bias of 5 V (red circles). The theoretical model agrees well with the case of 0 V (blue line: $\Delta k = 12$ cm^{-1}) and 5 V (red line: $\Delta k = 10.5$ cm^{-1}). (From K. K. Tsia et al. *Opt. Express* 9838–9843, 2008.)

(to the phase mismatch of ~12 cm^{-1} at zero bias). While the tunable phase match control provided by the piezoelectric transducer is sufficient to compensate for fabrication-induced error in waveguide dimensions, it is not enough to bring the waveguide to the zero phase-mismatch condition. To arrive at phase matching, one must engineer the waveguide dimensions such that they are closer to the phase matching condition. The piezoelectric technology can then correct the fabrication errors and also maintain the phase-matching condition in the presence of temperature or acoustic-induced fluctuations.

It is worthwhile to point out another important feature of the PZT-integrated silicon photonic devices: low static power consumption. As shown in Figure 3.4c, the integrated PZT capacitor exhibits negligible leakage current (~10^{-7} A/cm^2), resulting in a power consumption as small as 10 nW in our current device design. Hence, it can be regarded as a power-efficient approach to achieve the aforementioned active tuning of silicon's optical properties, which is of great importance for integrated optoelectronic devices. One example is wavelength-agile devices used in dense wavelength division multiplexing (DWDM) telecommunications. In these devices, the stability of the operation wavelength plays a pivotal role and continuous active trimming of wavelength is thus essential. While the common active trimming approach is based on the thermo-optics effect, it however suffers considerable power consumption due to the ohmic heat dissipation of the thermo-optics devices. The piezo-induced refractive index change of 10^{-4}–10^{-3} is comparable to that achieved by the thermo-optic devices. Meanwhile, PZT-integrated silicon photonic devices have much lower power consumption, and hence could represent an attractive candidate for realizing green integrated photonics.

3.5 Periodically Poled Silicon (PePSi)

3.5.1 Concepts and Designs

Bulk silicon has a centrosymmetric crystal structure that forbids second-order $\chi^{(2)}$ optical nonlinearity. This property limits silicon's range of applications in nonlinear optics. As a result of this fundamental limitation, nonlinear processes in silicon photonics have been based on the third-order $\chi^{(3)}$ susceptibility such as the Raman and Kerr effects.

However, the disruption of crystal symmetry at the surface can enable $\chi^{(2)}$ processes such as second-harmonic generation (SHG) [27,28]. In addition, the crystal symmetry can also be perturbed by applying inhomogeneous mechanical stress (strain gradient) on the crystal [10]. Recently, strained silicon has been used to demonstrate linear electro-optic modulators [10]. The stress can be induced by depositing silicon nitride on top of waveguides, and a $\chi^{(2)}$ value of ~15 pm/V has been measured with 1 GPa stress. Motivated by these large values of $\chi^{(2)}$, we have recently proposed a new concept: *periodically poled silicon* (PePSi) [11,12]. PePSi creates alternating stress gradients along a silicon waveguide using a periodic arrangement of strained films. The structure creates appreciable $\chi^{(2)}$ and simultaneously achieves quasi-phase matching (QPM). Based on $\chi^{(2)}$ values reported in [10], our simulations have shown efficient midwave infrared (MWIR) generation through the quasi-phase-matched difference frequency generation process (QPM-DFG). The PePSi concept is meant to broaden the capabilities of silicon as a nonlinear optical medium.

The PePSi structure considered here is a channel waveguide integrated with two types of silicon nitride (SiN) stressed films: one with tensile stress and another with compressive

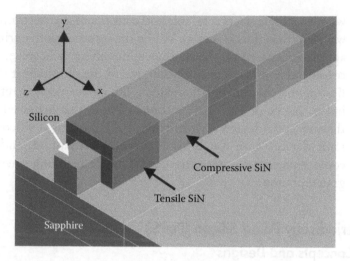

FIGURE 3.9 An example of the PePSi waveguide formed by covering a silicon channel waveguide with two types of silicon nitride (SiN) stressed films periodically along the waveguide: one induces compressive stress, whereas the other produces tensile stress. (From N. K. Hon et al., *Appl. Phys. Lett.*, 091116, 2009.)

stress, periodically deposited along the waveguide, as shown in Figure 3.9. Hence, the sign of stress induced within the silicon core alternates along the waveguide and results in alternating dipole regions: a new type of periodically poled structure.

SiN was chosen because it achieves high film stress [29] and is compatible with CMOS processing. In addition, the stress in SiN films can be readily tailored from compressive to tensile stress depending on the deposition conditions [29]. It thus permits flexible engineering of stress, and hence $\chi^{(2)}$, in silicon. In practice, the design shown in Figure 3.9 can be realized by two different SiN deposition steps, giving rise to alternating stresses along the waveguide. For MWIR applications, SOI is not the desirable platform, because of the high losses of silicon dioxide at MWIR wavelengths. Instead, silicon-on-sapphire (SOS) is employed here as sapphire is transparent to MWIR.

The PePSi waveguide we consider has the silicon core dimensions of 2×2 µm, which supports optical waveguiding in both MWIR and near-infrared (NIR) regimes. We assume the in-plane stresses in the two different SiN films (with 1 µm thickness) to be +1 GPa (*tensile*) and –500 MPa (*compressive*) [29]. The SiN stressed film period is designed to be 8 µm in order to phase-match the interacting waves in the DFG process for MWIR generation (discussed in the next section). To estimate the stress-induced $\chi^{(2)}$

in the PePSi waveguide, we simulated the stress distribution in the waveguide by ANSYS. In principle, all the stress components (i.e., the normal and shear stresses in all directions) should be considered for evaluating the stress-induced $\chi^{(2)}$ because it is the anisotropy of the stress resulting in breaking the original crystal symmetry. Nevertheless, it was found that, compared to all other stress components, the y-component normal stress (σ_{yy}) gives rise to a highly uniform stress distribution in the silicon core with a considerably higher average stress magnitude in the present PePSi waveguide design. Hence, it is conceivable that stress-induced $\chi^{(2)}$ here is dominated by σ_{yy}. Simulated cross-sectional σ_{yy} distributions in the tensile and compressive SiN cladding regions are shown in Figures 3.10a and 3.10b, respectively. In one half of the period, when the SiN film exhibits tensile stress that deforms the waveguide, a compressive stress field is induced and confined inside the waveguide core (Figure 3.10a) in order to counteract the deformation under elastic equilibrium. Conversely, in another half of the period, the compressive SiN cladding induces tensile stress within the silicon core (Figure 3.10b). In addition, the silicon core, covered by the conformal SiN stressed cladding, displays a stress distribution, and hence the stress-induced $\chi^{(2)}$ distribution, with good uniformity (Figure 3.10a and 3.10b). This feature is important to ensure efficient $\chi^{(2)}$ interaction of the optical modes within the waveguide. The idea of periodic poling becomes more appealing when we observe the average stress (σ_{yy}) along the waveguide. As illustrated in Figure 3.10c, the periodic oscillation with –500 MPa peak average compressive stress and +200 MPa peak average tensile stress is evident.

Based on a classical anharmonic oscillator model [30], the stress-induced $\chi^{(2)}$ can be estimated as $\chi^{(2)} = 4q^3 S / (m^2 \varepsilon \omega^4 a^4)$, where q is the electron charge, m is the electron mass, ε is the dielectric permittivity, a is the lattice constant, ω is the angular frequency of light, and S is the strain, which is related to the stress by Young's modulus of the material and an element of the stress tensor. Thus, based on the simulated stress values, the induced $\chi^{(2)}$ is estimated to have oscillatory values from –15 pm/V to +6 pm/V in a period of 8 μm (Figure 3.10c). This oscillatory behavior is the desired property for QPM.

The estimated $\chi^{(2)}$ agrees qualitatively with the prior work for the same order of magnitude of stress [10]. It should be noted that the foregoing formula is meant as a heuristic tool, and should not be interpreted rigorously. A full calculation of the nonlinearity

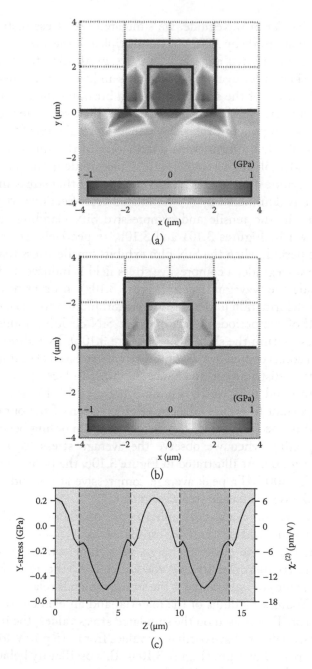

FIGURE 3.10 (a) Simulated cross-sectional *y*-component normal stress (σ_{yy}) distribution in which the PePSi waveguide is covered by silicon nitride films with 1 GPa (*tensile*) stress and (b) –500 MPa (*compressive*) stress. (c) Simulated average σ_{yy} and estimated second-order susceptibility ($\chi(2)$) induced along the waveguide (*z*-direction). (From N. K. Hon et al., *Appl. Phys. Lett.*, 091116, 2009.)

induced in silicon by applied stress remains an open question. The values measured by Jacobsen et al. [10] are the fundamental justification for the value of the nonlinear parameter used in our simulations. In fact, it was recently found that there appears to be a difference between the experimental [10,28] and theoretical [27] values of $\chi^{(2)}$ reported in the literature for strained silicon, with theoretical values (based on a two-dimensional Coulomb's law model [31] and a three-dimensional bond additivity model [27]) being much lower than their experimental counterparts [10,28]. The apparent disagreement suggests that there may be important missing physics in the existing theoretical calculations. Interestingly, the $\chi^{(2)}$ value obtained using a formula derived from a heuristic model based on the foregoing classical anharmonic oscillator [30] agrees with the measurement of Jacobson et al. In this simple formula, $\chi^{(2)}$ is proportional to the total stress (integrated stress gradient) rather than the stress gradient itself. This may provide a clue regarding the present gap between theory and experiments.

3.5.2 MWIR Generation in PePSi

The PePSi technology introduced here allows for efficient silicon photonic devices based on different $\chi^{(2)}$ processes. As an example, we consider the generation of MWIR radiation in the PePSi waveguide from two NIR beams using DFG. Such a capability has many applications, including remote sensing of chemical and biological agents and environmental monitoring [6]. We note that MWIR generation in silicon can also be implemented by first-order stimulated Raman scattering (SRS) pumped at shorter MWIR wavelengths [32], or by cascaded SRS pumped at NIR wavelengths [33,34]. In contrast, the present MWIR generation approach using PePSi is a single-step conversion from NIR via difference frequency generation (DFG). This is more favorable in terms of the wide availability of the NIR pump sources and also more efficient in terms of circumventing the cumulative two-photon absorption (TPA) and associated free-carrier absorption (FCA) that occurs in the cascaded SRS approach.

We numerically investigated the QPM-DFG process in the PePSi waveguide using the nonlinear Schrödinger equation (NLSE), which incorporates (1) stress-induced $\chi^{(2)}$ effects and (2) $\chi^{(3)}$ effects (Kerr effect, TPA). We simultaneously calculate the free-carrier concentration resulting from TPA and utilize known empirical relations, proposed by Soref and Bennet, to determine the associated FCA and free-carrier refraction (FCR) [35]. Taking the waveguide dispersion into account, the 8-μm-period of the SiN film pattern is designed to satisfy the QPM condition in the DFG process: pump at 1.3 μm, signal at 1.75 μm, and idler at 5.1 μm. We consider the fundamental TE-polarized modes of the three interacting waves. Although the waveguide is multimode at the pump and signal wavelengths, the higher-order modes are expected to have no significant effect on the QPM-DFG process if the pump and signal propagate predominantly in the fundamental modes, because higher-order modes have different mode profiles and dispersion properties, which lead to different phase-matching conditions.

In the model, we input transform-limited pump and signal pulses (both with a pulse width of 12 ps), which have peak intensities of 1.5 GW/cm^2 and 12.5 MW/cm^2, respectively, into a 2-cm-long PePSi waveguide. The peak intensities are chosen to obtain the highest conversion efficiency. Through QPM-DFG, the MWIR idler pulse can be efficiently generated at 5.1 μm at the waveguide output (Figure 3.11a). During the process, SHG and sum frequency generation (SFG), albeit without phase matching, can generate photons with energies above the silicon bandgap. This inevitably causes single-photon absorption (SPA), which introduces additional FCA and thus decreases the DFG efficiency. We incorporated both SHG and SFG in the model, but found that the efficiencies of both processes are negligibly low along the waveguide because both second-harmonic and sum-frequency waves are highly attenuated by SPA in the early stage of propagation. On the other hand, since the second-harmonic and sum-frequency waves remain weak throughout the waveguide, SPA is insignificant. We find that the average free-carrier concentration along the waveguide generated by SPA (due to SHG and SFG) is ~10^{12} cm^{-3}, whereas that generated by TPA is ~10^{15} cm^{-3}. Hence, the contribution of SPA to the overall FCA is still overwhelmed by that of TPA. We also remark that even though the idler wave suffers from larger FCA than the pump and signal waves due to the wavelength dependence of FCA, only the trailing edge of the idler pulse is suppressed by the FCA since more free carriers are generated toward the trailing edge [36]. As a result, high peak conversion efficiency can still be obtained.

As depicted in Figure 3.11b, which shows the peak conversion efficiency as a function of the peak pump intensity, the PePSi

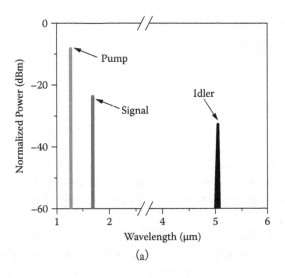

(a)

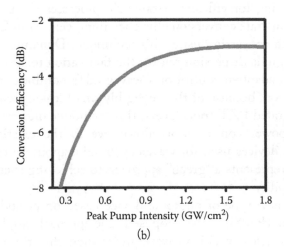

(b)

FIGURE 3.11 (a) Calculated output spectra in a 2-cm-long PePSi waveguide (period ~ 8 μm): a 12 ps pump pulse at 1.3 μm (input peak intensity = 1.5 GW/cm²), a 12 ps signal pulse at 1.75 μm (input peak intensity = 12.5 MW/cm²), idler at 5.1 μm. (b) Calculated conversion efficiency of MWIR generation. (From N. K. Hon et al., *Appl. Phys. Lett.*, 091116, 2009.)

waveguide is able to achieve a maximum conversion efficiency of the MWIR generation by QPM-DFG as high as ~–3 dB (~50%) under a peak pump intensity of 1.5 GW/cm². This intensity level is readily achievable under experimental conditions. The slight decrease of conversion efficiency at high pump intensities observed in Figure 3.11b is primarily due to TPA and FCA in the pump, signal, and idler waves, as well as FCR, which introduces an additional phase mismatch in the DFG process.

3.6 Conclusions

In summary, stress engineering is of immense value in silicon photonics as it makes it possible to alter both the linear and nonlinear optical properties of silicon. In particular, two new types of stress-engineered silicon photonic devices, piezoelectric-transducer-integrated silicon waveguides and PePSi, are introduced.

Using the integrated PZT transducers, we are able to achieve dynamic tuning of a silicon waveguide's birefringence. This capability has many promising applications requiring phase matching for efficient parametric processes. In particular, we demonstrate conversion efficiency improvement in CARS wavelength conversion using this technique. Dynamic tunability of waveguide dispersion relaxes the fabrication tolerances and also offers adaptive control of a waveguide's nonlinear response. In addition, because of the negligibly small leakage current in the integrated PZT transducers, this technique offers a much more low-power consumption alternative to the existing thermo-optic devices used for wavelength-agile applications. Hence, it also represents a "green" approach to achieving such dynamical tunability.

A new type of photonic device based on periodically poled silicon (PePSi) is also proposed. The approach employs periodic stress fields in silicon waveguides such that the crystal symmetry of silicon is broken in a periodically alternating fashion. Introducing, what is in effect, the functional equivalent of periodic poling technology into silicon photonics offers a path to realizing efficient wave mixing devices based on second-order optical nonlinearity. As an example of the utility of PePSi technology, it was numerically shown that using QPM-DFG, efficient MWIR generation can be achieved. The use of PePSi for frequency conversion can also be extended to terahertz generation as silicon exhibits excellent transmission properties at these frequencies. When combined with piezoelectric stressed layers, this technology will offer the capability to dynamically control $\chi^{(2)}$ in silicon.

References

1. T. Ghani, M. Armstrong, C. Auth et al., A 90 nm high volume manufacturing logic technology featuring novel 45 nm gate length strained silicon CMOS transistors, IEEE International Electron Devices Meeting, 2003 Technical Digest, pp. 11.6.1–11.6.3, 2003.

2. A. Arnau (Ed.), *Piezoelectric Transducers and Applications*, Springer, 2004.

3. B. E. A. Saleh, *Fundamentals of Photonics*, 1st edition, Wiley-Interscience, 1991.

4. W. N. Ye, D.-X. Xu, S. Janz, P. Cheben, M.-J. Picard, B. Lamontagne, and N. G. Tarr, Birefringence control using stress engineering in silicon-on-insulator (SOI) waveguides, *J. Lightwave Technol.* **23**, pp. 1308–1317, 2005.

5. V. Raghunathan and B. Jalali, Stress-induced phase matching in silicon waveguides, Conference of Lasers and Electro-Optics (CLEO), Long Beach, CA (2006) Paper CMK5.

6. K. K. Tsia, S. Fathpour, and B. Jalali, Electrical control of parametric processes in silicon waveguides, *Opt. Express* **16**, pp. 9838–9843, 2008.

7. K. K. Tsia, S. Fathpour, and B. Jalali, Electrical tuning of birefringence in silicon waveguides, *Appl. Phys. Lett.* **92**, 061109, 2008.

8. B. Jalali, Making *silicon* lase, *Sci. Am.* **296**, 58, 2007.

9. Q. Lin, O. J. Painter, and G. P. Agrawal, Nonlinear optical phenomena in silicon waveguides: Modeling and applications, *Opt. Express* **15**, 16604–16644, 2007.

10. R. S. Jacobsen, K. N. Andersen, P. I. Borel, J. Fage–Pedersen, L. H. Frandsen, O. Hansen, M. Kristensen, A. V. Lavrinenko, G. Moulin, H. Oul, C. Peucheret, B. Zsigri, A. Bjarklev, W. Wu, J. Sun, K. Cao, S. Wang, and B. Shi, Strained silicon as a new electro-optic material, *Nature* **441**, pp. 199–202, 2006.

11. N. K. Hon, K. K. Tsia, D. R. Solli, and B. Jalali, Periodically-poled silicon, *Appl. Phys. Lett.* **94**, 091116, 2009.

12. N. K. Hon, K. K. Tsia, D. R. Solli, and B. Jalali, Addendum: Periodically poled silicon [*Appl. Phys. Lett.* **94**, 091116 (2009)], *Appl. Phys. Lett.* **94**, 159902, 2009.

13. C. W. Wong, Peter T. Rakich, S. G. Johnson, M. Qi, H. I. Smith, E. P. Ippen, and L. C. Kimerling, Strain-tunable silicon photonic band gap microcavities in optical waveguides, *Appl. Phys. Lett.* **84**, 1242, 2004.

14. N. Setter, *Electroceramic-Based MEMS: Fabrication-Technology and Applications*, Springer, New York, 2005.

15. S. Donati, L. Barbieri, and G. Martini, Piezoelectric actuation of silica-on-silicon waveguide devices, *IEEE Photon. Technol. Lett.* **10**, pp. 1428, 1998.

16. H. Ishiwara, M. Okuyama, and Y. Arimoto, *Ferroelectric Random Access Memories: Fundamentals and Applications*, Springer, New York, 2004.

17. S. T. Kim, H. H. Kim, M. Y. Lee, W. J. Lee, Investigation of Pt/ Ti bottom electrodes for $Pb(Zr,Ti)O_3$ films, *Jpn. J. Appl. Phys.* **36**, pp. 294–300, 1997.

18. R. Guo, L. E. Cross, S-E. Park, B. Noheda, D. E. Cox, and G. Shirane, Origin of the high piezoelectric response in $PbZr_{1-x}Ti_xO_3$, *Phys. Rev. Lett.* **84**, pp. 5423–5426, 2000.

19. S. M. Koo, D. P. Kim, K. T. Kim, S. H. Song, and C. I. Kim, Etching properties of lead-zirconate-titanate thin films in Cl_2/Ar and BCl_3/Ar gas chemistries, *J. Vac. Sci. Technol. A* **22**, pp. 1519–1523, 2004.

20. K. Lefki and G. J. M. Dormans, Measurement of piezoelectric coefficients of ferroelectric thin films, *J. Appl. Phys.* **76**, pp. 1764–1767, 1994.

21. J. M. Liu, B. Pan, H. L. W. Chan, S. N. Zhu, Y. Y. Zhu, and Z. G. Liu, Piezoelectric coefficient measurement of piezoelectric thin films: An overview, *Mat. Chem. and Phys.*, **75**, pp. 12–18, 2002.

22. M. A. Foster, A. C. Turner, J. E. Sharping, B. S. Schmidt, M. Lipson, and A. L. Gaeta, Broad-band optical parametric gain on a silicon photonic chip, *Nature* **441**, 960–962, 2006.

23. M. A. Foster, A. C. Turner, R. Salem, M. Lipson, and A. L. Gaeta, Broad-band continuous-wave parametric wavelength conversion in silicon nanowaveguides, *Opt. Express* **15**, 12949–12958, 2007.

24. D. Dimitropoulos, V. Raghunathan, R. Claps, and B. Jalali, Phase-matching and nonlinear optical processes in silicon waveguides, *Opt. Express* **12**, 149–160, 2004.

25. V. Raghunathan, R. Claps, D. Dimitropoulos, and B. Jalali, Parametric Raman wavelength conversion in scaled silicon waveguides, *J. Lightwave Technol.* **23**, 2094–2102, 2005.

26. K. K. Tsia, S. Fathpour, and B. Jalali, Energy harvesting in silicon wavelength converters, *Opt. Express* **14**, 12327–12333, 2006.

27. J. Y. Huang, Probing inhomogeneous lattice deformation at interface of Si(111)/SiO2 by optical second-harmonic reflection and Raman spectroscopy, *Jpn. J. Appl. Phys.* **33**, 3378, 1994.

28. S. V. Govorkov, V. I. Emel'yanov, N. I. Koroteev, G. I. Petrov, I. L. Shumay, V. V. Yakovlev, and R. V. Khokhlov, Inhomogeneous deformation of silicon surface layers probed by second-harmonic generation in reflection, *J. Opt. Soc. Am. B* **6**, 1117–1124, 1989.

29. P. Temple–Boyer, C. Rossi, E. Saint–Etienne, and E. Scheid, Residual stress in low pressure chemical vapor deposition SiN films deposited from silane and ammonia, *J. Vac. Sci. Technol. A* **16**, 2003, 1998.

30. R. W. Boyd, *Nonlinear Optics*, Academic Press, New York, 1994.

31. N. K. Hon, K. K. Tsia, D. R. Solli, J. B. Khurgin, and B. Jalali, Periodically-Poled Silicon, presented at SPIE Photonics West, January 2010

32. V. Raghunathan, D. Borlaug, R. R. Rice, and B. Jalali, Demonstration of a mid-infrared silicon Raman amplifier, *Opt. Express* **15**, 14355, 2007.

33. H. Rong, S. Xu, O. Cohen, O. Raday, M. Lee, V. Sih and M. Paniccia, A cascaded silicon Raman laser, *Nat. Photonics* **2**, 170, 2008.

34. M. Krause, R. Draheim, H. Renner, and E. Brinkmeyer, Cascaded silicon Raman lasers as mid-infrared sources, *Electron. Lett.* **42**, 1224, 2006.

35. R. A. Soref and B. R. Bennett, Electrooptical effects in silicon, *IEEE J. Quantum Electron.* **23**, 123, 1987.

36. P. Koonath, D. R. Solli, and B. Jalali, Limiting nature of continuum generation in silicon, *Appl. Phys. Lett.* **93**, 091114, 2008.

CHAPTER 4

Pulse Shaping and Applications of Two-Photon Absorption

Ozdal Boyraz

Contents

4.1 Two-Photon Absorption and Free Carriers

Two-photon absorption (TPA) is a nonlinear optical effect that is particularly strong in semiconductors. This effect results in pump depletion and generation of free carriers that give rise to an additional broadband absorption and free carrier plasma effect. Strictly speaking, TPA has been shown to be negligible from the point of view of pump depletion at moderate peak intensity levels [1]. This is plausible since the TPA coefficient in silicon, β, is relatively small, ~0.7 cm/GW at 1.55 μm wavelength. This value is about an order of magnitude smaller than the Raman effect. On the other hand, TPA-induced free-carrier absorption (FCA) is a broadband process that competes with the gain in silicon nonlinear amplifiers, especially when carrier densities approach 10^{16} cm^{-3}. This effect has been identified as a limiting factor in all-optical switching in III-V semiconductor waveguides

[2–6]. It has also been discussed as a potential limit to achievable Raman gain in GaP waveguides [7], even though a Raman gain of 24 dB was demonstrated in these waveguides [8]. In particular, TPA-induced FCA has been studied in silicon waveguides in the context of the Raman process [9,10] and in transmission of ultra-short pulses in silicon waveguides [11,12]. The physics of TPA and free-carrier effects has been investigated thoroughly, and several excellent books and papers can be found on the subject [13,14]. In this chapter, we will start with a brief summary of the TPA process and free-carrier accumulation, as well as the pulse shaping mechanism induced by these effects.

4.1.1 Physics of TPA and Free Carrier Plasma Effect

Because of its crystal inversion symmetry, the lowest-order non-linearity in crystalline silicon originates from third-order optical susceptibility. In the presence of a single and two input signals, the third nonlinear polarization vector, P, related to TPA in the silicon waveguide can be described as [15]:

$$P^{(3)}(\omega) = \varepsilon_0 \frac{3}{4}\chi^{(3)}(-\omega;\omega,-\omega,\omega)|E_\omega|^2 E_\omega \qquad (4.1)$$

$$P^{(3)}(\omega_2) = \varepsilon_0 \frac{3}{2}\chi^{(3)}(-\omega_2;\omega_1,-\omega_1,\omega_2)|E_{\omega_1}|^2 E_{\omega_2} \qquad (4.2)$$

where ε_0 is the permittivity of free space and $P^{(3)}$ is the third-order susceptibility tensor. The first equation describes the Kerr effect and the degenerate TPA, and the latter corresponds to nondegenerate Kerr and TPA effects. TPA and Kerr effects are related to the imaginary part and the real part of $P^{(3)}$ that correspond to the intensity dependence of the optical absorption and the refractive index perturbation, respectively. As illustrated in Figure 4.1, the TPA process in silicon couples two photons when the total energy of two input photons is greater than the bandgap energy of silicon ($E_g = E_c - E_v \approx 1.12$ eV), and excites an electron from the valence band to the conduction band, resulting in generation of a free-carrier (electron-hole) pair. Depending on the parity difference in the initial and final sates, three types of two-photon transitions can occur, which are referred to as allowed-allowed

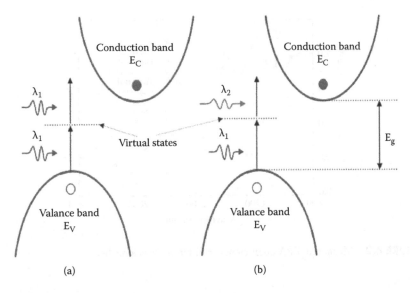

FIGURE 4.1 Schematic illustration of TPA in silicon: (a) degenerate TPA, (b) nondegenerate TPA.

(a-a), allowed-forbidden (a-f), and forbidden-forbidden (f-f). The TPA coefficient

$$\beta(\omega) = \frac{3\pi}{c\varepsilon_0 \lambda n_0^2} \operatorname{Im}\left\{\chi^{(3)}\right\}$$

is a fundamental parameter to evaluate TPA. Considering the three types of two-photon processes, the total TPA coefficient can be given as $\beta(\omega) = \sum_{n=0}^{2} \beta^{(n)}(\omega)$, where n = 0, 1, and 2 for a-a, a-f, and f-f transitions [16–18].

Because $\beta^{(n)}(\omega) = 2CF_2^{(n)}(h\upsilon / E_g)$ incorporates a material-dependent constant C, the contribution of each type process is determined by the fundamental photon energy, $h\upsilon$. In the expression of the foregoing equation, $F_2^{(n)}(x) = \left[\pi(2n+1)! / 2^{n+2}(n+2)!\right](2x)^{-5}(2x-1)^{n+2}$, $x = h\upsilon/E_g$, for parabolic electron and hole bands. Assuming that phonon energies are $\ll E_g$, the f-f process is weaker than the two other processes, but it gets its maximum at $h\upsilon \approx E_g/2$ [18] . The other two processes become stronger when photon energies are near E_g. In the degenerate TPA, two photons with the same wavelength are absorbed, and the power of the input light will be depleted along the silicon waveguide.

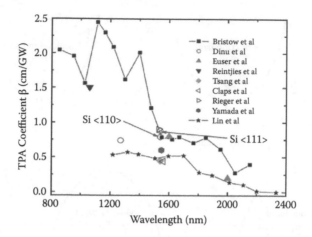

FIGURE 4.2 Measured TPA coefficients at different wavelengths.

A lot of experimental and theoretical work related to degenerate TPA in semiconductors has been carried out. It should be noted that due to the off-diagonal elements in the third-order susceptibility tensor, the TPA exhibits optical polarization dependence, and the presence of TPA anisotropy in silicon was observed [19]. So far, the z-scan technique and pulse transmission method are used to measure or elevate the TPA coefficient of silicon. The measured TPA coefficients are summarized in Figure 4.2 [18–23].

In the nondegenerate TPA process, a nonlinear absorption is induced on a signal by a second pump signal instead of by itself. In other words, a signal photon is absorbed when a pump photon is present and leads to cross-absorption modulation. From Equation 4.2, the nondegenerate TPA process is two times stronger than the degenerate TPA process. It should be pointed out that the nondegenerate TPA is polarization dependent, and the linearly polarized beams show stronger polarization dependence than the circularly polarized ones [24–26]. Assuming that there is an isotropic coupling between momentum space of the semiconductor, the effective TPA coefficient will reduce by 2/3. For instance, the effective TPA coefficient between two linearly polarized light signals with an angle ϕ of separation between them will see an angle-dependent effective TPA coefficient of $\beta(\phi) = \beta_{Linear}(1 - \frac{2}{3}\sin^2\phi)$. In the case of two-photon interaction of two optical beams with elliptical polarizations or in an anisotropic media, the TPA coefficient will be strongly dependent on the local normalized Stokes parameters of an individual beam [26].

Ignoring the free-carrier effect, TPA alone is an intensity-dependent attenuation. Analytically, this attenuation due to the TPA process can be described as

$$\frac{dI_P}{dz} = -\beta I_P^2 \tag{4.3}$$

where β is the net TPA coefficient in silicon and I_p is the pump intensity. As described in Figure 4.2, the measured value of the TPA coefficient varies between 0.5 to 1 cm/GW at 1.55 μm. Here in this book, we use 0.7 cm/GW to demonstrate the effect of TPA in different applications unless stated otherwise. To illustrate the amount of attenuation induced by TPA, the transmission function of a 1-cm-long silicon waveguide is presented in Figure 4.3. The modal area of the waveguide here is assumed to be 5 μm². The attenuation induced by TPA is intensity square dependent, and it is fairly small for moderate intensity levels. Only 40% attenuation is expected at 50 W (which corresponds to 1 GW/cm²) input power levels. Readers are advised to remember this value and compare it to the losses induced by free carriers as described next.

The TPA process in silicon leads to the generation of free carriers, whose density depends on the incident optical intensity, and then leads to FCA and free-carrier dispersion (FCD) as a consequence of these free carriers. However, diffusion will reduce the

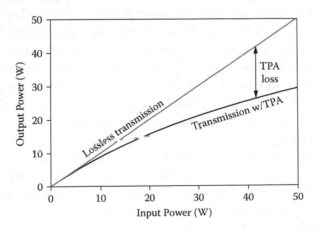

FIGURE 4.3 Transmission function of a 1-cm-long silicon waveguide.

carrier density at the center of the optical mode. The dynamics of free carriers induced by TPA can be described by the Drude model [13]. Since the TPA-induced free carriers in the silicon waveguide vary slowly along the propagation direction z, lateral diffusion (along the x axis) will dominate and impact the free-carrier density. The rate equation for the free-carrier density in the silicon waveguide can be given in terms of free-carrier generation and recombination as [13]:

$$\frac{dN}{dt} = \frac{\beta I^2}{2h\nu} - D'\frac{\partial^2 N}{\partial x^2} - \frac{N}{\tau_0} \tag{4.4}$$

where $h\nu$ is the photon energy, I is the optical intensity (time and space dependent), D' is the effective diffusion constant of the free carriers, and τ_0 is the carrier recombination time. For a given carrier gradient, Equation 4.4 can be written as

$$\frac{dN}{dt} = \frac{\beta I^2}{2h\nu} - \frac{N}{\tau_{eff}}$$

where τ_{eff} represents the effective carrier lifetime due to recombination process and diffusion. Free carriers are accumulative inside the silicon waveguide. For continuous-wave (CW) signals or pulsed signals whose pulse widths are much wider than τ_{eff}, free-carrier density will be stabilized at a local position with a value of $N(z) = \tau_{eff}\beta I(z)/(2h\nu)$ at steady state. For pulsed signals whose pulse widths are much narrower than τ_{eff}, on the other hand, one must consider the repetition rate R of the pulse [27,28]. When the pulse signals operate at a low repetition rate ($R\tau_{eff} \ll 1$), the free-carrier density is time dependent, and the peak density depends on the pulse width, just like a single pulse. When the pulse repetition rate increases, those free carriers generated by the former pulses may not have enough time to recombine, which will impact the later pulses.

Since the free carriers induced by TPA are inevitable in silicon at high optical intensities, the consequences of these free carriers should be outlined clearly. Two main effects induced by free carriers are the optical absorption and the plasma dispersion effect. Analytically, the absorption and the refractive index change induced by free carriers can be given as [13]:

$$\alpha^{FCA} = \frac{e^3 \lambda^2}{4\pi^2 c^2 \varepsilon_0 n} \left(\frac{\Delta N_e}{m_{ce}^2 \mu_e} + \frac{\Delta N_h}{m_{ch}^2 \mu_h} \right)$$

$$\approx 1.45 \times 10^{-17} \left(\frac{\lambda}{1550} \right)^2 N(z,t) \tag{4.5}$$

$$\Delta n_{FC} = \frac{e^2 \lambda^2}{8\pi^2 c^2 \varepsilon_0 n} \left(\frac{\Delta N_e}{m_{ce}} + \frac{\Delta N_h}{m_{ch}} \right)$$

$$\approx -8.2 \times 10^{-22} \lambda^2 N(z,t) \tag{4.6}$$

Here, $m_{ce(h)}$ is the electron (hole) effective mass, and $\mu_{e(h)}$ is the electron (hole) mobility. So, the dispersion induced by free carriers can be given by

$$\Delta D_{FC} = \frac{1}{c} \frac{\partial \Delta n_{FC}}{\partial \lambda} = -5.46 \times 10^{-30} \lambda N \tag{4.7}$$

Because the TPA in silicon is intensity dependent, both FCA loss and FCD will depend on the optical intensity, as well. As a result, during propagation inside the waveguide, the optical intensity $I(z,t)$ will decrease due to linear absorption and scattering, TPA and FCA as described by

$$\frac{dI(z,t)}{dz} = \alpha I(z,t) - \alpha^{FCA} I(z,t) - \beta I^2(z,t) \tag{4.8}$$

Here, α is the linear loss coefficient. Figure 4.4 illustrates the schematic diagram of pulse propagation in a silicon waveguide where free-carrier losses are significant. Upon entering the waveguide, the front end of the pulse will be slightly attenuated by TPA and, hence, the free carriers will start to accumulate wherever optical energy is present. As a result, the photons entering the waveguide will suffer from the free-carrier losses generated by the earlier photons. This process will lead to attenuation of the trailing edge of the pulse and, hence, pulse compression in the time domain (see Figure 4.4). Because of the linear and nonlinear absorptions, the FCA will be spatially varying and will be much

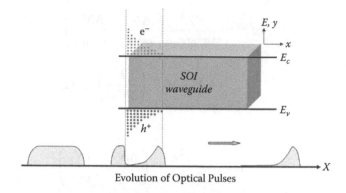

Evolution of Optical Pulses

FIGURE 4.4 Schematic description of free-carrier accumulation for pulse compression. A high-intensity optical pulse will leave paint of free carriers behind, which will contribute to pulse attenuation at the trailing edge and hence compression. (From E. K. Tien et al., *Opt. Express*, 6500, 2007. With permission.)

stronger at the input facet of the silicon waveguide. To have a fair comparison with the TPA-induced loss presented earlier, the attenuation constant induced by free carriers along the front facet of a 5 μm² waveguide is illustrated in Figure 4.5. As one can see, the consequence of a small attenuation due to TPA may lead to several orders of magnitude larger losses due to the FCA.

The fundamental parameter that governs the TPA mediated loss is the carrier lifetime, τ_{eff}. It is well known that the carrier lifetime in silicon-on-insulator (SOI) is much shorter than that in a bulk silicon sample with comparable doping concentration.

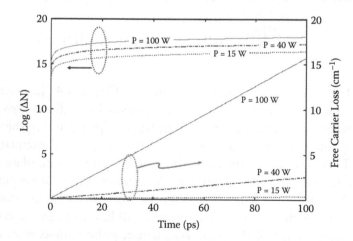

FIGURE 4.5 FCA evolution along time in a 5 μm² waveguide. (From E. K. Tien et al., *Opt. Express*, 15i 6500, 2009. With permission.)

This lifetime reduction is due to the presence of interface states at the boundary between the top silicon and the buried oxide layer. This effect depends on the method used for preparation of the SOI wafer and the film thickness, with measured and expected values ranging between 10 and 200 ns based on waveguide geometry [29,30]. In SOI waveguides, the lifetime is further reduced to a few nanoseconds, or even below in the case of submicron waveguides, due to the recombination at the etched waveguide facets and, in the case of rib waveguides, due to diffusion into the slab regions [31,32]. The lifetime can be further reduced by application of a reverse bias p-n junction [9,10,31] or by introduction of midgap states through high-energy irradiation, and gold or platinum doping. A modest amount of CW gain has been observed in deep submicron waveguides by Osgood's group at Columbia University [32], where the impact of surface and interface recombination plays a critical role in reducing the lifetime. As a matter of fact, net CW Raman gain was enabled by carrier lifetime reduction due to sweeping the free carriers using a reverse bias p-n junction diode [31].

4.1.2 Active Carrier Sweeping and Reduced Carrier Lifetime

In bulk silicon samples with comparable doping concentrations, the carrier lifetime is determined by the recombination processes. The radiative recombination process is phonon assisted and requires millisecond waiting times. Nonradiate recombination due to trap levels is relatively faster and results in microsecond free-carrier lifetimes. Planar waveguides fabricated on SOI wafers exhibit free-carrier lifetimes on the order of 10–200 ns. This lifetime reduction is due to the presence of interface states at the boundary between the top silicon and the buried oxide layer. The diffusion of free carriers to these recombination centers reduces the carrier lifetimes approximately three orders of magnitude. From a device point of view, the free-carrier lifetime is by no means limited by the recombination processes. In a waveguide geometry, instead of waiting for full recombination, removing the free carriers out of the optically active region by diffusion or carrier sweeping by using an electric field is more than adequate [33]. In this section, the physical limitations of carrier lifetime in Si are discussed. In particular, the intensity dependence of the carrier lifetime in a rib waveguide in which a reverse-biased p-n junction is used to sweep out the carriers is studied.

The magnitude of the nonlinear optical loss is proportional to the carrier density that has been created in the waveguide, which is in turn proportional to the time it takes for carriers to be removed from the waveguide core. The carrier lifetime is therefore the critical parameter for CW operation, or for pulsed operation when the pulse repetition period is shorter than the carrier lifetime.

The lifetime is determined by the combination of diffusion and interface/surface recombination currents. In a bare SOI waveguide, the geometry of the waveguide plays a significant role in determining the carrier lifetime, as an order-of-magnitude reduction can be achieved by varying the slab height and the rib-to-slab-height ratio [33]. The carrier lifetime also depends on the generation rate in the waveguide and therefore on the intensity of the pump beam. As a result, one obtains a complex coupling between gain and free-carrier concentrations and hence loss, which cannot be solved analytically. However, there are numerical methods that can solve the problem.

One can understand the essential features by examining the case of a p-n junction, where there is a uniform generation rate in the depletion region. When no generation is taking place, the applied field, E_o, is proportional to applied voltage, V, and is inversely proportional to the width of the depletion region. Now, if an electron-hole pair is generated, the particles will drift under the action of the applied electric field. We note that the carrier drift velocity in silicon has a nonlinear dependence on the applied electric field due to velocity saturation. For electric fields smaller than the saturation field, $E_{SAT} \approx 2 \times 10^4 V / cm$, the carrier velocities increase linearly with the applied field as $v_e = \mu_e E$ and $v_h = \mu_h E$ ($\mu_e = 1500 \ cm^2 \cdot s^{-1} \cdot V^{-1}$ and $\mu_h = 450 \ cm^2 \cdot s^{-1} \cdot V^{-1}$ for intrinsic Si). For higher electric fields, the carrier velocity reaches the thermal velocity and it is saturated. For electrons, this saturation velocity is $v_{e,SAT} \approx 10^7 cm/s$. The electric field can exceed E_{SAT}, but it cannot exceed the material breakdown field $E_B \sim 3 \times 10^5 V / cm$. When the generation rate is low enough, the generated carriers are swept out of the waveguide core, and the carrier lifetime for electron or holes is therefore the mean transit time through the rib region with width, w, that is, $\tau_{e(h)} = w / (2\mu_{e(h)} E_o)$ when $E_o < E_{SAT}$ and $\tau_{e(h)} = w / (2v_{e(h),SAT})$ when $E_o > E_{SAT}$.

Furthermore, since the electrons and holes drift in opposite directions, a secondary electric field will be set up in the depletion region with direction opposite to the applied field. The applied

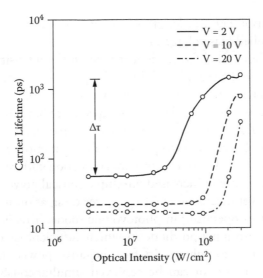

FIGURE 4.6 Dependence of the effective carrier lifetime on the optical intensity for various values of the applied voltage. (From D. Dimitropoulos et al., *Appl. Phys. Lett.*, vol. 87, pp. 261108-3, 2005.)

electric field will be screened by the secondary field, leading to an increase of the carrier lifetime as observed in Figure 4.6 [34]. The screening of the applied field by the carriers is the same effect that is responsible for the saturation of responsivity in high-power photodetectors. In the regime in Figure 4.6 where saturation has set in, drift is negligible, and carrier transport is governed by diffusion and recombination. In this case, the lifetime should approach the lifetime of carriers in a waveguide with no junction. Therefore, the difference in lifetime at the low- and high-intensity regimes in Figure 4.6, $\Delta\tau$, is the lifetime reduction enabled by the junction. Note that the generation rate along the waveguide varies slowly in comparison to that in the transverse direction. Therefore, in calculations of carrier lifetime, longitudinal carrier diffusion can be neglected.

The reduction in lifetime and the concomitant reduction in optical loss that is achieved with the p-n junction come at the expense of on-chip electrical power dissipation [35]. The power dissipation per unit length is a strong function of pump intensity, which varies along the waveguide length. Neglecting the recombination of carriers (this is permissible when the dominant current is the drift current), the current (per unit length of the waveguide) is simply determined by the generation rate as $j = e(G \times w \times H)$. The power dissipation per unit length of the waveguide is $W_{DISS} = jV$ under the applied bias voltage of V.

In short, the lifetime dependence on the intensity is due to the screening of the applied field by the photo-generated carriers, and can result in an increase in the lifetime by more than an order of magnitude. Hence, this effect is expected to have significant implications for the operation of silicon Raman lasers and amplifiers that employ carrier sweep-out via a p-n junction. In addition, the benefits gained by free-carrier loss reduction with reverse bias come at the expense of increased on-chip electrical power dissipation.

A novel means of avoiding power dissipation is demonstrated in devices operating in photovoltaic mode. This is done by realizing a p-n junction diode in which sufficient carrier sweep-out occurs at zero voltage. In fact, negative power dissipation and CW Raman gain can be achieved simultaneously in the same structure. This approach invokes the photovoltaic effect caused by TPA, and is applicable to any photonic device based on nonlinear interactions in silicon. These include Raman amplifiers/lasers, as well as devices that exploit Raman- or Kerr-based four wave mixing [36–38].

In addition to reverse-biased p-i-n diode waveguides, free-carrier lifetime in silicon waveguides can be reduced by increasing the recombination centers. This is a well-known approach in electronics that comes at the expense of increasing impurity levels. In classical semiconductor theory, Au, Pt, Cu, Mn, Cr, and Fe are considered to be common impurities to introduce trap levels near silicon mid-bandgap. For photonic applications, the large absorption induced by these metals are not desirable. In photonics and high-voltage electronics helium and hydrogen (He^{++}, H^+) implantation of silicon is also emerging as an alternative approach to lifetime reduction in high-voltage electronic devices for photonic applications [39,40]. In this process, a high-energy ion penetrating the silicon generates defect levels along its whole trajectory. The most stable defect levels are produced near the end of the particle range [27]. Hence, local defect levels generated by high-energy ions, where defect levels are controlled by ion energy, are being utilized as high-power bipolar diodes.

It is obvious that helium implantation is an effective tool for carrier lifetime reduction for photonic devices. However, the size of lifetime reduction and optical losses highly depends on defect concentrations and, hence, the implantation dose. Subnanosecond free-carrier lifetimes are achievable at the expense of optical losses. For given loss values, the reverse-biased p-i-n diode approach appears more attractive. However, improved subbandgap photon absorption can be utilized as an inline detector. Recently, this has been proposed as a low-loss inline optical detector to monitor the optical power flowing through the fibers [41]. The dose of the helium implantation can be optimized for low-loss transmission and optimum photo current detection. Helium doses less than 10^{13} cm^2 can produce nearly 0.5 mA of photo current with 5 dB insertion loss [41].

4.2 Two-Photon Absorption and Free Carriers in Nonlinear Silicon Optics

4.2.1 Optical Pulse Shaping Applications of TPA and FCA

When optical pulses enter the silicon waveguide, the transient behaviors of TPA and TPA-induced free carriers will have a significant effect on the final output pulses. Upon entering the silicon waveguide, the front edge of the pulse will be slightly attenuated by TPA and hence the free carriers will start to accumulate. Since the free-carrier recombination time is much larger than the optical pulse widths, the generated free carriers accumulate and introduce time-dependent attenuation along the pulse. As a result, the trailing edge of the pulse entering the silicon waveguide is attenuated by both the TPA and the free carriers generated by the front edge of the same pulse. This transient behavior of free-carrier concentration will build up over the pulse duration every time the pulse propagates through the silicon waveguide and facilitate a self-compression and mode-locking inside a laser cavity. To further understand the pulse shaping, the pulse envelope evolution in the silicon waveguide can be analyzed through the following nonlinear Schrödinger equation [12,42]:

$$\frac{\partial E(z,t)}{\partial z} = -\frac{1}{2}\left[\alpha + \alpha_{FC}(z,t) + \alpha_{TPA}(z,t)\right]E(z,t)$$

$$-i\gamma|E(z,t)|^2 E(z,t) + i\frac{2\pi}{\lambda}\Delta n_{FC}(z,t)E(z,t)$$

(4.9)

where $\alpha_{TPA}(z,t) = \ln(1+\beta I_0 z)/z$ and α_{TPA}, γ are attenuation coefficients by TPA and Kerr nonlinear coefficients, respectively.

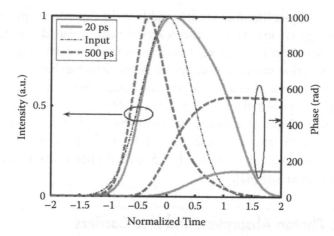

FIGURE 4.7 Estimated output pulses and the chirp induced by the free-carrier plasma effect. (From E. K. Tien et al., *Appl. Phys. Lett.*, 201115, 2007, American Institute of Physics. With permission.)

Based on the peak pulse intensity and pulse energy, the nonlinear response in silicon can be divided into a TPA-dominated regime and an FCA-dominated regime [42,43]. In the TPA-dominated regime, the attenuation at the center of the pulse is larger and, hence, net pulse broadening is expected. In the FCA-dominated regime, on the other hand, the attenuation at the trailing edge is larger than losses at any part and, hence, pulse compression is expected. Figure 4.7 shows the calculated results of two regimes for 20 ps and 500 ps optical pulses with 1 kW peak power. The waveguide is 1 cm long with a 5 μm^2 effective area, and the free-carrier recombination time is 16 ns. The fixed 1 kW peak power facilitates the same level of TPA (90%) at the center for comparison, in Figure 4.8. The 20 ps pulses are broadened due to TPA dominance, and 500 ps pulses are compressed to 350 ps by FCA dominance. The FCD causes a linear phase change across the pulse.

We expect that the total length of the waveguide, input pulse width, and the pulse energy will have a prominent influence on the final outcome after propagation. Figure 4.9 illustrates the compression behavior of 10 to 1000 ps optical pulses in a 10-mm-long silicon waveguide. At low peak powers, <10^{16} cm^{-3} carriers are generated to facilitate attenuation of the trailing edge and to provide self-compression. However, stronger TPA at the center of the pulse dominates at these intensities and induces a net pulse broadening at the 3 dB point. Free-carrier dominance is expected

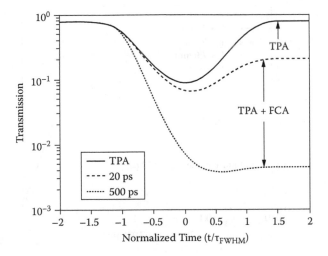

FIGURE 4.8 TPA and free-carrier loss profiles of 20- and 500-ps-wide input pulses. (From E. K. Tien et al., *Appl. Phys. Lett.*, 201115, 2007, American Institute of Physics. With permission.)

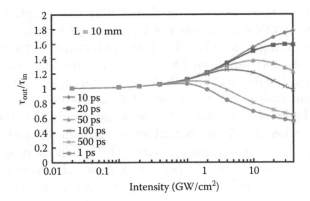

FIGURE 4.9 Predicted pulse compression ratio in 10-mm-long silicon waveguide for different initial pulse widths. (From E. K. Tien et al., *Appl. Phys. Lett.*, 201115, 2007, American Institute of Physics. With permission.)

at higher intensities. For example, 1 ns pulses with 2 GW/cm² intensities generate 4×10^{18} cm⁻³ free carriers, which dominate over TPA at the center of the pulse and facilitate net self-compression. For 500- and 100-ps-wide optical pulses, it requires 3 GW/cm² (75 nJ) and 32 GW/cm² (160 nJ) to initiate the same type of compression, respectively. For input pulses shorter than 50 ps, TPA-induced pulse broadening surpasses the self-compression even at 40 GW/cm² and, hence, net pulse broadening

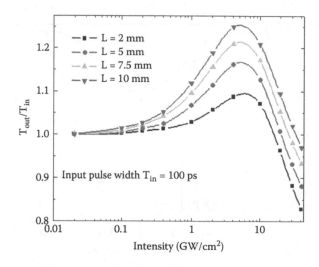

FIGURE 4.10 Predicted pulse compression ratio in silicon waveguides with different lengths.

is expected. However, we should note that simple free-carrier density calculations may not be reliable at such high intensities due to Auger recombination [44]. For instance, as the free-carrier density increases from 10^{18} to 10^{19} cm^{-3}, the free-carrier lifetime due to the Auger effect will decrease from ~1.3 to ~20 ns, which is comparable to the free-carrier lifetime of our waveguides.

Figure 4.10 illustrates the pulse compression for 100 ps optical pulses at different optical intensities with different waveguide lengths. At low intensity, strong TPA-induced pulse broadening dominates over FCA-induced pulse self-compression. FCA dominance can be achieved at higher intensities and facilitate pulse compression.

Large pulse compression factors require large intensity values. This is a practical challenge. However, one can place this scheme inside a laser cavity to obtain mode-locked short pulse lasers. Inside the laser cavity, a small compression factor will be amplified by circulation until a very short pulse is obtained at steady state. This scheme can be tested in the experimental configuration presented in Figure 4.11. Here, the silicon waveguide has a p-i-n diode structure to inject carriers and, hence, a modulation capability. The output of the waveguide is connected to a 3/97 tap coupler where 3% used as an output and 97% is fed into the gain medium, a high-power erbium-doped fiber amplifier (EDFA). The resonator is formed by launching the EDFA output back

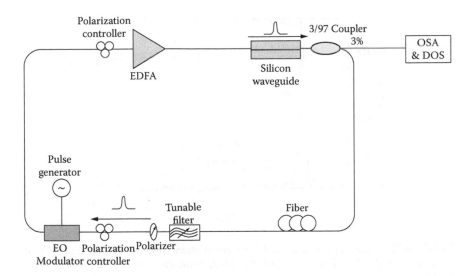

FIGURE 4.11 Experimental setup used for TPA-based modelocking and pulse compression scheme. (From E. K. Tien et al., *Appl. Phys. Lett.*, 201115, 2007, American Institute of Physics. With permission.)

into the silicon waveguide input end. An optical bandpass filter is inserted to ensure lasing at a desired wavelength and suppress undesired amplified spontaneous emission (ASE) accumulation. Since the pulse shaping by free carriers requires a time-varying optical signal circulating inside the laser cavity, an electro-optic (EO) modulator is connected to the waveguide to start initial pulsation. The output pulse width is expected to be minimum when the frequency of the function generator matches the fundamental cavity frequency. Figure 4.12 shows that the 1.5 ns pulse can be compressed to 60 ps by FCA. Figure 4.13 illustrates the pulse widths measured at the laser output for different peak powers and different rise/fall times. The shortest pulse width is measured to be 60 ps by using RF signals with 140 ps rise/fall time. Figure 4.13 shows the tendency of the pulse compression with respect to different peak powers and different rise/fall times. At fixed pulse energy, slower rise time results in less free-carrier generation. As a result, the pulse width increases to 85 ps if the rise time is tuned to 650 ps for the same pulse energy. These measurements indicate that free-carrier concentration increases more rapidly for modulation signals with sharper rise times. To compensate the slope effect, we need to increase pulse energy to achieve the same compression. This tendency agrees with the theoretical results presented in Figure 4.13, which shows the higher peak power

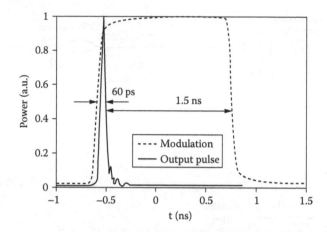

FIGURE 4.12 Laser output and the 1.5 ns modulation with 140 ps rise time. (From E. K. Tien et al., *Appl. Phys. Lett.*, 201115, 2007, American Institute of Physics. With permission.)

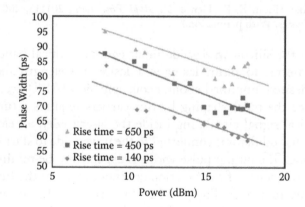

FIGURE 4.13 Variation of output pulse width with respect to pulse energy. (X. Z. Sang et al., *IEEE Photonics Technol. Lett.*, vol. 20, pp. 1184–1186, July–August 2008. With permission.)

requirement for low-energy pulses. For instance, to achieve 80 ps output pulses, the average power entering the silicon waveguide has to be maintained at 9, 13, and 16 dBm levels for 140, 450, and 650 ps rise and fall times, respectively. The peak power corresponding to an 80 ps pulse with 9 dBm average power is ~120 W (2.4 GW/cm²). The total loss at this setting is measured to be <6 dB, of which 2 dB is the linear loss and the rest originates from the nonlinear effects.

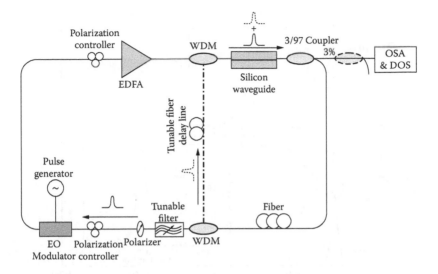

FIGURE 4.14 Experimental setup used for dual wavelength laser. (X. Z. Sang et al., *IEEE Photonics Technol. Lett.*, vol. 20, pp. 1184–1186, July–August 2008. With permission.)

Once the short pulses are generated inside the laser cavity, they can be used to stimulate other nonlinear effects in silicon. For instance, mode-locked pulses generated by silicon can induce stimulated Raman scattering and hence provide dual wavelength lasing. A slight adjustment in the laser cavity presented in Figure 4.14 is required to facilitate dual wavelength lasing. After the modification of the laser cavity and alignment of pump and Stokes pulses temporally, one can observe lasing at both wavelengths [45]. The spectra of the dual wavelength laser are shown in Figure 4.15. The pump wavelength is selected by the bandpass filter inside the laser cavity at 1540 nm, and the expected Raman Stokes signal is at a wavelength of 1675 nm.

4.3 Conclusion

We review the physics behind TPA and the free-carrier effect in this chapter. TPA and TPA-induced free-carrier absorption are major obstacles for chip-scale high-efficiency silicon-based photonics devices and ultimately limit the operation optical power and speed. Free-carrier absorption can be reduced by decreasing the lifetime of the carrier. This can be done by ion implantation or using a p-i-n structure to sweep the free carrier inside the waveguide.

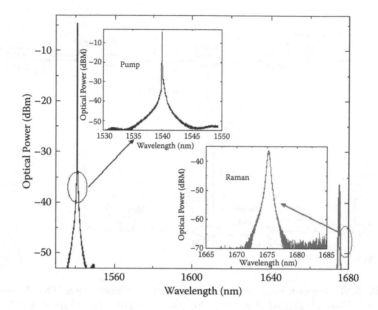

FIGURE 4.15 Spectra of the pulse-compressed pump and Raman laser.

Because of the varying property of free-carrier concentrations generated by short pulses, free-carrier absorbtion induces an asymmetric attenuation on optical pulses when they are propagating in silicon waveguides. We present our work on utilizing this transient loss on pulse compression and pulse shaping. The nonlinear loss in silicon is an interesting topic, and those who design fast and high-power silicon photonic devices should familiarize themselves with it.

References

1. G. G. L. Pavesi, *Optical Interconnects: The Silicon Approach*, Springer-Verlag, Netherlands 2006.
2. J. H. Yee and H. H. M. Chau, 2-Photon indirect transition in gap crystal, *Opt. Commun.*, vol. 10, pp. 56–58, 1974.
3. K. W. Delong and G. I. Stegeman, 2-Photon absorption as a limitation to all-optical wave-guide switching in semiconductors, *Appl. Phys. Lett.*, vol. 57, pp. 2063–2064, November 12, 1990.
4. A. Villeneuve, C. C. Yang, G. I. Stegeman, C. N. Ironside, G. Scelsi, and R. M. Osgood, Nonlinear absorption in a GaAs wave-guide just above half the band-gap, *IEEE J. Quantum Electron.*, vol. 30, pp. 1172–1175, May 1994.

5. A. M. Darwish, E. P. Ippen, H. Q. Le, J. P. Donnelly, and S. H. Groves, Optimization of four-wave mixing conversion efficiency in the presence of nonlinear loss, *Appl. Phys. Lett.*, vol. 69, pp. 737–739, August 1996.

6. Y. H. Kao, T. J. Xia, M. N. Islam, and G. Raybon, Limitations on ultrafast optical switching in a semiconductor laser amplifier operating at transparency current, *J. Appl. Phys.*, vol. 86, pp. 4740–4747, Nov 1999.

7. K. Suto, T. Kimura, T. Saito, and J. Nishizawa, Raman amplification in GaPAlxGa1-xP waveguides for light frequency discrimination (vol 145, p. 105, 1998), *IEE Proceedings-Optoelectronics*, vol. 145, pp. 247–247, August 1998.

8. S. Saito, K. Suto, T. Kimura, and J. I. Nishizawa, 80-ps and 4-ns pulse-pumped gains in a GaP-AlGaP semiconductor Raman amplifier, *IEEE Photonics Technol. Lett.*, vol. 16, pp. 395–397, February 2004.

9. T. K. Liang and H. K. Tsang, Role of free carriers from two-photon absorption in Raman amplification in silicon-on-insulator waveguides, *Appl. Phys. Lett.*, vol. 84, pp. 2745–2747, April 2004.

10. R. Claps, V. Raghunathan, D. Dimitropoulos, and B. Jalali, Influence of nonlinear absorption on Raman amplification in Silicon waveguides, *Opt. Express*, vol. 12, pp. 2774–2780, June 2004.

11. A. R. Cowan, G. W. Rieger, and J. F. Young, Nonlinear transmission of 1.5 μm pulses through single-mode silicon-on-insulator waveguide structures, *Opt. Express*, vol. 12, pp. 1611–1621, April 19, 2004.

12. O. Boyraz, P. Koonath, V. Raghunathan, and B. Jalali, All optical switching and continuum generation in silicon waveguides, *Opt. Express*, vol. 12, pp. 4094–4102, August 2004.

13. R. A. Soref and B. R. Bennett, Electrooptical effects in silicon, *IEEE J. Quantum Electron.*, vol. 23, pp. 123–129, Jan 1987.

14. P. Y. A. M. Cardona, *Fundamentals of Semiconductors*, 3rd ed., Springer, Germany 2001.

15. P. Dumon, G. Priem, L. R. Nunes, W. Bogaerts, D. Van Thourhout, P. Bienstman, T. K. Liang, M. Tsuchiya, P. Jaenen, S. Beckx, J. Wouters, and R. Baets, Linear and nonlinear nanophotonic devices based on silicon-on-insulator wire waveguides, *Jpn. J. Appl. Phys Part 1—Regular Papers Brief Communications and Review Papers*, vol. 45, pp 6589–6602, August 2006.

16. M. Sheikbahae, D. J. Hagan, and E. W. Vanstryland, Dispersion and band-gap scaling of the electronic kerr effect in solids associated with 2-photon absorption, *Phys. Rev. Lett.*, vol. 65, pp. 96–99, July 2 1990.

17. H. Garcia and R. Kalyanaraman, Phonon-assisted two-photon absorption in the presence of a dc-field: the nonlinear Franz-Keldysh effect in indirect gap semiconductors, *J. Phys. B—Atomic Mol. Opt. Physics*, vol. 39, pp. 2737–2746, June 28, 2006.

18. A. D. Bristow, N. Rotenberg, and H. M. van Driel, Two-photon absorption and Kerr coefficients of silicon for 850-2200 nm, *Appl. Phys. Lett.*, vol. 90:91104, May 7, 2007.

19. M. Dinu, F. Quochi, and H. Garcia, Third-order nonlinearities in silicon at telecom wavelengths, *Appl. Phys. Lett.*, vol. 82, pp. 2954–2956, May 5, 2003.

20. H. K. Tsang, C. S. Wong, T. K. Liang, I. E. Day, S. W. Roberts, A. Harpin, J. Drake, and M. Asghari, Optical dispersion, two-photon absorption and self-phase modulation in silicon waveguides at 1.5 μm wavelength, *Appl. Phys. Lett.*, vol. 80, pp. 416–418, January 2002.

21. T. G. Euser, H. Wei, J. Kalkman, Y. Jun, A. Polman, D. J. Norris, and W. L. Vos, Ultrafast optical switching of three-dimensional Si inverse opal photonic band gap crystals, *J. Appl. Phys.*, vol. 102, p. 6, September 2007.

22. T. G. Euser and W. L. Vos, Spatial homogeneity of optically switched semiconductor photonic crystals and of bulk semiconductors, *J. Appl. Phys.*, vol. 97:043102, February 15, 2005.

23. J. F. Reintjes and J. C. Mcgroddy, Indirect 2-photon transitions in Si at 1.06 μm, *Phys. Rev. Lett.*, vol. 30, pp. 901–903, 1973.

24. M. Sheikbahae, J. Wang, and E. W. Vanstryland, Nondegenerate optical kerr-effect in semiconductors, *IEEE J. Quantum Electron.*, vol. 30, pp. 249–255, February 1994.

25. E. K. Tien and O. Boyraz, Polarization state of silicon Raman lasers dictated by non-degenerate two-photon absorption, in *3rd IEEE International Conference on Group IV Photonics, 2006*, 2006, pp. 58–60.

26. R. Salem and T. E. Murphy, Polarization-insensitive cross correlation using two-photon absorption in a silicon photodiode, *Opt. Lett.*, vol. 29, pp. 1524–1526, July 2004.

27. Y. Liu and H. K. Tsang, Time dependent density of free carriers generated by two photon absorption in silicon waveguides, *Appl. Phys. Lett.*, vol. 90, May 2007.

28. A. J. Sabbah and D. M. Riffe, Femtosecond pump-probe reflectivity study of silicon carrier dynamics, *Phys. Rev. B*, vol. 66:165217, Oct. 15, 2002.

29. M. A. Mendicino, Comparison of properties of available SOI materials, in *Properties of Crystalline Silicon*, vol. 18, R. Hull, Ed., 1998, pp. 992–1001. London.

30. J. L. Freeouf and S. T. Liu, Minority carrier lifetime results for SOI wafers, in *SOI Conference, 1995. Proceedings, 1995 IEEE International*, 1995, pp. 74–75.

31. A. S. Liu, H. S. Rong, and M. Paniccia, Net optical gain in a low loss silicon-on-insulator waveguide by stimulated Raman scattering, *Opt. Express*, vol. 12, pp. 4261–4268, September 2004.

32. R. L. Espinola, J. I. Dadap, R. M. Osgood, S. J. McNab, and Y. A. Vlasov, Raman amplification in ultrasmall silicon-on-insulator wire waveguides, *Opt. Express*, vol. 12, pp. 3713–3718, August 9, 2004.

33. D. Dimitropoulos, R. Jhaveri, R. Claps, J. C. S. Woo, and B. Jalali, Lifetime of photogenerated carriers in silicon-on-insulator rib waveguides, *Appl. Phys. Lett.*, vol. 86, February 2005.

34. D. Dimitropoulos, S. Fathpour, and B. Jalali, Limitations of active carrier removal in silicon Raman amplifiers and lasers, *Appl. Phys. Lett.*, vol. 87, pp. 261108-3, 2005.

35. D. Dimitropoulos, S. Fathpour, and B. Jalali, Limitations of active carrier removal in silicon Raman amplifiers and lasers, *Appl. Phys. Lett.*, vol. 87, December 2005.

36. S. Fathpour and B. Jalali, Energy harvesting in silicon optical modulators, *Opt. Express*, vol. 14, pp. 10795–10799, Oct 2006.

37. K. M. Tsia, S. Fathpour, and B. Jalali, Energy harvesting in silicon Raman amplifiers, in *3rd IEEE International Conference on Group IV Photonics, 2006*, 2006, pp. 231–233.

38. K. K. Tsia, S. Fathpour, and B. Jalali, Energy harvesting in silicon wavelength converters, *Opt. Express*, vol. 14, pp. 12327–12333, December 2006.

39. P. Spirito, S. Daliento, A. Sanseverino, L. Gialanella, M. Romano, B. N. Limata, R. Carta, and L. Bellemo, Characterization of recombination centers in Si epilayers after He implantation by direct measurement of local lifetime distribution with the AC lifetime profiling technique, *IEEE Electron Device Lett.*, vol. 25, pp. 602–604, September 2004.

40. V. K. Khanna, Carrier lifetimes and recombination-generation mechanisms in semiconductor device physics, *Eur. J. Phys.*, vol. 25, pp. 221–237, 2004.

41. Y. Liu, C. W. Chow, W. Y. Cheung, and H. K. Tsang, In-line channel power monitor based on helium ion implantation in silicon-on-insulator waveguides, *IEEE Photonics Technol. Lett.*, vol. 18, pp. 1882–1884, September–October 2006.

42. E. K. Tien, F. Qian, N. S. Yuksek, and O. Boyraz, Influence of nonlinear loss competition on pulse compression and nonlinear optics in silicon, *Appl. Phys. Lett.*, vol. 91, November 2007.

43. R. Dekker, A. Driessen, T. Wahlbrink, C. Moormann, J. Niehusmann, and M. Forst, Ultrafast Kerr-induced all-optical wavelength conversion in silicon waveguides using 1.55 μm femtosecond pulses, *Opt. Express*, vol. 14, pp. 8336–8346, September 4, 2006.

44. M. J. Kerr and A. Cuevas, General parameterization of Auger recombination in crystalline silicon, *J. Appl. Phys.*, vol. 91, pp. 2473–2480, February 15, 2002.

45. X. Z. Sang, E. K. Tien, N. S. Yuksek, F. Qian, Q. Song, and O. Boyraz, Dual-wavelength mode-locked fiber laser with an intracavity silicon waveguide, *IEEE Photonics Technol. Lett.*, vol. 20, pp. 1184–1186, July–August 2008.

CHAPTER 5

Theory of Silicon Raman Amplifiers and Lasers

Michael Krause, Hagen Renner, and Ernst Brinkmeyer

Contents

5.1 Introduction

Stimulated Raman scattering (SRS) is a nonlinear optical process in which energy is transferred from an optical pump wave to a signal (or Stokes) wave at a longer wavelength. This energy transfer is mediated by atomic vibrations of the material; thus, it is strongest when the pump and Stokes waves are spectrally separated by an amount corresponding to a vibrational frequency (15.6 THz in silicon). SRS is an attractive optical amplification process since the indirect band structure of silicon that prohibits radiative electronic transitions is no obstacle here; furthermore, no doping of the material is required. Finally, Raman amplification is possible at any wavelength inside the transparency range of silicon ($\lambda = 1.1$–5 µm) provided a pump source with a suitably smaller wavelength is available. The effects of spontaneous and stimulated Raman scattering in silicon waveguides have been experimentally observed for the first time in 2002 and 2003, respectively (Claps et al. 2002; Claps et al. 2003). Since then the topic has been picked up by various groups, which has led to the demonstration of pulsed (Boyraz and Jalali 2004), continuous-wave (Rong et al. 2005), and cascaded (Rong et al. 2008) silicon Raman lasers

(SRLs), as well as silicon Raman amplifiers (SRAs) with a total gain of up to 3.7 dB (Jones et al. 2005), such that SRS can today be considered a practical optical gain mechanism for silicon photonics (Jalali 2008).

Unfortunately, SRS in silicon waveguides competes with two nonlinear absorption effects at the standard telecom wavelengths around 1.5 μm. First, even though silicon is nominally transparent, the effect of two-photon absorption (TPA) becomes noticeable at the high optical powers necessary for Raman amplification. In the TPA process, two photons are absorbed simultaneously, generating an electron–hole pair. The absorption caused by this effect alone does not impair the Raman-amplification process significantly (Claps et al. 2002). However, the increased conductivity of the silicon waveguide due to TPA-generated charge carriers leads in turn to additional optical losses called free-carrier absorption (FCA), which can indeed be substantial (Liang and Tsang 2004) All these effects must be properly balanced in designing Raman-based components in order to achieve optimum performance. The aim of this chapter is to give an overview over the known theoretical results that can help achieve this goal.

In Section 5.2, starting from Maxwell's equations, we derive a full-vectorial model for the propagation of the pump and Stokes waves in a silicon waveguide under the influence of nonlinear effects such as SRS, TPA, and FCA. Readers not interested in the details of the derivation may skip directly to Section 5.2.7, which summarizes the final result. This model is then used in Sections 5.3 and 5.4 to discuss general properties and advanced designs for amplifiers and lasers, respectively. We restrict our discussion to the steady-state (continuous-wave, CW) operation of these devices; literature that discusses their transient behavior will be summarized in Section 5.3.8.

5.2 Fundamentals

In the following, we derive from Maxwell's equations the propagation equations for co- and counter-propagating pump and Stokes waves in a silicon waveguide, where (in contrast to a scalar or semivectorial formulation) all three cartesian mode-field components are taken into account. Parts of the following derivation parallel that of Sipe et al. (2002), Chen et al. (2006), Krause et al. (2007), and Krause et al. (2008). For clarity, we state here the Fourier relationships between a time-domain function $\tilde{\Psi}(t)$ and its frequency-domain counterpart $\Psi(\omega)$ as used throughout this chapter:

$$\Psi(\omega) = \mathcal{F}[\tilde{\Psi}(t)] = \int_{-\infty}^{+\infty} \tilde{\Psi}(t)e^{-j\omega t}\,dt \qquad (5.1)$$

$$\tilde{\Psi}(t) = \mathcal{F}^{-1}[\Psi(\omega)] = \frac{1}{2\pi}\int_{-\infty}^{+\infty} \Psi(\omega)e^{j\omega t}\,d\omega \qquad (5.2)$$

5.2.1 Spectral Amplitudes

The temporal and spatial evolution of the electromagnetic field inside the silicon waveguide is described by Maxwell's equations for a nonmagnetic medium:

$$\nabla \times \tilde{\mathbf{E}} = -\mu_0 \frac{\partial \tilde{\mathbf{H}}}{\partial t}, \qquad \nabla \times \tilde{\mathbf{H}} = \frac{\partial}{\partial t}(\varepsilon_0 n^2 \tilde{\mathbf{E}} + \tilde{\mathbf{P}}) + \tilde{\mathbf{J}}, \qquad (5.3)$$

$$\nabla \cdot (\varepsilon_0 n^2 \tilde{\mathbf{E}} + \tilde{\mathbf{P}}) = \tilde{\rho}, \qquad \nabla \cdot \tilde{\mathbf{H}} = 0 \qquad (5.4)$$

The refractive-index profile $n(x, y)$ of the waveguide does not vary along the waveguide axis z, and $\tilde{\mathbf{P}}(\mathbf{r},t)$ [where $\mathbf{r} = (x, y, z)$] represents a perturbation that will remain completely general in this section. In the ambipolar approximation (which will be used in Section 5.2.6 to incorporate the effect of FCA), the excess hole and electron densities are equal at each point such that the electric charge and current densities may be assumed to be zero in the following; that is, $\tilde{\rho} = 0$ and $\tilde{\mathbf{J}} = 0$. Fourier-transforming Maxwell's equations (Equations 5.3 and 5.4), we therefore obtain

$$\nabla \times \mathbf{E} = -j\omega\mu_0 \mathbf{H}, \qquad \nabla \times \mathbf{H} = j\omega(\varepsilon_0 n^2 \mathbf{E} + \mathbf{P}), \qquad (5.5)$$

$$\nabla \cdot (\varepsilon_0 n^2 \mathbf{E} + \mathbf{P}) = 0, \qquad \nabla \cdot \mathbf{H} = 0 \qquad (5.6)$$

We now assume that light is propagating in the waveguide only in the form of strictly monochromatic waves at the pump and Stokes frequencies (ω_p and ω_s), such that the total electric field may be written in the form

$$E(r,\omega) = E^*(r,-\omega) = E_p(r)\delta(\omega - \omega_p) + E_s(r)\delta(\omega - \omega_s) \quad \text{for } \omega > 0$$

$$(5.7)$$

The nonlinear polarization **P** generated in the waveguide by the electric field (5.7) will also consist of discrete spectral lines, some of which coincide with the $\omega_{p,s}$, and some of which do not. We are interested only in the contributions coinciding with the $\omega_{p,s}$ and therefore write the nonlinear polarization in the form

$$P(r,\omega) = P^*(r,-\omega) = P_p(r)\delta(\omega - \omega_p) + P_s(r)\delta(\omega - \omega_s)$$

$$+ P_{\text{remainder}}(r,\omega) \quad \text{for } \omega > 0$$

$$(5.8)$$

At each frequency ($q = p$ or s) the transverse (x and y) field components of the solution of (5.5) and (5.6) (for any perturbing polarization **P**) can be expanded in the complete set of forward- (+) and backward-propagating (−) normal modes of the linear (**P** = 0) waveguide (Snyder and Love 1983):

$$E_q^t(r) = \sum_m \left[A_{q,m}^+(z)e^{-j\beta_{q,m}z} + A_{q,m}^-(z)e^{j\beta_{q,m}z} \right] f_{q,m}^{+,t}(x,y) \quad (5.9)$$

where the $f_{q,m}^+$ are the electric fields of the forward-propagating mode m of the linear waveguide at frequency ω_q with propagation constants $\beta_{q,m}$. The amplitudes $A_{q,m}^\pm(z)$ are dimensionless. The superscript t denotes the transverse part of the corresponding vector, and we are using the convention that the transverse mode fields are chosen real and independent of the propagation direction. Then the forward- and backward-propagating mode fields are related as (Snyder and Love 1983)

$$f^+ = f^{+,t} + f^{+,z}\hat{z}, \qquad f^- = f^{+,t} - f^{+,z}\hat{z} \quad (5.10)$$

The summations over m in (5.9) are understood to represent the summation over the finite number of guided modes and the integration over all propagating and evanescent radiation modes.

Using the conjugated reciprocity theorem (Snyder and Love 1983), one can show that the amplitudes $A_{q,m}^\pm(z)$ occurring in (5.9) evolve due to the perturbing polarization $P_q(r)$ according to

$$\frac{\mathrm{d}A_{q,m}^{\sigma}(z)}{\mathrm{d}z} = -j\sigma\frac{\omega_q}{4N_{q,m}}e^{j\sigma\beta_{q,m}z}\int \mathbf{f}_{q,m}^{\sigma*}(x,y)\cdot \mathbf{P}_q(x,y,z)\mathrm{d}A \quad (5.11)$$

where σ denotes propagation directions (+ or − when used as superscripts, corresponding to +1 and −1 when written as a variable) and $N_{q,m}$ is the mode normalization

$$N_{q,m} = \frac{1}{2}\int\left[\mathbf{f}_{q,m}^{+}(x,y)\times\mathbf{h}_{q,m}^{+*}(x,y)\right]\cdot\hat{z}\,\mathrm{d}A \quad (5.12)$$

where $\hat{z}$ is the unit vector in the z (propagation) direction, and the $h_{q,m}$'s are the magnetic mode fields. Strictly speaking, (5.11) is only valid for modes with a real propagation constant. The corresponding expression for complex and evanescent modes will not be required in the remainder of this chapter.

5.2.2 Nonlinear Perturbation

As the perturbing polarization $\tilde{\mathbf{P}}$ that occurs in (5.3) and (5.4), we now consider a third-order nonlinearity of the form (Butcher and Cotter 1990)

$$\tilde{P}^i(\mathbf{r},t) = \varepsilon_0\iint\int_0^\infty \tilde{\chi}_{ijkl}^{(3)}(\tau_1,\tau_2,\tau_3)\tilde{E}^j(\mathbf{r},t-\tau_1)\tilde{E}^k(\mathbf{r},t-\tau_2)$$

$$\tilde{E}^l(\mathbf{r},t-\tau_3)\,d\tau_1\,d\tau_2\,d\tau_3 \quad (5.13)$$

where superscripts denote cartesian components x,y,z, and a summation over $j, k, l = x, y, z$ is implicit on the right-hand side. The polarization at a given time t thus depends on the electric field at all earlier times and along all cartesian directions according to the response function $\tilde{\chi}_{ijkl}^{(3)}(\tau_1,\tau_2,\tau_3)$. Expression (5.13) is sufficiently general that it can in principle describe effects such as SRS, TPA, self-phase modulation (SPM), cross-phase modulation (XPM), and four-wave mixing (FWM). FCA will be incorporated later in Section 5.2.6.

The frequency-domain polarization $\mathbf{P}(\mathbf{r},\omega)$ occurring in (5.11) is obtained upon Fourier-transforming (5.13) as

$$P^i(\mathbf{r},\omega) = \frac{\varepsilon_0}{4\pi^2} \int \int_{-\infty}^{+\infty} \chi_{ijkl}^{(3)}(\omega_1,\omega_2,\omega-\omega_1-\omega_2) \tag{5.14}$$

$$\cdot E^j(\mathbf{r},\omega_1)E^k(\mathbf{r},\omega_2)E^l(\mathbf{r},\omega-\omega_1-\omega_2)\,d\omega_1\,d\omega_2$$

where the frequency-dependent $\chi^{(3)}$ tensor is the Fourier-transformed response function

$$\chi_{ijkl}^{(3)}(\omega_1,\omega_2,\omega_3) = \int \int \int_{-\infty}^{+\infty} \tilde{\chi}_{ijkl}^{(3)}(\tau_1,\tau_2,\tau_3)e^{-j(\omega_1\tau_1+\omega_2\tau_2+\omega_3\tau_3)}\,d\tau_1\,d\tau_2\,d\tau_3 \tag{5.15}$$

To evaluate (5.14), we need all three cartesian electric-field components $E^{x,y,z}$. However, (5.9) only gives an expansion of the transverse electric field $E^{x,y}$ in the waveguide. The remaining longitudinal component E^z can be calculated from the transverse components through Maxwell's equations (5.5) and (5.6). As a result, the complete (transverse and longitudinal) field can be written in the form

$$E(\mathbf{r},\omega) = \bar{E}(\mathbf{r},\omega) - \frac{P^z(\mathbf{r},\omega)}{\varepsilon_0 n^2}\hat{\mathbf{z}} \tag{5.16}$$

where we have defined an effective electric field $\bar{E}$ (with all three cartesian components) as

$$\bar{E}(\mathbf{r},\omega) = \bar{E}*(\mathbf{r},-\omega) = \bar{E}_p(\mathbf{r})\delta(\omega-\omega_p) + \bar{E}_s(\mathbf{r})\cdot\delta(\omega-\omega_s) \quad \text{for } \omega > 0 \tag{5.17}$$

where

$$\bar{E}_q(\mathbf{r}) = \sum_m \left[A_{q,m}^+(z)e^{-j\beta_{q,m}z}\mathbf{f}_{q,m}^+(x,y) + A_{q,m}^-(z)e^{j\beta_{q,m}z}\mathbf{f}_{q,m}^-(x,y) \right]$$

for $q = p,s$ \hfill (5.18)

Note that the polarization **P** is given by (5.14) as a function of the electric field **E**, but the electric field itself depends on the polarization through (5.16). An approximate explicit expression for the polarization can be obtained from a perturbation expansion of the polarization (5.14) and the electric field (5.16), keeping only the first-order term:

$$P^i(\omega) = \frac{\varepsilon_0}{4\pi^2} \int\int_{-\infty}^{+\infty} \chi_{ijkl}^{(3)}(\omega_1, \omega_2, \omega - \omega_1 - \omega_2)\bar{E}^j(\omega_1)\bar{E}^k(\omega_2)$$

$$\bar{E}^l(\omega - \omega_1 - \omega_2)\,d\omega_1\,d\omega_2 \qquad (5.19)$$

where we have simplified the notation by suppressing the explicit dependence on **r**, and $\bar{E}$ is the *effective* electric field defined in (5.17).

Finally, for the cases of interest in this chapter, the frequencies of the light propagating in the structure span at most one octave. Physically, this means neglecting the possibility of sum-frequency generation, which is usually not phase matched (Agrawal 2001; Boyd 2003). Equation (5.19) can then be further simplified so as to include only integrations over positive frequencies by making use of the intrinsic permutation symmetry of the nonlinear susceptibility tensor $\chi^{(3)}$ (Boyd 2003) and the reality of the time-domain functions, such that for $\varepsilon > 0$ we have

$$P^i(\omega) = \frac{3\varepsilon_0}{4\pi^2} \int\int_0^{\infty} \chi_{ijkl}^{(3)}(\omega_1, \omega_2, \omega - \omega_1 - \omega_2)\bar{E}^j(\omega_1)\bar{E}^k(\omega_2)$$

$$\bar{E}^{l*}(\omega_1 + \omega_2 - \omega)\,d\omega_1\,d\omega_2 \qquad (5.20)$$

and $P^i(-\omega) = [P^i(\omega)]^*$. Equations (5.17) and (5.20) now explicitly give the nonlinear polarization in terms of the spectral envelope functions $A_{q,m}^{\pm}(z)$, as desired.

In order to find the two nonlinear polarizations $\mathbf{P}_{p,s}(\mathbf{r})$ defined in (5.8) and required for (5.11), we insert our effective electric field of (5.17) into (5.20), yielding

$$P_p^i = \frac{3\varepsilon_0}{4\pi^2}\left[\chi_{ijkl}^{(3)}(\omega_p, \omega_p, -\omega_p)\bar{E}_p^j\bar{E}_p^k\bar{E}_p^{l*} + 2\chi_{ijkl}^{(3)}(\omega_p, \omega_s, -\omega_s)\bar{E}_p^j\bar{E}_s^k\bar{E}_s^{l*}\right]$$

$$(5.21)$$

$$P_s^i = \frac{3\varepsilon_0}{4\pi^2}\left[\chi_{ijkl}^{(3)}(\omega_s,\omega_s,-\omega_s)\bar{E}_s^j\bar{E}_s^k\bar{E}_s^{l*} + 2\chi_{ijkl}^{(3)}(\omega_s,\omega_p,-\omega_p)\bar{E}_s^j\bar{E}_p^k\bar{E}_p^{l*}\right]$$

$$(5.22)$$

where the explicit **r** dependence has been suppressed and the intrinsic permutation symmetry of $\chi^{(3)}$ has been used. The first terms on the right-hand sides of (5.21) and (5.22) represent the nonlinear actions of each wave on itself, while the second terms represent the interaction of the pump and Stokes waves. The induced polarizations at the other two frequencies ($2\omega_s - \omega_p$ and $2\omega_p - \omega_s$) are of no relevance here; by making the two-frequency ansatz (5.7) in the first place, we have implicitly assumed that they are unable to significantly drive the electric field due to phase mismatch.

5.2.3 Single-Mode Approximation

We now make the further assumption that the pump and Stokes fields $\bar{E}_{p,s}$ of (5.18) are well described by a single waveguide mode of the linear waveguide; that is, we assume that the waveguide nonlinearity changes only the amplitude and the phase of the mode during propagation but does not significantly excite other normal modes:

$$\bar{E}_p(\mathbf{r}) = A_p^+(z)e^{-j\beta_p z}\mathbf{f}_p^+(x,y) + A_p^-(z)e^{j\beta_p z}\mathbf{f}_p^-(x,y) \quad (5.23)$$

$$\bar{E}_s(\mathbf{r}) = A_s^+(z)e^{-j\beta_s z}\mathbf{f}_s^+(x,y) + A_s^-(z)e^{j\beta_s z}\mathbf{f}_s^-(x,y) \quad (5.24)$$

The pump and Stokes modes with propagation constants β_p and b_s may be the same or entirely different modes of the structure.

Inserting (5.23) and (5.24) into (5.21) and (5.22) and the result into (5.11), we obtain differential equations for the longitudinal evolution of the pump and Stokes amplitudes $A_{p,s}^\pm(z)$ in terms of just these amplitudes, the corresponding mode fields, and the $\chi^{(3)}$ tensor:

$$\frac{dA_p^\pm(z)}{dz} = \mp j\frac{3\varepsilon_0\omega_p}{16\pi^2 N_p}\sum_{\sigma_1,\sigma_2,\sigma_3\in\{+,-\}}\left(Q_{p,\pm,\sigma_1,p,\sigma_2,\sigma_3} + 2Q_{p,\pm,\sigma_1,s,\sigma_2,\sigma_3}\right)$$

$$(5.25)$$

$$\frac{\mathrm{d}A_s^\pm(z)}{\mathrm{d}z} = \mp j \frac{3\varepsilon_0 \omega_s}{16\pi^2 N_s} \sum_{\sigma_1,\sigma_2,\sigma_3 \in \{+,-\}} \left(Q_{s,\pm,\sigma_1,s,\sigma_2,\sigma_3} + 2Q_{s,\pm,\sigma_1,p,\sigma_2,\sigma_3} \right)$$

$$(5.26)$$

with contributions of the form

$$Q_{q,\sigma_0,\sigma_1,r,\sigma_2,\sigma_3} = A_q^{\sigma_1} A_r^{\sigma_2} A_r^{\sigma_3*} \exp\left(-j\Delta\beta_{q,\sigma_0,\sigma_1,r,\sigma_2,\sigma_3} z\right) I_{q,\sigma_0,\sigma_1,r,\sigma_2,\sigma_3}$$

$$(5.27)$$

where we have defined a nonlinearity parameter

$$I_{q,\sigma_0,\sigma_1,r,\sigma_2,\sigma_3} = \int \chi_{ijkl}^{(3)}(\omega_q,\omega_r,-\omega_r) f_q^{-i\sigma_0} f_q^{j\sigma_1} f_r^{k\sigma_2} f_r^{-l\sigma_3} \, \mathrm{d}A \quad (5.28)$$

and a phase mismatch

$$\Delta\beta_{q,\sigma_0,\sigma_1,r,\sigma_2,\sigma_3} = (\sigma_1 - \sigma_0)\beta_q + (\sigma_2 - \sigma_3)\beta_r \qquad (5.29)$$

Most of the contributions to the right-hand sides of (5.25) and (5.26) include exponential terms that are highly oscillatory along z, as quantified by (5.29). Adopting the usual slowly varying amplitude approximation, we assume that the effect of these contributions. Only those contributions with $\Delta\beta = 0$ will be taken into account. Of these contributions, some turn out to be identical due to the intrinsic permutation symmetry of the $\chi^{(3)}$ tensor, since the latter implies $Q_{q,\sigma_0,\sigma_1,q,\sigma_2,\sigma_3} = Q_{q,\sigma_0,\sigma_2,q,\sigma_1,\sigma_3}$. For example, the longitudinal evolution of forward-propagating Stokes envelopes has four phase-synchronous contributions:

$$\frac{\mathrm{d}A_s^+(z)}{\mathrm{d}z} = T_{s,+,s,+} + 2T_{s,+,s,-} + 2T_{s,+,p,+} + 2T_{s,+,p,-} \qquad (5.30)$$

where $T_{q,\sigma_0,r,\sigma_2} = -j(3\varepsilon_0\omega_q)/(16\pi^2 N_q) \cdot Q_{q,\sigma_0,\sigma_0,r,\sigma_2,\sigma_2}$.

5.2.4 Material Parameters and Effective Areas

The goal of this section is to specialize (5.30) for the specific tensorial structure of the nonlinearities in silicon and write the result in a form that is consistent with the usual notation in fiber optics, that is, by separating the material and waveguide contributions through the introduction of bulk nonlinear constants and effective areas (Agrawal 2001).

5.2.4.1 General Structure of Susceptibility Tensor $\chi^{(3)}$

First we note that in silicon the $\chi^{(3)}$ tensor has two contributions, which, in the Born–Oppenheimer approximation, may be treated independently (Hellwarth 1977): one describing the electronic (or nonresonant) contribution that leads to FWM, SPM, XPM, and TPA; and another one describing the nuclear (resonant) contribution responsible for Raman scattering: $\chi^{(3)} = \chi^{(3),\text{el}} + \chi^{(3),\text{Raman}}$. Since (5.28) is linear in $\chi^{(3)}$, the two contributions may be separated in the amplitude-evolution equation (5.30), yielding

$$\frac{dA_s^+(z)}{dz} = T_{s,+,s,+}^{\text{el}} + 2T_{s,+s,-}^{\text{el}} + 2T_{s,+p,+}^{\text{el}} + 2T_{s,+p,+}^{\text{Raman}} + 2T_{s,+p,-}^{\text{el}} + 2T_{s,+p,-}^{\text{Raman}}$$

$$(5.31)$$

where the Raman contributions to $T_{s,+s,+}$ and $T_{s,+s,-}$ have been neglected, which is permissible since in silicon $\chi^{(3),\text{Raman}}(\omega_s,\omega_s,-\omega_s)$ is small due to the small Raman bandwidth.

Usually, silicon waveguides are oriented along a $\langle 011 \rangle$ direction on the substrate instead of along the crystallographic axes; see Figure 5.1. To make the following discussion more convenient, we evaluate the tensor-field product under the integral in (5.28) in a coordinate system $(\tilde{x}, \tilde{y}, \tilde{z})$ coinciding with the crystallographic axes; that is, a summation over $i,j,k,l = \tilde{x}, \tilde{y}, \tilde{z}$ is implicit on the right-hand side of (5.28). Since silicon is a cubic material belonging to the m3m class, all 81 components of a nonlinear

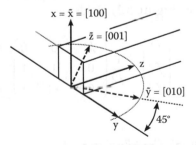

FIGURE 5.1 Definition of coordinate systems: $(\tilde{x}, \tilde{y}, \tilde{z})$ denote the crystallographic axes, while (x, y, z) is a coordinate system aligned with the waveguide. Silicon waveguides are usually fabricated along a $\langle 011 \rangle$ direction, such that (5.34) describes the transformation between the two coordinate systems. (Reprinted, with permission, from M. Krause et al., *Appl. Phys. Lett.* 95, 261111. © 2009, American Institute of Physics.)

susceptibility tensor $\chi^{(3)}$ can be fully expressed in terms of the four elements $\chi_{\tilde{x}\tilde{x}\tilde{x}\tilde{x}}$, $\chi_{\tilde{x}\tilde{x}\tilde{y}\tilde{y}}$, $\chi_{\tilde{x}\tilde{y}\tilde{x}\tilde{y}}$ and $\chi_{\tilde{x}\tilde{y}\tilde{y}\tilde{x}}$ as (Boyd 2003)

$$\chi_{ijkl}^{(3)} = \chi_{\tilde{x}\tilde{x}\tilde{y}\tilde{y}}\delta_{ij}\delta_{kl} + \chi_{\tilde{x}\tilde{y}\tilde{x}\tilde{y}}\delta_{ik}\delta_{jl} + \chi_{\tilde{x}\tilde{y}\tilde{y}\tilde{x}}\delta_{il}\delta_{jk} + \chi_{\text{aniso}}\delta_{ijkl} \quad (5.32)$$

In (5.32), δ denotes the Kronecker symbol, which is unity only when all of its subscripts are equal; the indices i,j,k,l can take the values $\tilde{x}, \tilde{y}, \tilde{z}$, and $\chi_{\text{aniso}} = \chi_{\tilde{x}\tilde{x}\tilde{x}\tilde{x}} - \chi_{\tilde{x}\tilde{x}\tilde{y}\tilde{y}} - \chi_{\tilde{x}\tilde{y}\tilde{x}\tilde{y}} - \chi_{\tilde{x}\tilde{y}\tilde{y}\tilde{x}}$. Using (5.32), we now vcast (5.28) in a more explicit form. Note that for (5.31) the evaluation of (5.28) will only be required for $\sigma_0 = \sigma_1$ and $\sigma_2 = \sigma_3$. We therefore abbreviate $I_{q,\sigma_0,r,\sigma_2} = I_{q,\sigma_0,\sigma_0,r,\sigma_2,\sigma_2}$ and obtain (Dvorak et al. 1994)

$$I_{q,\sigma_0,r,\sigma_2} = \int_{si} \chi_{\tilde{x}\tilde{x}\tilde{y}\tilde{y}} \, |\mathbf{f}_q|^2 |\mathbf{f}_r|^2 + \chi_{\tilde{x}\tilde{y}\tilde{y}\tilde{x}} \, |\mathbf{f}_q^{\sigma_0} \cdot \mathbf{f}_r^{\sigma_2}|^2 + \chi_{\tilde{x}\tilde{y}\tilde{x}\tilde{y}} \, |\mathbf{f}_q^{\sigma_0} \cdot \mathbf{f}_r^{-\sigma_2}|^2$$

$$+ \chi_{\text{aniso}} \sum_{i=\tilde{x},\tilde{y},\tilde{z}} |f_q^i|^2 |f_r^i|^2 \, dA \quad (5.33)$$

where the frequency arguments $(\omega_q, \omega_r, -\omega_r)$ of the nonlinear susceptibilities are implicit. The propagation directions σ_0 and σ_2 of the modes q and r, respectively, do not have an influence on the first and last terms in (5.33), so the corresponding superscripts have been dropped there. Finally, to evaluate the last term in (5.33) we need a relation between the mode-field components along the crystallographic coordinates $(\tilde{x}, \tilde{y}, \tilde{z})$ and those along the original waveguide coordinates (x, y, z). A rotation about the x axis by $45°$ (see Figure 5.1) yields

$$\begin{pmatrix} f^{\tilde{x}} \\ f^{\tilde{y}} \\ f^{\tilde{z}} \end{pmatrix} = \frac{1}{\sqrt{2}} \begin{pmatrix} \sqrt{2} & 0 & 0 \\ 0 & 1 & 1 \\ 0 & -1 & 1 \end{pmatrix} \begin{pmatrix} f^x \\ f^y \\ f^z \end{pmatrix} \quad (5.34)$$

5.2.4.2 Electronic Contribution to $\chi^{(3)}$

As far as the electronic contribution to the nonlinear susceptibilities occurring in (5.33) is concerned, we assume that the four elements sufficient to specify the full $\chi^{(3)}$ tensor according to (5.32) are related as

$$\chi_{\bar{x}\bar{x}\bar{y}\bar{y}}^{el}(\omega_q,\omega_r,-\omega_r) = \chi_{\bar{x}\bar{y}\bar{x}\bar{y}}^{el}(\omega_q,\omega_r,-\omega_r) = \chi_{\bar{x}\bar{y}\bar{y}\bar{x}}^{el}(\omega_q,\omega_r,-\omega_r) =$$

$$\frac{1}{3}\chi_{\bar{x}\bar{x}\bar{x}\bar{x}}^{el}(\omega_q,\omega_r,-\omega_r)$$

$$(5.35)$$

such that $\chi_{aniso}^{el}(\omega_q,\omega_r,-\omega_r) = 0$. The assumption expressed by (5.35) is that the electronic third-order susceptibility in silicon is isotropic and fulfills the Kleinman symmetry condition (Boyd 2003). This is consistent with the experimental results available in the 1.55 μm region from Dinu et al. (2003), Salem and Murphy (2004), and Bristow et al. (2007). However, slight deviations from isotropy or Kleinman symmetry have been reported, too (Zhang et al. 2007; Kagawa and Ooami 2007).

The real and imaginary parts of $\chi_{\bar{x}\bar{x}\bar{x}\bar{x}}^{el}(\omega_q,\omega_r,-\omega_r)$ are related to the bulk Kerr and TPA coefficients $n_{2,qr}$ and b_{qr} (customarily quoted for collinear polarizations of the two involved beams q and r) according to Butcher and Cotter (1990):

$$\chi_{\bar{x}\bar{x}\bar{x}\bar{x}}^{el}(\omega_q,\omega_r,-\omega_r) = \frac{2n_qn_r}{3}\left(\frac{2n_{2,qr}}{Z_0} - j\frac{\beta_{qr}}{\mu_0\omega_q}\right) \qquad (5.36)$$

where Z_0 is the vacuum impedance and $n_{q,r}$'s are linear refractive indices. Using (5.35) and (5.36) in (5.33), the electronic contributions to the right-hand side of the amplitude-propagation equation (5.31) can be written in the desired form

$$T_{q,\sigma_0,r,\sigma_2}^{el} = -\left(\frac{jk_qn_{2,qr}}{A_{qr}^{TPA}} + \frac{\beta_{qr}}{2A_{qr}^{TPA}}\right)\frac{N_r}{\pi^2}\mid A_r^{\sigma_2}\mid^2 A_q^{\sigma_0} \qquad (5.37)$$

where $k_q = 2\pi/\lambda_q$ is the wavenumber, and

$$A_{qr}^{TPA} = \frac{4Z_0^2N_qN_r}{n_qn_r}\left\{\frac{1}{3}\int_{si}\mid f_q\mid^2\mid f_r\mid^2 + \mid f_q\cdot f_r\mid^2 + \mid f_q\cdot f_r^*\mid^2 dA\right\}^{-1} \quad (5.38)$$

is the effective area for the electronic contribution to the $\chi^{(3)}$ non-linearity in silicon waveguides, characterizing the strength of both the TPA and the Kerr processes. It is identical to the effective area derived for the single-beam Kerr effect in tapered fibers by Tzolov et al. 1995. It does not depend on the propagation directions $\sigma_{0,2}$ of the modes; furthermore, $A_{qr}^{TPA} = A_{rq}^{TPA}$.

5.2.4.3 Raman Contribution to $\chi^{(3)}$

The second contribution to the nonlinear susceptibilities occurring in (5.33) is the Raman contribution $\chi^{(3),\text{Raman}}$, which in a classical polarizability model can be written in the form (Shen and Bloembergen 1965; Dimitropoulos et al. 2003)

$$\chi_{ijkl}^{(3),\text{Raman}}(\omega_1,\omega_2,\omega_3) = \chi_{\text{max}}^R H(\omega_1 + \omega_3) R_{ik}^m R_{jl}^m \qquad (5.39)$$

where

$$H(\omega) = \frac{2\omega_v \Gamma}{\omega_v^2 - \omega^2 + j2\omega\Gamma} \qquad (5.40)$$

is the normalized Raman gain spectrum, and R is the Raman tensor of the Brillouin-zone-center optical phonon in silicon. Its components along the crystallographic axes $\tilde{x}, \tilde{y}, \tilde{z}$ are given by

$$R^{\tilde{x}} = \begin{pmatrix} 0 & 0 & 0 \\ 0 & 0 & 1 \\ 0 & 1 & 0 \end{pmatrix}, R^{\tilde{y}} = \begin{pmatrix} 0 & 0 & 1 \\ 0 & 0 & 0 \\ 1 & 0 & 0 \end{pmatrix}, R^{\tilde{z}} = \begin{pmatrix} 0 & 1 & 0 \\ 1 & 0 & 0 \\ 0 & 0 & 0 \end{pmatrix} \qquad (5.41)$$

The Raman shift in silicon, which is given by $\omega_v = 2\pi \times 15.6$ THz, and $\Gamma = 2\pi \times 50$ GHz is the half-width of the Raman gain spectrum (Claps et al. 2003). Equation (5.39) is valid for the cases of interest in this chapter, where it needs to be evaluated with frequency arguments, either $(\omega_1, \omega_2, \omega_3) \approx (\omega_s, \omega_p, -\omega_p)$ or $\approx (\omega_p, \omega_s, -\omega_s)$ while $\omega_p - \omega_s \approx \omega_v$. Raman contributions to other frequency combinations [such as those leading to coherent anti-Stokes Raman scattering (CARS)] are not covered by (5.39).

The four elements sufficient to specify the full $\chi^{(3)}$ tensor according to Equation (5.32) may be obtained from (5.39) and (5.41). The χ_{max}^R occurring in (5.39) is a real constant, such that the on-resonance Raman susceptibility $\chi_{ijkl}^{(3),\text{Raman}} \omega_s, \omega_p, -\omega_p$ is purely imaginary when $\omega_p - \omega_s = \omega_v$. It is related to the bulk Raman-gain constant g_{max} of silicon (customarily quoted for linear pump and Stokes polarizations along orthogonal crystallographic axes) according to (Butcher and Cotter 1990)

$$g_{\max} = \frac{3\mu_0\omega_s}{n_s n_p} \operatorname{Im}\chi_{\bar{x}\bar{x}\bar{y}\bar{y}}^{(3),\text{Raman}}(\omega_s,\omega_p,-\omega_p)\Big|_{\omega_p-\omega_p=\omega_v} = \frac{3\mu_0\omega_s}{n_s n_p}\chi_{\max}^{R} \quad (5.42)$$

Using (5.39) and (5.42) in (5.33), we find that the Raman contribution to the amplitude-propagation equation (5.31) can be written as [note that this contribution only exists when $(q, r) = (p, s)$ or (s, p)]

$$T_{q,\sigma_0,r,\sigma_2}^{\text{Raman}} = j\frac{\lambda_s}{\lambda_q}\frac{g_{\max}H(\omega r-\omega_q)}{4A_{s\sigma_0,p\sigma_2}^{\text{Raman}}}\frac{N_r}{\pi^2}\mid A_r^{\sigma_2}\mid^2 A_q^{\sigma_0} \quad (5.43)$$

The effective area for Raman amplification is given by

$$A_{s\sigma_0,p\sigma_2}^{\text{Raman}} = \frac{4Z_0^2 N_p N_s}{n_p n_s}\left\{\int_{\text{Si}}\mid f_p\mid^2\mid f_s\mid^2 + \mid f_p^{\sigma_0}\cdot f_s^{\sigma_2}\mid^2\right.$$

$$\left. -2(f_s^x)^2(f_p^x)^2 - \left[(f_s^y)^2-(f_s^z)^2\right]\cdot\left[(f_p^y)^2-(f_p^z)^2\right]dA\right\}^{-1} \quad (5.44)$$

where we have made use of the convention that the transverse (i.e., x and y) components of the pump and Stokes electric mode fields $f_{p,s}$ are real, while the z components are imaginary (Snyder and Love 1983).

5.2.4.4 Nonreciprocal Raman Gain in Silicon Waveguides

Note that, in contrast to the "electronic" effective area (5.38), the "Raman" effective area $A_{s\sigma_s,p\sigma_p}^{\text{Raman}}$ given by (5.44), depends on the relative propagation direction of the pump and Stokes waves, that is, in general $A_{s+,p+}^{\text{Raman}} \neq A_{s+,p-}^{\text{Raman}}$. This Raman nonreciprocity results from the existence of longitudinal (i.e., along the propagation direction z) mode-field components that are out of phase by 90° with respect to the transverse field. The sign of this relative phase is inverted by reversing the propagation direction, which mathematically corresponds to a complex conjugation of the corresponding electric mode field; see (5.10). This will affect the second contribution to the integral in (5.44), since in general $\mid f_p\cdot f_s\mid^2 \neq \mid f_p^*\cdot f_s\mid^2$. The co- and counterpropagating modal Raman-gain coefficients may therefore be different.

The longitudinal mode field components can be particularly strong in highly confining submicron silicon photonic wires with their large refractive-index contrast. As an example, we consider the case of a silica-clad strip waveguide 500 nm wide and 220 nm high, where both the pump and the Stokes waves propagate in the quasi-TE mode. Using a full-vectorial mode solver (Krause et al. 2006) for pump and Stokes wavelengths of 1455 and 1574 nm, respectively, we obtain from (5.44) effective areas of 0.12 and 0.080 μm² for co- and counterpropagation, respectively. This result demonstrates that the efficiency of stimulated Raman scattering can depend significantly on whether the pump and Stokes waves propagate co- or counter-directionally—here, the counterdirectional gain is 50% larger than the co-directional gain. SRS in silicon waveguides may therefore be used for realizing all-optical nonreciprocal components (see Sections 5.3.7 and 5.4.4). In Krause et al. (2009b), we further show that the nonreciprocity depends strongly on the waveguide orientation in the substrate plane, that is, on the angle between the y and $\tilde{y}$ axes in Figure 5.1, which in this section was assumed to be fixed to the most commonly used value of 45°. An experimental confirmation of Raman-induced nonreciprocity has been reported by Krause et al. (2010).

Finally, note that since the Raman-induced nonreciprocity is caused by the presence of longitudinal mode-field components, the nonreciprocity is much smaller (<2%) for the larger rib waveguides with dimensions on the order of microns such as those of Rong et al. (2005), where the mode fields do not exhibit strong longitudinal components.

5.2.5 Power-Evolution Equations

In Section 5.2.1, we have assumed strictly monochromatic wave propagation. Furthermore, in Section 5.2.3, we have neglected non-phase-matched terms in the amplitude-propagation equations (5.23) and (5.24). This makes it possible to derive equations for the longitudinal evolution of the *powers* (as opposed to the amplitudes A_q^σ) of the pump and Stokes waves.

From (5.9) and a similar expansion of the transverse magnetic field inside the waveguide, the energy flow of the electromagnetic field can be calculated by integrating the Poynting vector over the waveguide cross section. The expression

$$P_q^\sigma(z) = \frac{N_q}{\pi^2} \mid A_q^\sigma(z) \mid^2 \qquad (5.45)$$

is then seen to represent the longitudinally varying power of the wave $q = p, s$ in the direction $\sigma = +, -$ on a time scale long compared to $2\pi/ \mid \omega_p - \omega_s \mid$. By differentiating (5.45) with respect to z and inserting (5.31), (5.37), and (5.43), we obtain

$$\frac{1}{P_s^+}\frac{dP_s^+}{dz} = -\frac{\beta_{ss}}{A_{ss}^{\text{TPA}}}(P_s^+ + 2P_s^-) + \left(\frac{g}{A_{s+,\,p+}^{\text{Raman}}} - \frac{2\beta_{sp}}{A_{sp}^{\text{TPA}}}\right)P_p^+$$

$$+ \left(\frac{g}{A_{s+,\,p-}^{\text{Raman}}} - \frac{2\beta_{sp}}{A_{sp}^{\text{TPA}}}\right)P_p^- \tag{5.46}$$

where $g = g_{\text{max}} \operatorname{Im} H(\omega_s - \omega_p)$ is the bulk Raman gain corresponding to the given pump and Stokes frequencies $\omega_{p,s}$. Differential equations similar to (5.46) for the remaining powers P_s^- and $P_p^{\pm}$ are summarized in Section 5.2.7.

5.2.6 Incorporation of Free-Carrier Absorption

In this section, we will extend the wave-propagation model derived so far (see Section 5.2.5) to include the effect of FCA. As before, we restrict all derivations to the stationary case, that is, $\partial P_{p,s}^{\pm} / \partial t = 0$.

5.2.6.1 Optical Absorption due to Free Carriers

Excess charge carriers are generated by TPA, diffuse through the silicon, and recombine both inside the silicon and at the interfaces between silicon and the surrounding cladding materials. The steady-state carrier density is determined by an equilibrium between these processes. As we are dealing only with optically generated charge carriers without any externally applied electric fields, the diffusion process is well described in the ambipolar approximation (McKelvey 1966; Seeger 1991); that is, the excess electron and hole densities are equal at each position due to the Coulomb attraction of electrons and holes, and the joint diffusion of electrons and holes is characterized by an ambipolar diffusion constant D.

The dependence of optical losses on the amount of excess charge carriers in silicon has been extensively investigated by Soref and Bennett (1986) and Soref and Bennett (1987). Their results are usually formulated in a compact form as [see, e.g., pages 289 and 367 in Pavesi and Lockwood (2004)]

$$\Delta\alpha = 8.5 \times 10^{-18} \text{cm}^2 \cdot \Delta N_e + 6.0 \times 10^{-18} \text{cm}^2 \cdot \Delta N_h \qquad (5.47)$$

where $\Delta N_{e,h}$ are the densities of excess electrons and holes, and $\Delta\alpha$ is the optical absorption at 1.55 μm. Soref and Bennett (1986) and Soref and Bennett (1987) are often cited as the source of (5.47), even though this formula is not explicitly stated there, and the experimental results reported in those publications deviate somewhat from a linear relationship as given by (5.47). According to the Drude theory of free-carrier absorption, the absorption (5.47) should vary with the square of the wavelength (Soref and Bennett 1987). Setting $\Delta N_e = \Delta N_h = N$ (since we consider ambipolar carrier processes), we thus write (5.47) as (Claps et al. 2004)

$$\Delta\alpha(x, y, z, \lambda) = \bar{\varphi}\lambda^2 N(x, y, z) \qquad (5.48)$$

where N is the free-carrier density, and we have defined the FCA efficiency as

$$\bar{\varphi} = 6.0 \times 10^{-10} \qquad (5.49)$$

The effect of FCA can now be included in the model of Section 5.2.1 by expressing the absorption as the imaginary part of a dielectric constant, $\Delta\varepsilon = j(nc/\omega)\Delta\alpha$ (Agrawal 2001), and then using this to add the additional perturbing polarization $P_{FCA} = \varepsilon_0\Delta\varepsilon\bar{E}$ to the P in (5.5). The result is an additional contribution to the equations describing the longitudinal evolution of the pump and Stokes powers,

$$\pm \frac{1}{P_p^{\pm}}\frac{dP_p^{\pm}}{dz} = -\alpha_p^{FCA}(z), \qquad \pm \frac{1}{P_s^{\pm}}\frac{dP_s^{\pm}}{dz} = -\alpha_s^{FCA}(z) \qquad (5.50)$$

where $\alpha_{p,s}^{FCA}$ is the FCA loss determined by the overlap of the mode with the steady-state carrier density,

$$\alpha_k^{FCA}(z) = \frac{\bar{\varphi}\lambda_k^2 n_k}{2Z_0 N_k}\int_{Si} N(x, y, z)|f_k(x, y)|^2 \, dA \qquad (5.51)$$

which corresponds to the standard perturbation result for the modal attenuation caused by an absorbing waveguide profile (Snyder and Love 1983). To proceed, we need to relate the TPA-generated carrier density $N(x,y,z)$ to the powers of the waves propagating inside the waveguide.

5.2.6.2 Constant-Carrier-Density Approximation

We now assume that the optical powers in the waveguide vary slowly enough along the propagation direction z such that the carrier-generation rate and thus the steady-state carrier density vary with z only on a scale large compared to the carrier's diffusion length. For example, in a simple effective one-dimensional model of a silicon slab of thickness $h = 1$ μm that is sandwiched between two interfaces with recombination velocity of $S = 8000$ cm/s (Dimitropoulos et al. 2005), the effective lifetime is $\tau = h/(2S) = 6.3$ ns. Assuming an ambipolar diffusion constant of $D = 20$ cm²/s (Zhao 2008), we obtain a diffusion length of $L_D = \sqrt{D\tau} = 3.5\mu m$, which is much shorter than the typical lengths over which the pump and Stokes powers will change significantly. Therefore, we can reduce the carrier-diffusion problem to a locally two-dimensional process taking place only in the transverse cross-sectional plane (x,y) of the waveguide. The steady-state carrier density $N(x,y)$ then obeys the two-dimensional diffusion equation (McKelvey 1966)

$$DV^2 N + G - \frac{N}{\tau_b} = 0 \qquad (5.52)$$

where $G(x,y)$ is the local carrier-generation rate, and τ_b is the bulk carrier lifetime. At the interfaces between silicon and the surrounding waveguide materials, N fulfills the boundary conditions

$$D\mathbf{n} \cdot \nabla N = -SN \qquad (5.53)$$

where $\mathbf{n}$ is a unit vector normal to the interface and directed outward, and S is the recombination velocity at the interface. In order to solve (5.52) and (5.53) for the carrier density $N(x,y)$, we need to know the local carrier generation rate $G(x,y)$. The spatial distribution and the magnitude of the generation rate $G(x,y)$ will depend on the mode-field shapes and powers of the various forward- and backward-propagating waves. However, a simple approximate solution can be obtained as follows.

Since the diffusion equation and boundary condition (5.52) and (5.53) are linear in N, their solution can always be expressed as

$$N(x, y) = \int_{Si} G(x', y') M(x, y, x', y') \, dA' \qquad (5.54)$$

where M is the Green's function; that is, $M(x, y, x', y')$ represents the carrier density for a point source located at (x', y'). Inserting (5.54) into (5.51), we obtain the FCA loss seen by mode k in the form

$$\alpha_k^{FCA} = \frac{\overline{\varphi} \lambda_k^2 n_k}{2 Z_0 N_k} \int \int_{Si} G(x', y') |f_k(x, y)|^2 \, M(x, y, x', y') \, dA' dA \qquad (5.55)$$

To proceed, we assume that the diffusion process is so fast that the Green's function $M(x, y, x', y')$ may be considered approximately constant in those regions where the electric mode fields f_k and thus also the carrier generation rate G are concentrated [this assumption is often justified (Dimitropoulos et al. 2005)]. Consequently, we may neglect the variation of M in that region and pull it out of the integral to obtain

$$\alpha_k^{FCA}(z) = \overline{\varphi} \lambda_k^2 \eta \overline{N}(z) \qquad (5.56)$$

where

$$\eta_k = \frac{n_k}{2 Z_0 N_k} \int_{Si} |f_k(x, y)|^2 \, dA \qquad (5.57)$$

is a confinement factor, and

$$\overline{N}(z) = M G_{tot}(z) \qquad (5.58)$$

is an effective free-carrier density. In (5.58)

$$G_{tot} = \int_{\infty} G(x, y) \, dA \qquad (5.59)$$

is the total number of carriers generated through TPA in the entire cross section of the waveguide per unit time and per unit waveguide length (to be derived in Section 5.2.6.3), and M is a constant of proportionality with unit s/m^2 that depends only on the waveguide geometry and the two material parameters, bulk recombination lifetime τ_b, and interface recombination velocity S. The constant M can be related to an "effective carrier lifetime" that is often used in the literature; see Section 5.2.6.4.

For rib waveguides, an explicit expression for M can be derived by assuming that the mode fields and thus the generation rate $G(x,y)$ are confined to the rib (i. e., they do not extend significantly into the adjacent slabs) and that the diffusion in the adjacent slabs is effectively one-dimensional (Dimitropoulos et al. 2005):

$$M = \left\{ \frac{wH}{\tau_b} + wS + \left[w + 2(H-h)\right]S' + 2h\sqrt{D\left(\frac{1}{\tau_b} + \frac{S+S'}{h}\right)} \right\}^{-1} \quad (5.60)$$

where w, h, and H are the width, adjacent-slab height, and total silicon rib height, respectively, and S and S' are the recombination velocities at the bottom and top interfaces, respectively.

5.2.6.3 Total Carrier Generation Rate

An expression for the total carrier generation rate $G_{tot}(z)$ required in (5.58) is obtained by rewriting the differential equations (5.46) in terms of the photon fluxes

$$F_{p,s}^{\pm}(z) = \frac{P_{p,s}^{\pm}(z)}{h\nu_{p,s}} \quad (5.61)$$

which represent the number of photons flowing through the entire waveguide cross section at z per unit time and h in Planck's constant. In terms of these, the TPA contributions to (5.46) and the corresponding equations for the three other waves read as follows:

$$\pm \frac{1}{F_p^{\pm}} \frac{dF_p^{\pm}}{dz} = \frac{\beta_{pp}}{A_{pp}^{TPA}}(F_p^{\pm} + 2F_p^{\mp})h\nu_p - \frac{\beta_{ps}}{A_{ps}^{TPA}}(2F_s^+ + 2F_s^-)h\nu_s \quad (5.62)$$

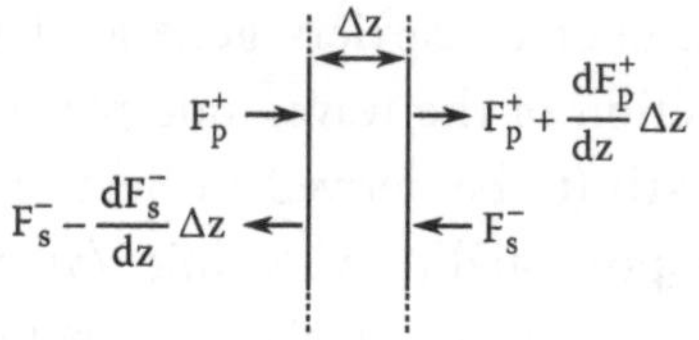

$$\Delta z$$

$$F_p^+ \rightarrow \quad \rightarrow F_p^+ + \frac{dF_p^+}{dz}\Delta z$$

$$F_s^- - \frac{dF_s^-}{dz}\Delta z \leftarrow \quad \leftarrow F_s^-$$

FIGURE 5.2 Illustration of the photon fluxes into and out of a slice Δz of the waveguide (analogous contributions of F_s and F^-_p are not shown). (Reprinted, with permission, from Krause, M., Efficient Raman amplifiers and lasers in optical fibers and silicon waveguides: New concepts, Ph.D. thesis, Technische Universität Hamburg-Harburg, Hamburg, Germany. Cuvillier Verlag, Göttingen, 2007, ©2007 Cuvillier Verlag, Göttingen.)

$$\pm \frac{1}{F_p^\pm}\frac{dF_s^\pm}{dz} = \frac{\beta_{ss}}{A_{ss}^{TPA}}(F_s^\pm + 2F_s^\mp)h\nu_s - \frac{\beta_{sp}}{A_{ps}^{TPA}}(2F_p^+ + 2F_p^-)h\nu_p \quad (5.63)$$

Consider now a thin slice Δz of the waveguide. Figure 5.2 illustrates the photon fluxes entering and leaving this slice. By subtracting the outgoing photon flux from the photon flux going into that slice, one obtains the total number B of photons that are absorbed inside that slice per unit time by TPA:

$$B = \Delta z\left(\frac{dF_p^-}{dz} + \frac{dF_s^-}{dz} - \frac{dF_p^+}{dz} - \frac{dF_s^+}{dz}\right) \quad (5.64)$$

For each two absorbed photons, an electron–hole pair is generated, so that the total carrier-generation rate in that slice is given by $B/2$, and the desired total carrier-generation rate per unit length is $G_{tot} = B/(2\Delta z)$, which in terms of powers reads as follows:

$$G_{tot}^{(z)} = \frac{1}{2}\left[\begin{array}{c} \dfrac{\beta_{pp}}{h\nu_p A_{pp}^{TPA}}\left(P_p^{+2} + P_p^{-2} + 4P_p^+ P_p^-\right) + \dfrac{\beta_{pp}}{h\nu_p A_{pp}^{TPA}}\left(P_p^{+2} + P_p^{-2} + 4P_p^+ P_p^-\right) \\ \\ + \dfrac{4\beta_{sp}}{h\nu_s A_{sp}^{TPA}}\left(P_p^+ + P_p^-\right)\left(P_s^+ + P_s^-\right) \end{array}\right] \quad (5.65)$$

where we have used $\beta_{ps} = (\lambda_s / \lambda_p)\beta_{sp}$ (Shen 1984).

Note the occurrence of a factor of four in the mixed contributions in (5.65), where some authors use a "more intuitive" factor of two. Physically, the factor of four is due to interference. For example, consider the two counterpropagating pump beams P^+ and P^-. When they are ideally monochromatic, there will be a standing-wave interference pattern inside the waveguide. In the valleys, no TPA will occur, while in the peaks TPA will be especially strong since TPA is proportional to the fourth power of the electric field. As a result, on the average more carriers are generated than expected from the square of the sum of the individual intensities (the spatial averaging over the interference pattern is implicitly performed by the slowly varying amplitude approximation during the derivation of the propagation equations in Section 5.2.3).

5.2.6.4 Relation to Conventional Models

Here we show how the constant M introduced in Section 5.2.6.2 may be related to the effective carrier lifetime τ_{eff} that occurs in other models in the literature. Let us consider the propagation of a single wave propagating in the waveguide. Dropping the respective subscripts "p" or "s," in our model [(5.46) and (5.56)], the wave is attenuated due to TPA and FCA according to

$$\frac{1}{P}\frac{dP}{dz}\bigg|_{\text{our model}} = -\frac{\beta}{A^{\text{TPA}}}P - = -\frac{\overline{\varphi}\lambda^2\eta M\beta}{2h\nu A^{\text{TPA}}}P^2 \qquad (5.66)$$

where P is the power of the wave, A^{TPA} is the TPA effective area of (5.38), and η is the confinement factor of (5.57). On the other hand, in conventional models (Claps et al. 2004; Rong et al. 2004; Liu et al. 2006), the corresponding propagation equation is

$$\frac{1}{P}\frac{dP}{dz}\bigg|_{\text{conventional model}} = -\frac{\beta}{A_{\text{eff}}P}P - \frac{\overline{\varphi}\lambda^2\beta\tau_{\text{eff}}}{2h\nu A_{\text{eff}}^2}P^2 \qquad (5.67)$$

where A_{eff} is an effective area that might be slightly different from ours, that is, $A_{\text{eff}} \neq A^{\text{TPA}}$. We therefore cannot expect an exact agreement between the predictions of the conventional model and the more detailed one derived in this chapter. However, we can obtain an order-of-magnitude estimate for the M by ignoring the difference between the two areas and assuming $A_{\text{eff}} = A^{\text{TPA}}$. Then the TPA contributions to (5.66) and (5.67) are identical, and by requiring that the FCA contributions to (5.66) and (5.67) are also identical we obtain the desired relation between M and the effective lifetime τ_{eff} occurring in the conventional model:

$$\tau_{\text{eff}} = \eta A^{\text{TPA}} M \qquad (5.68)$$

As an example of the application of (5.68), an estimate for the constant M required for our simulations is now obtained as follows. In a standard silica-clad strip waveguide (500×220 nm^2), a typical lifetime seen by the fundamental mode would be $\tau_{\text{eff}} = 1$ ns (Yamada et al. 2005). The effective area and the confinement factor of the fundamental quasi-TE mode are numerically calculated at our wavelength of 1455 nm to be $A^{\text{TPA}} = 0.086$ μm^2 and $\eta = 1.1$, respectively, yielding an M of 1.1×10^4 s/m^2. For larger rib waveguides—for example, those of Rong et al. (2005)—the confinement factor is near-unity, and (5.68) can further be simplified to $M = \tau_{\text{eff}}/A^{\text{TPA}}$, yielding a value of $M = 625$ s/m^2 for the corresponding $A^{\text{TPA}} = 1.6$ μm^2 (compare Section 5.4.2) and a lifetime of 1 ns.

5.2.7 Summary: Model for Continuous-Wave Operation

We now summarize the model derived so far. It is the main result of this section and forms the basis for the treatment of amplifiers and lasers in Sections 5.3 and 5.4.

Adding the FCA contribution (5.50) to the power-evolution equation (5.46) and incorporating the linear waveguide losses $\alpha_{p,s}$ phenomenologically, we obtain the final CW model in the following form. The longitudinal evolution of the powers of forward- (+) and backward-propagating (−) CW pump ("p") and Stokes ("s") waves is governed by the differential equations

$$\pm \frac{1}{P_p^{\pm}} \frac{\mathrm{d}P_p^{\pm}}{\mathrm{d}z} = -\alpha_p - \frac{\lambda_s}{\lambda_p} g \left(\frac{P_s^{\pm}}{A_{s+,p+}^{\text{Raman}}} + \frac{P_s^{\mp}}{A_{s+,p-}^{\text{Raman}}} \right)$$

$$- \beta_{pp} \frac{P_p^{\pm} + 2P_p^{\mp}}{A_{pp}^{\text{TPA}}} - \frac{\lambda_s}{\lambda_p} \beta_{sp} \frac{2P_s^{+} + 2P_s^{-}}{A_{sp}^{\text{TPA}}} - \eta_p \overline{\varphi} \lambda_p^2 \overline{N}$$

$$\text{(5.69)}$$

$$\pm \frac{1}{P_s^{\pm}} \frac{dP_s^{\pm}}{dz} = -\alpha_s + g \left(\frac{P_p^{\pm}}{A_{s+,p+}^{\text{Raman}}} + \frac{P_p^{\mp}}{A_{s+,p-}^{\text{Raman}}} \right)$$

$$\text{(5.70)}$$

$$- \beta_{ss} \frac{P_s^{\pm} + 2P_s^{\mp}}{A_{ss}^{\text{TPA}}} - \beta_{sp} \frac{2P_p^{+} + 2P_p^{-}}{A_{sp}^{\text{TPA}}} - \eta_s \overline{\varphi} \lambda_s^2 \overline{N}$$

where the five terms on the right-hand sides of (5.69) and (5.70) represent, respectively, linear (scattering) losses given by

α_p and α_s, SRS, frequency-degenerate TPA, nondegenerate TPA, and FCA. The effective free-carrier density $\bar{N}$ occurring in Eqs. (5.69) and (5.70) is given by

$$\bar{N} = \frac{M}{2}\left[\frac{\beta_{pp}}{h\nu_p A_{pp}^{\text{TPA}}}(P_p^{+2} + P_p^{-2} + 4P_p^+ P_p^-) + \frac{\beta_{ss}}{h\nu_s A_{ss}^{\text{TPA}}}(P_s^{+2} + P_s^{-2} + 4P_s^+ P_s^-) \right.$$

$$\left. + \frac{4\beta_{sp}}{h\nu_s A_{sp}^{\text{TPA}}}(P_p^+ + P_p^-)(P_s^+ + P_s^-) \right] \qquad (5.71)$$

where M characterizes the steady state of the carrier diffusion process. It can be related to the effective carrier lifetime τ_{eff} used in conventional models; see Section 5.2.6.4. The constants g and β_{ab} represent the bulk coefficients for SRS and TPA in silicon. Tsang and Liu (2008) give an overview of the various measurement results that are available for these constants.

The effective areas occurring in (5.69)–(5.71) and defined in (5.38) and (5.44) have a different form than in fiber optics. Only for weakly guiding waveguides and strictly y-polarized mode fields do the effective areas derived here become identical and reduce to the one used in fiber optics (Agrawal 2001). This is a good approximation for the quasi-TE modes of silicon rib waveguides, for example. While many basic properties of SRAs and SRLs can be understood with models less detailed than (5.69)–(5.71) (for example, by treating all of the effective areas as approximately equal), the full model as described here will be used to analyze cladding-pumped silicon Raman amplifiers (Section 5.3.5) and cascaded silicon Raman lasers (Sections 5.4.3 and 5.4.4). Finally, we note that the simplified version of equations (5.69)–(5.71) as published by us earlier (Krause et al. 2004) is not consistent since it neglects the difference between the pump and Stokes wavelengths for the TPA process, while allowing for it in the FCA and SRS contributions to the propagation equations.

5.3 Amplifiers

In this section, we discuss the performance of *unsaturated* silicon Raman amplifiers, that is, we assume signal or Stokes waves that propagate much less power than the pump wave and thus the Stokes does not deplete the pump. Analytical results for *saturated* amplifiers, however, are scarce in the literature. On the one hand, Rukhlenko et al. (2009) obtain explicit results for saturated SRAs, after performing several mathematical simplifications of the model presented in Section 5.2.7. On the other hand, Renner (2010) derives explicit optimal pump conditions and an upper limit of the amplifiable Stokes power for saturated SRAs without needing any analytical solution of the pump and Stokes evolution along the amplifier.

We define the total gain G of an amplifier as the ratio of the signal power at the output to that at the input:

$$G = \frac{S_{\text{out}}}{S_{\text{in}}} \tag{5.72}$$

Gains are usually quoted in decibel units, that is, $G_{\text{dB}} = 10 \log_{10} G$.

5.3.1 Single-Pass Pumping in the Absence of Nonlinear Absorption

Concise and explicit design rules for silicon Raman amplifiers can be derived for the case that the nonlinear absorption effects TPA and FCA are negligible. This is the case in the mid-infrared (mid-IR) spectral region (Soref et al. 2006; Raghunathan et al. 2007), where the energy of two photons is not sufficient to excite an electron from the valence band into the conduction band. The only physical effects we take into account here are stimulated Raman scattering and linear waveguide losses. In the most simple scheme considered in this section, we assume that the pump and Stokes waves pass the amplifying waveguide exactly once (single-pass pumping, SPP; see Figure 5.3a) without seeing any resonator structures or reflectors. Simplifying the model of Section 5.2.7, the evolution of the pump and Stokes powers P and S, respectively, is therefore governed by the differential equations

$$\frac{1}{P} \frac{dP}{dz} = -\alpha - \frac{\lambda_s}{\lambda_p} \frac{g}{A} S \tag{5.73}$$

$$\frac{1}{S} \frac{dS}{dz} = -\alpha + \frac{g}{A} P := \gamma(z) \tag{5.74}$$

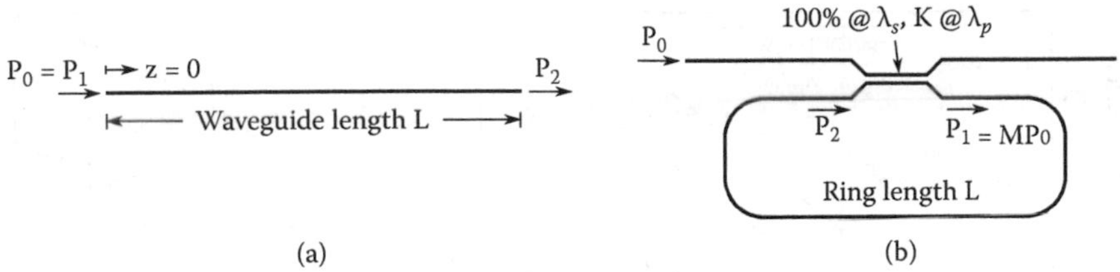

FIGURE 5.3 Silicon Raman amplifiers (*a*) with single-pass pumping (SPP) and (*b*) with ring-resonator (RR) enhancement of the pump power. The signal, not indicated here, passes the amplifying waveguide only once without seeing any resonator. (Reprinted, with permission, from Krause (2010), ©2010 IEEE.)

where A is the effective modal area (5.44), g is the bulk Raman-gain constant of silicon, α represents the linear waveguide losses that we assume to be equal for the pump and Stokes waves, and $\gamma(z)$ is the local gain experienced by the signal. Furthermore, we assume unsaturated operation of the amplifier; that is, the signal powers are so small that the depletion of the pump wave by the SRS process, represented by the second term on the right-hand side of (5.73), can be neglected.

By integrating the local gain along the waveguide of length L, one obtains the total gain G of the amplifier,

$$G = \frac{S(L)}{S(0)} = \exp\left[\int_0^L \gamma(z)\,dz\right] = \exp(-\alpha L)e^{F[1-\exp(-\alpha L)]} \quad (5.75)$$

where

$$F = \frac{gP}{\alpha A} \quad (5.76)$$

io a normalized pump power. Equating to zero the derivative of (5.75) with respect to L shows that by choosing the optimal length

$$L_{\text{opt}} = \tfrac{1}{\alpha}\ln F \quad (5.77)$$

the amplifier gain reaches its maximum

$$G_{\text{max}} = \frac{e^F}{e \cdot F} \quad (5.78)$$

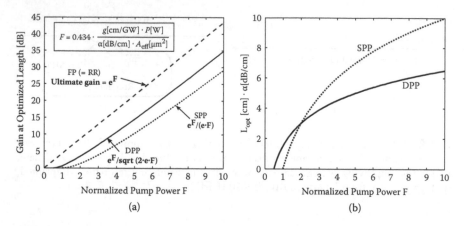

FIGURE 5.4 (a) Maximum possible gain and (b) corresponding optimal lengths of silicon Raman amplifiers for various pumping schemes. The ultimate limit for the achievable gain (dashed curve) is given by $\exp(F)$ (see Section 5.3.4.2), where F is the normalized pump power defined in the inset. (Reprinted, with permission, from Krause (2008), ©2008 IEEE.)

The dotted curves in Figure 5.4 show the maximum SPP gain and the corresponding optimal length as a function of the normalized pump power F. Realization of an SPP amplifier with $G > 0$ dB is only possible when $F > 1$. For example, using a waveguide with an effective area of A = 1.0 μm², assuming g = 7.5/GW (Raghunathan et al. 2007) and a rather large α = 2 dB/cm in order to take into account the increased attenuation of the surrounding silica at mid-IR wavelengths (Soref et al. 2006), an amplifier can be built provided at least 616 mW of pump power are available. If $F < 1$, it is impossible to build an SPP amplifier. By using a double-pass scheme (DPP), the limit can be lowered to $F_{min,DPP}$ = 1/2 (Krause and Renner 2008), and by using resonant pumping, it is even possible to obtain amplifiers with a gain >0 dB for every pump power, however small it is; see Section 5.3.4.

5.3.2 Single-Pass Pumping in the Presence of TPA

While the previous section (Section 5.3.1) ignored all nonlinear absorption effects, we now include TPA (Renner 2007). As mentioned in the introduction to Section 5.3, we again assume the signal Stokes power $S(z)$ to be small enough not to deplete the pump wave $P(z)$. In the absence of FCA, the equations for propagation along the waveguide distance z, (5.69)–(5.71), can be written as follows:

$$\frac{1}{P}\frac{dP}{dz} = -\alpha - \frac{\beta}{A}P \tag{5.79}$$

$$\frac{1}{S}\frac{dS}{dz} = -\alpha + \frac{g-2\beta}{A}P =: \gamma(z) \tag{5.80}$$

Here, A is the effective modal area (assumed equal for TPA and SRS, as in micron-sized rib waveguides; see the discussion in Section 5.2.7), α is the linear attenuation constant due to absorption and mainly scattering, while β and g are the TPA and Raman-gain coefficients of silicon, respectively. The local Raman gain for the Stokes wave is given by $\gamma(z) = -\alpha + (g - 2\beta)$ $P(z)/A$ on the right-hand side of (5.80). In the following calculations, we assume the same parameters as in Krause et al. (2004): g = 20 cm/GW, α = 1.0 dB/cm, β = 0.7 cm/GW, and use A = 1 μm².

With an input pump power $P(0) = P_0$ assumed to be launched at the waveguide input $z = 0$, (5.79) is solved by

$$P(z) = \frac{\alpha A P_0}{(\beta P_0 + \alpha A)\exp(\alpha z) - \beta P_0} \tag{5.81}$$

We derive from (5.80) the total gain G of an SRA of length L:

$$G = \frac{S(L)}{S(0)} = \exp\left\{-\alpha L + \frac{g-2\beta}{A}\int_0^L P(z)dz\right\} \tag{5.82}$$

In order to achieve $G \geq 1$ (= 0 dB) at all, the input pump power P_0 has to be larger than $P_{0,\min} = \alpha A/(g - 2\beta)$ to ensure a positive local gain $\gamma(z)$ at least at the waveguide input $z = 0$. Using (5.81) in (5.82), we obtain the total Raman gain taking into account TPA but neglecting FCA:

$$G = \exp(-\alpha L)\left\{1 + \frac{\beta P_0}{\alpha A}\left[1 - \exp(-\alpha L)\right]\right\}^{(g-2\beta)/\beta} \tag{5.83}$$

In Figure 5.5a, the total gain is shown as a function of the waveguide length L and the input pump power P_0. For each L, the total gain G increases monotonously with P_0. With $g > 2\beta$, we conclude

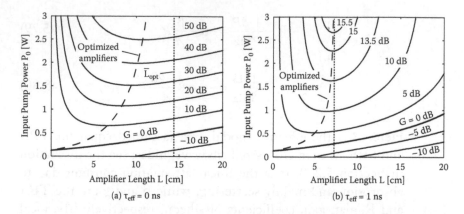

FIGURE 5.5 (a) Total gain G of SRAs with TPA, but without FCA (reprinted, with permission, from Renner (2007), ©2007 IEEE.); (b) same, but with FCA. The dashed lines designate the optimal amplifier lengths from (5.84) and (5.94) (in the latter case, P_0 replaced for P_U), respectively. The optimal lengths asymptotically approach the limits (5.86) and (5.94) (in the latter, ∞ replaced for P_U, respectively, indicated by vertical lines).

from (5.83) that any value of total gain G can theoretically be reached simply by increasing the input pump power P_0, because G increases monotonously with P_0 for any length L of the amplifier.

Under the restriction that only a certain limited input pump power $\hat{P}_0 > P_{0,\min}$ is available, the optimum length $L_{opt} = L_{opt}(\hat{P}_0)$ of the amplifier waveguide providing the largest gain $\hat{G} = G(L_{opt}, \hat{P}_0)$ possible follows from maximizing G in (5.83) over L. We obtain the optimal amplifier length

$$L_{opt} = \frac{1}{\alpha} \ln \left[\frac{\hat{P}_0 (g - \beta)}{\alpha A + \beta \hat{P}_0} \right] \tag{5.84}$$

and the corresponding maximum possible total gain is

$$\hat{G} = \frac{(\alpha A + \beta \hat{P}_0)}{\hat{P}_0 (g - \beta)} \left[\frac{(g - 2\beta)(\alpha A + \beta \hat{P}_0)}{(g - \beta)\alpha A} \right]^{\frac{g - 2\beta}{\beta}} \tag{5.85}$$

In order to realize an amplifier with $G = 10$ dB total gain, for example, one finds from Figure 5.5a an optimum length of $L = 6.6$ cm, requiring an input pump power of $\hat{P}_0 = 0.65$ W.

When trying to increase the total gain further and further by increasing the input pump power $\hat{P}_0 \gg \alpha A / \beta$, one observes the optimum length to become less and less dependent of $\hat{P}_0$ and to tend toward an asymptotic limit $\overline{L}_{opt}$,

$$\overline{L}_{opt} = L_{opt}(P_0 \gg \alpha A / \beta) = \frac{1}{\alpha} \ln\left(\frac{g-\beta}{\beta}\right) \qquad (5.86)$$

Assuming linear attenuation coeffeicients α of 0.2 dB/cm, or 1.0 dB/cm, (5.86) yields optimum high-power amplifier lengths $\overline{L}_{opt}$ of 72.0 cm or 14.4 cm, respectively. The optimum length L_{opt} never exceeds $\overline{L}_{opt}$ independently of the input pump power, and any length in excess of $\overline{L}_{opt}$ would necessarily add undesired attenuation. This is in contrast to the "TPA-less" case studied in Section 5.3.1, where the optimal length increases without limit as the available pump power increases. Moreover, while decreasing either α or A in (5.83) affects the total gain of an arbitrary amplifier in different ways, the maximum total gain (5.85) possible for a fixed input pump power $\hat{P}_0$ can be increased only by decreasing the product αA. When designing the amplifier, it should be kept in mind that a reduction of the modal dimensions often goes along with an increase of the scattering losses (Sparacin et al. 2005).

Figure 5.5b shows the same data as Figure 5.5a, but including the effect of FCA with an effective carrier lifetime of $\tau_{eff} = 1$ns. As discussed in the following section (Section 5.3.3), FCA has a significant impact on the characteristics of an SRA.

5.3.3 Single-Pass Pumping in the Presence of TPA and FCA

In this section, we finally analyze the full amplifier model, including linear losses, SRS, and TPA as well as FCA. For simplicity, we here assume equal effective areas A_{eff} for TPA and SRS, as in micron-sized rib waveguides; see the discussion in Section 5.2.7. The evolution of the pump (P) and signal (S) powers inside the waveguide is then governed by the differential equations (Rong et al. 2006)

$$\frac{1}{P}\frac{dP}{dz} = -\alpha - b_p P - c_p P^2 = \phi(P) \qquad (5.87)$$

$$\frac{1}{S}\frac{dS}{dz} = -\alpha + b_s P - c_s P^2 = \gamma(P) \qquad (5.88)$$

where we abbreviate $b_p = \beta/A_{\text{eff}}$, $b_s = (g - 2\beta)/A_{\text{eff}}$, $c_p = \sigma_p \beta \tau_{\text{eff}} / (2E_p A_{\text{eff}}^2)$, and $c_s = (\lambda_s / \lambda_p)^2 c_p$. The first and last terms on the right-hand sides of (5.87)–(5.88) represent linear (α) and free-carrier losses, respectively, while the second terms represent degenerate TPA losses and the combined action of Raman gain and nondegenerate TPA, respectively. Furthermore, g and β are the SRS and TPA coefficients of bulk silicon, respectively, σ_p is the FCA cross section at the pump wavelength, τ_{eff} is the effective free-carrier lifetime, E_P is the pump-photon energy, and λ_p and λ_s are the pump and signal wavelengths, respectively. As in the preceding sections, here we assume unsaturated operation of the amplifier; that is, the Stokes powers are much smaller than the pump powers (S << P) such that the depletion of the pump by SRS may be neglected.

5.3.3.1 Numerical Simulation

Given $P_1 = P(z = 0)$ is the injected pump power at the waveguide input $z = 0$ and L is the amplifier length, the total gain of the amplifier is

$$G(P_1, L) = \frac{S(L)}{S(0)} = \exp \int_0^L \gamma[P(z)]dz \qquad (5.89)$$

where the right-hand side follows from (5.88). Unfortunately, there is no known explicit expression for the pump-power evolution $P(z)$ as a function of arbitrary P_1, L and material parameters, so (5.89) cannot be made more explicit. A numerical method has to be used to calculate $P(z)$, and the integration in (5.89) can then be performed numerically.

The typical form of the characteristics [i.e., gain G versus pump power $P(0)$] of SRAs is shown in Figure 5.6b. The curves are numerical simulations of the gain G using the models (5.87) and (5.88), while the symbols are the corresponding experimental results of Liu et al. (2006). This waveguide was equipped with a p-i-n structure in order to extract the TPA-generated free carriers, different bias voltages thus corresponding to different effective carrier lifetimes τ_{eff}. Note that the amplifier gain in Figure 5.6b does not increase monotonously with an increase of the pump power; instead, there is always a roll-over point beyond which the gain decreases again. This is due to FCA, which starts to dominate over the Raman gain at large pump powers (a similar

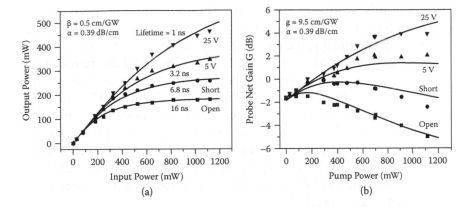

FIGURE 5.6 Characteristics of a p-i-n silicon waveguide for various bias voltages. (a) Output pump power versus input pump power; (b) total Stokes gain *G* versus input pump power. Symbols are experimental results, and curves are numerical calculations based on (5.87) and (5.88). (Reprinted, with permission, from Liu A. et al., *J. Lightwave Technol.* **24**(3): 1440–1455, 2006, ©2006 IEEE.)

behavior is exhibited by the characteristics of silicon Raman lasers; see Section 5.4.2).

Finally, Figure 5.6a shows the pump-transmission characteristics $P(L)$ versus $P(0)$. As in Figure 5.6b, the agreement between theory and experimental results is very good, which confirms the practical applicability of the models (5.87) and (5.88). We note in passing that for a different set of experimental data (Rong et al. 2004), we reported a similarly good agreement between theory and experiment in Figure 5.2 of Krause et al. (2004). Unfortunately, this comparison is not valid since we misinterpreted the experimental data of Rong et al. (2004).

To quantitatively illustrate the results in the remainder of this section, we assume the following material and waveguide parameters: λ_p = 1550 nm, λ_s = 1686 nm, β = 0.5 cm/GW, g = 9.5 cm/GW, α = 0.6 dB/cm, A_{eff} = 1.6 μm², and σ_P = 1.45 × 10⁻¹⁷ cm². For example, the observed threshold powers of the various silicon Raman lasers characterized by Rong et al. (2006) are accurately predicted by numerical simulations of (5.87) and (5.88) when using these parameters, as shown by Rong et al. (2006).

5.3.3.2 Explicit Analytical Optimization

While in general the gain of a given amplifier has to be calculated numerically (see Section 5.3.3.1), it is nevertheless possible to derive fully explicit formulas for the maximum possible gain as well as for the optimal length and pump power required to achieve the latter (Renner and Krause 2006; Krause et al. 2010).

In Figure 5.7, the solid curve illustrates the distribution of the pump power $P(z)$ inside the amplifier. Since all the terms in the differential equation (5.87) for P contribute to an attenuation, that is, $0(P) < 0$, P decays strictly monotonously. Also, (5.87) is a first-order differential equation; thus, the shape of the function $P(z)$ depends only on the material parameters and on the initial value at any position z_0, for example at the waveguide input. Thus, increasing or decreasing the input pump power simply shifts the rest of the solid curve in Figure 5.7 toward larger or shorter distances z, respectively.

Corresponding to the pump power $P(z)$ at each point z is a local gain $\gamma[P(z)]$ experienced by the signal wave, which is determined by $P(z)$ according to (5.88). The dashed curve in Figure 5.7 shows the distribution of the local gain $\gamma(z)$. For large local pump powers (left-hand side of Figure 5.7), the local signal gain is negative due to the dominance of the FCA term $c_s P^2$ in (5.88). For small local pump powers (right-hand side of Figure 5.7), the local

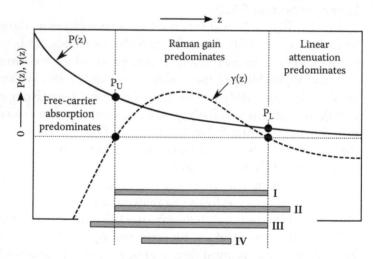

FIGURE 5.7 Schematic view of the longitudinal distribution of the pump power $P(z)$ (solid) and the corresponding local signal gain $\gamma(z)$ (dashed). In the optimal single-pass-pumped amplifier (I), the signal experiences the largest possible amount of only positive local gain γ; the corresponding maximum possible total gain is represented by the area under the positive part of the dashed curve. (II) SRA too long, total gain reduced by linear attenuation near the output. (III) SRA too long, total gain reduced by FCA near the input. (IV) SRA too short, only part of the total gain exploited. (Reprinted, with permission, from Krause, M. et al., *IEEE J. Sel. Top. Quantum Electron.* **16**(1): 216–225, 2010, ©2010 IEEE.)

signal gain is negative due to the dominance of the linear losses α. Between these two regimes there is a pump-power range $[P_L, P_U]$ in which Raman gain predominates and the local signal gain is positive. The limits of this regime are obtained by finding the zeros of $\gamma(P)$ of (5.88):

$$P_L = \frac{b_s - w_s}{2c_s} \quad \text{and} \quad P_U = \frac{b_s + w_s}{2c_s} \tag{5.90}$$

where $w_s = (b_s^2 - 4\alpha c_s)^{1/2}$. When the material parameters are such that $\alpha c_s > b_s^2 / 4$ (i.e., either the linear losses α or the free-carrier losses τ_{eff} are too large), the local signal gain cannot become positive for any pump power. No amplifier can be realized with this waveguide technology then. Throughout the remainder of this section, we assume the more useful case

$$\alpha c_s < \frac{b_s^2}{4} \quad (\text{i.e.}, w_s = \text{real}) \tag{5.91}$$

We are here interested in the total gain of (5.89), which is related to the area under a certain portion of the dashed curve in Figure 5.7. The pump power P_1 and the amplifier length L determine exactly which portion that is. Intuitively, it is clear that there is a maximum possible gain that corresponds to the integral exclusively over the positive part of the dashed curve. The pump power and the amplifier length have to be arranged such that only this part of the curve contributes to the integral in (5.89).

Explicit expressions for the optimal length, the optimal pump power, and the maximum possible gain can be derived by exploiting the monotonicity of $P(z)$ in order to express the gain (5.89) not in terms of the pump power P_1 and L, but instead in terms of the pump power P_1 and the residual pump power P_2 leaving the waveguide (Renner and Krause 2006; Krause et al. 2010). The maximum possible total gain $\hat{G}$ is given explicitly by

$$\hat{G} = \frac{P_L}{P_U} \left(\frac{\alpha + b_p P_L + c_p P_L^2}{\alpha + b_p P_U + c_p P_U^2} \right)^{(c_s / c_p - 1)/2}$$

$$\cdot \left(\frac{2c_p P_L + b_p + w_p}{2c_p P_L + b_p - w_p} \cdot \frac{2c_p P_U + b_p - w_p}{2c_p P_U + b_p + w_p} \right)^{[2b_s + b_p(1 + c_s / c_p)]/(2w_p)} \tag{5.92}$$

where P_U and P_L are given in (5.90), and $w_p = (b_p^2 - 4\alpha c_p)^{1/2}$. The optimal pump power required to achieve the maximum possible gain $\hat{G}$ is

$$\hat{P} = P(0) = P_U \qquad (5.93)$$

and the required optimal length $\hat{L}$ is that which makes the pump power decay from $P(0) = P_U$ at the waveguide input to precisely the value $P(\hat{L}) = P_L$ at the waveguide output, given explicitly by

$$\hat{L} = \frac{1}{2\alpha} \ln\left[\frac{P_U^2}{P_L^2} \cdot \frac{\alpha + b_p P_L + c_p P_L^2}{\alpha + b_p P_U + c_p P_U^2} \cdot \left(\frac{2c_p P_U + b_p + w_p}{2c_p P_U + b_p - w_p} \cdot \frac{2c_p P_L + b_p - w_p}{2c_p P_L + b_p + w_p} \right)^{b_p/w_p} \right].$$

$$(5.94)$$

If the optimal pump power P_U is not available, the amplifier gain is maximized by replacing whatever pump power is available (P_{avail}) for P_U (Krause et al. 2010). Finally, in case the available pump power is too low ($P_{avail} \le P_L$), no SPP amplifier with total gain >0 dB can be realized.

For the parameters chosen in Section 5.3.3.1, we show the maximum possible SPP gain $G_{SPP,max}$ in Figure 5.8a as a function of the effective free-carrier lifetime τ_{eff} and the available pump power P_{avail}. For example, a gain of more than 3 dB can be achieved with a pump power of 1 W when the lifetime is reduced to values below 1 ns. The corresponding optimal length $L_{spp,opt}$ is on the order of 7 cm (Figure 5.8b). The thick dashed curve in Figure 5.8a represents the minimum required pump power P_L. If the available pump power is lower than P_L, it is impossible to realize an SPP amplifier. Even when reducing the lifetime τ_{eff} to zero, a minimum pump power of 260 mW is necessary to realize an amplifier in order to compensate for the linear losses α. On the other hand, the thick dotted curve in Figure 5.8a represents the maximum useful pump power P_U: larger pump powers would inevitably decrease the amplifier gain due to the predominance of FCA. Therefore, the optimal pump power is clamped at P_U, and above the thick dotted curve in Figures 5.8a and 5.8b the gains and lengths remain independent of P_{avail}.

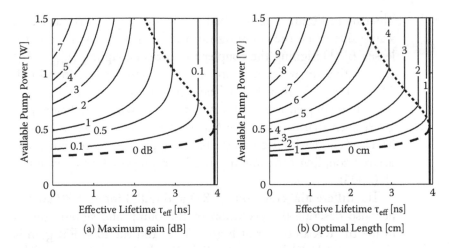

FIGURE 5.8 Optimal single-pass-pumped (SPP) silicon Raman amplifiers as a function of the effective lifetime τ_{eff} and the available pump power P_{avail}: (a) maximum gain, (b) optimal length. Assumed waveguide parameters are $\alpha = 0.6$ dB/cm and $A_{eff} = 1.6$ μm^2. (Reprinted, with permission, from Krause, M. et al., *IEEE J. Sel. Top. Quantum Electron.* **16**(1): 216–225, 2010, ©2010 IEEE.)

5.3.4 Ring-Resonator Enhancement of the Pump Power

In conventional single-pass-pumped SRAs such as those treated in Sections 5.3.1 through 5.3.3, a certain minimum Raman pump power is always required in order to make the waveguide at least locally transparent. A qualitatively new behavior is exhibited by ring-resonator-based SRAs, first proposed by Kuo et al. (2006), in which the externally applied pump power is resonantly enhanced. We now show that this reduces the threshold pump power of optimally designed amplifiers to zero; that is, RR-SRAs with a positive gain >0 dB can be realized with any pump power, however small (Krause et al. 2010).

5.3.4.1 Optimal RR-SRAs in the Presence of TPA and FCA

An RR amplifier consists of a silicon waveguide ring of length L and a coupler connecting this ring to an access waveguide; see Figure 5.3b. The coupler cross-couples a fraction K of the externally applied pump power P_0 into the ring. The pump powers at the beginning and at the end of the silicon waveguide are called P_1 and P_2, respectively. The ring and the coupler form a resonant cavity. When the pump wavelength has been tuned to a cavity resonance, the power circulating inside the cavity can become larger than the externally injected power P_0 (Siegman 1986).

The ring in Figure 5.3b is called "critically coupled" to the access waveguide when the coupling ratio K is equal to the round-trip losses L_{rt}. Designing the coupler for critical coupling ($K_{crit} = L_{rt}$) ensures that the resonant pump enhancement becomes maximal, $M_{max} = 1/L_{rt}$ (Siegman 1986). The signal, in contrast to the pump, should be coupled into and out of the ring with an efficiency of 100% such that it passes the entire ring exactly once. A tunable Mach–Zehnder-interferometer-based approach for realizing the desired coupling functionality has been proposed by Doylend et al. (2008).

The differential equations (5.87) and (5.88) for an SPP amplifier also describe the evolution of the pump and signal powers inside the RR amplifier of Figure 5.3b, and the total RR gain is given, as in the SPP case, by (5.89). The distinctive feature of the RR amplifier is that the resonant pump coupling can arrange for an effective enhanced pump power $P_1 = MP_0$. The total signal gain G_{RR} of the RR amplifier is equal to that of an SPP amplifier with its length equal to the ring length L and an enhanced pump power P_1. Consequently, when freely optimizing the coupler, the ring length, and the injected pump power, the maximum possible gain of an RR amplifier is the same as the maximum possible gain (5.92) of an SPP amplifier. The practical advantage is in the pump-power requirements: while the optimal SPP amplifier requires a pump power of P_U [see (5.93)], the optimal RR amplifier needs only $P_U - P_L$; that is, we save the amount of P_L (Krause et al. 2010).

On the other hand, when the optimal pump power $P_U - P_L$ is not available ($P_{avail} < P_U - P_L$), finding the optimal RR-SRA under these restricted circumstances is almost explicit except for the necessity of numerically finding a zero of the following equation:

$$\gamma(\bar{P}_1 - P_{avail})\phi(\bar{P}_1)\bar{P}_1 = \gamma(\bar{P}_1)\phi(\bar{P}_1 - P_{avail})(\bar{P}_1 - P_{avail}) \qquad (5.95)$$

where ϕ and γ have been defined in (5.87) and (5.88), respectively. Equation (5.95) is a polynomial equation of fifth order in $\bar{P}_1$ whose desired solution is in the interval $[P_L + P_{avail}, P_U]$. This zero can be quickly found numerically—a unique solution always exists (Krause et al. 2010). Once $\bar{P}_1$ has been found, the optimal length and best possible gain are given explicitly by (5.94) and (5.92), where P_U and P_L need to be replaced with $\bar{P}_1$ and $\bar{P}_1 - P_{avail}$, respectively. The optimal coupling ratio is $K_{RR,opt} = P_{avail} / \bar{P}_1$ (Krause et al. 2010).

Using the same technological parameters as for the SPP case of Figure 5.8, we now consider the maximum RR gain $G_{RR,max}$ as a function of the lifetime τ_{eff} and the available pump power P_{avail}; it is shown in Figure 5.9a. Comparing this result with the SPP case of Figure 5.8a, the most obvious difference is that the RR amplifier has no minimum required pump power. For example, assuming a lifetime of τ_{eff} = 0.5 ns, an SPP amplifier (Figure 5.8a) that is only just transparent (G = 0 dB) requires a pump power of at least 270 mW. On the other hand, with this pump power we could realize an RR amplifier (Figure 5.9a) with a gain of as much as 2 dB. To achieve this gain, the ring length and pump-coupling ratio have to be chosen as L = 2 cm and K = 30%, respectively (see Figures 5.9b–5.9c).

We show in Krause et al. (2010) by scaling the differential equations (5.87) and (5.88) that the maximum possible gain of SPP and RR amplifiers is determined by only two parameters, namely, the product $\alpha \cdot \tau_{eff}$, and the normalized available pump power $p_{avail} = P_{avail} / (\alpha A_{eff})$.

5.3.4.2 Optimal RR-SRAs in the Absence of Nonlinear Absorption

In the presence of TPA and FCA, finding the optimal ring-resonator SRA requires numerically finding a zero of the fifth-order polynomial (5.95). However, in the limit that TPA and FCA are negligible (i.e., in the mid-IR regime discussed in Section 5.3.1), the result can be made explicit as follows. An RR amplifier with critical coupling for the pump wavelength can be considered an SPP amplifier with an effective enhanced pump power of $P(0) = M_{max}P$, where M_{max} = 1/[1 − exp(−αL)]. The total gain is obtained from the SPP gain G of (5.75) by simply replacing F → $M_{max}F$, such that G_{RR} = $e^{-\alpha L}e^F$. When the length L approaches zero, the increasing enhancement M_{max} = 1/[1 − exp(−αL)] ≈ 1/(αL) precisely balances the decreasing available Raman interaction length L, while the linear losses seen by the signal vanish, with the overall effect that the total gain increases. In the limit of zero length, one obtains the ultimately achievable gain (see the dashed curve in Figure 5.4) (Krause and Renner 2008),

$$G_{ultimate} = e^F \quad \text{or, in numbers: } G_{ultimate}[dB] = 1.89 \frac{g[cm/GW].P[W]}{\alpha[dB/cm].A[\mu m^2]}$$

$$(5.96)$$

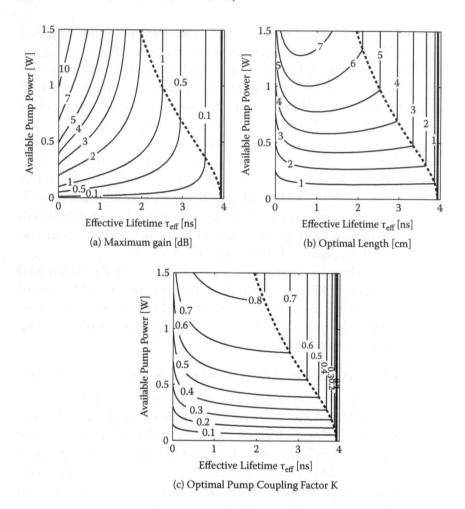

(a) Maximum gain [dB]

(b) Optimal Length [cm]

(c) Optimal Pump Coupling Factor K

FIGURE 5.9 Optimal ring-resonator (RR) silicon Raman amplifiers as a function of the effective lifetime τ_{eff} and the available pump power P_{avail}: (a) maximum gain, (b) optimal length, (c) optimal pump coupling ratio. Waveguide parameters are as in Figure 5.8. [Reprinted, with permission, from from Krause, M. et al., *IEEE J. Sel. Top. Quantum Electron.* **16**(1): 216–225, 2010, ©2010 IEEE.]

5.3.5 Cladding-Pumped Amplifiers

In this section, we show that the maximum achievable total gain of silicon Raman amplifiers can be significantly increased by injecting the pump power into a surrounding cladding instead of directly into the silicon core (Krause et al. 2006; Krause et al. 2008b).

The basic structure for a cladding-pumped silicon Raman amplifier is shown in Figure 5.10. It consists of a rectangular

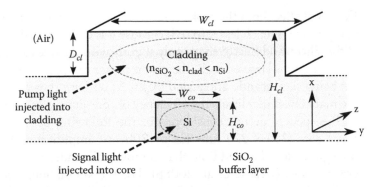

FIGURE 5.10 Waveguide cross section for a cladding-pumped silicon Raman amplifier. (Reprinted, with permission, from Krause, M. et al. *IEEE J. Quantum Electron.* **44**(7): 692–704, 2008b, ©2008 IEEE.)

silicon waveguide core on top of a silica buffer layer. The top cladding has a refractive index that lies between that of silica and silicon, that is, $n_{SiO_2} < n_{clad} < n_{Si}$. The signal light to be amplified is guided in the fundamental mode of the entire structure (Figure 5.11a), which is highly confined to the silicon core. The pump light, however, is injected into a higher-order mode, the power of which is mainly guided in the cladding (Figure 5.11b).

To understand the advantage of this structure, note that a small part of the pump mode extends into the silicon core (see Figure 5.11b). It thus amplifies the Stokes mode through SRS. It also generates free carriers through TPA, as in the conventional core-pumped amplifiers discussed in Section 5.3.3. The resulting

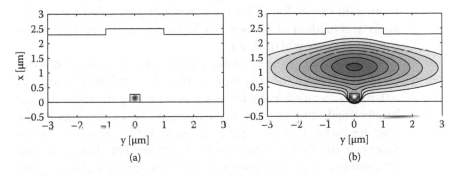

FIGURE 5.11 Illustration of typical mode-intensity profiles of the (a) Stokes and (b) pump in a cladding-pumped silicon Raman amplifier. While the Stokes light to be amplified is guided in the silicon core, the pump mode is guided mainly in the (nonabsorbing) cladding and is thus scarcely attenuated by FCA. (Reprinted, with permission, from Krause, M. et al. *IEEE J. Quantum Electron.* **44**(7): 692–704, 2008b, ©2008 IEEE.)

FCA will affect both the Stokes and the pump mode. However, the pump mode is much less affected, because it is concentrated outside of the silicon and thus overlaps only slightly with the free carriers. If the cladding is large enough, the pump power can propagate without being significantly attenuated by FCA; we have thus mitigated the main effect that limits the efficiency of core-pumped amplifiers.

The specific design considered in the following consists of a small core (350×270 nm^2 for a 1300 nm amplifier) and a rib waveguide for the cladding that removes all undesired higher-order modes (Krause et al. 2008b). For the simulations, we use a refractive index of $n_{\mathrm{clad}} = 2.0$, corresponding to silicon nitride, which can be integrated relatively easily with the processing of the rest of the structure. We assume fixed Stokes and cladding-mode losses of $\alpha_s = 2.0$ dB/cm and $\alpha_p = 5.0$ dB/cm, respectively, and the width of the cladding is kept at $W_{\mathrm{cl}} = 1.5\,\mu$m. Using a full-vectorial mode solver (Krause et al. 2006), we calculate the modes of the double-clad waveguide. We choose the predominantly y-polarized fundamental mode for the Stokes mode, and the next-higher-order predominantly y-polarized mode for the pump (cladding) mode. Using (5.38), (5.44), and (5.57), we then calculate the effective areas and confinement factors required for the simulation of the amplifier characteristics on the basis of (5.69)–(5.71). Note that in this waveguide structure, the effective areas for TPA and SRS may differ significantly. For example, for the design with $H_{\mathrm{cl}} = 2.5\,\mu$m and $D_{\mathrm{cl}} = 1.0\,\mu$m, the pump-Stokes-TPA effective area is $A_{sp}^{\mathrm{TPA}} = 0.41\,\mu$m^2, while the SRS effective area is only $A^{\mathrm{(SRS)}} = 0.23\,\mu$m^2. The optimal amplifier length and maximum gain can be found for a given pump-power limit using the analytical method of Section 5.3.3, which can be generalized for the present case where $\alpha_s \neq \alpha_p$.

Figure 5.12a gives an example for the maximum unsaturated gain under the assumption that $P_{\mathrm{pump}} = 300$ mW and $\tau_{\mathrm{eff}} = 1$ ns. With suitable cladding parameters, a gain of 18 dB can be obtained, and possibly even more if the dimensions of the core are included in the optimization and if the linear loss coefficient of the Stokes wave can be made smaller than 2 dB/cm. In contrast, when core-pumping the structure and choosing the optimal length and pump power, we can achieve a gain of only 2 dB. Note that the optimal cladding size depends on the available pump power. The larger the cladding (i.e., the further the pump mode is extracted from the silicon), the more pump power must be injected in order to keep the pump intensity in the silicon

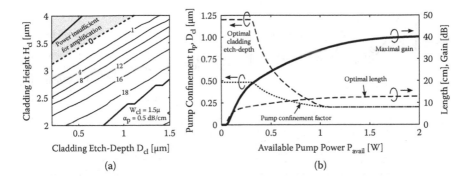

FIGURE 5.12 (*a*): Maximum unsaturated gain of cladding-pumped amplifiers for P_{pump} = 300 mW versus dimensions of the top cladding; (*b*): maximum possible gain (solid) versus available pump power, corresponding optimized amplifier length L (dashed), optimized cladding etch depth D_{cl} (dashed-dotted) and pump confinement factor η_p (dotted); cladding height was fixed at H_{cl} = 2.5 μm . (Reprinted, with permission, from Krause, M. et al. *IEEE J. Quantum Electron.* **44**(7): 692–704, 2008b, ©2008 IEEE.)

core high enough to enable Raman amplification. Vice versa, for large available pump powers, the cladding may be made larger than for small available pump powers. This is illustrated by the results shown in Figure 5.12b, where (in addition to the amplifier length) the cladding etch-depth D_{cl} (dashed-dotted) has also been optimized for various available pump powers. It can be seen that the larger the available pump power, the lower the optimized cladding etch depth and thus the pump-confinement factor η_p (dotted) becomes, leading to larger possible total gains (solid).

The aforementioned simulations have been performed for fixed Stokes and pump-mode losses of α_s = 2.0 dB/cm and α_p = 5.0 dB/cm, respectively, while the effective free-carrier lifetime has been assumed to be τ_{eff} = 1 ns. In contrast, Figure 5.13 shows the maximum gain for a wider range of linear pump and Stokes losses as well as free-carrier lifetimes in order to illustrate that the cladding-pumped SRA performs much better than the core-pumped SRA in a wide range of these parameters. For each parameter set, we have optimized the amplifier length for a maximum pump power of 300 mW such that the total gain is maximized. Figure 5.13 indeed shows that the cladding-pumped SRAs (solid curves) can deliver significantly more gain than the core-pumped SRAs (dotted curves). On the one hand, decreasing the linear pump-mode losses α_p and thus the total pump attenuation is advantageous since this enables longer amplifiers and thus a larger total gain. On the other hand, low linear Stokes losses α_s increase

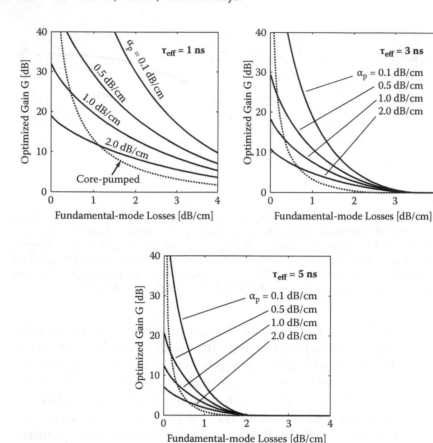

FIGURE 5.13 Maximum possible gain of SRAs with cladding dimensions $W_{cl} = 1.5\,\mu m$, $H_{cl} = 2.5\,\mu m$, and $D_{cl} = 1.2\,\mu m$ as a function of the fundamental-mode losses α_s for three different free-carrier lifetimes τ_{eff}, and maximum available pump power of 300 mW. The dotted curves show the maximum possible gain of core-pumped SRAs, while the solid curves correspond to cladding-pumped SRAs with varying pump-mode losses α_p. (Reprinted, with permission, from Krause, M. et al. *IEEE J. Quantum Electron.* **44**(7): 692–704, 2008b, ©2008 IEEE.)

the local Stokes gain, thus also contributing to a higher total gain. Finally, short lifetimes τ_{eff} result in both decreased pump attenuation (longer amplifiers) and a larger local Stokes gain. Further improvements are expected by including the cladding and core dimensions in the optimization, which we have kept fixed during the optimization for Figure 5.13.

5.3.6 Approaches for Reducing FCA

The preceding sections have shown that silicon Raman amplifiers at standard telecom wavelengths around 1.5μm are limited in their performance by FCA. Cladding pumping, as discussed in Section 5.3.5, can significantly reduce the problem. Further courses that have been pursued to mitigate the deterioration caused by FCA include, on the one hand, reducing the lifetime of the TPA-generated charge carriers by implanting helium (Liu and Tsang 2006), argon (Tanabe et al. 2007), or oxygen (Först et al. 2007). Also, an electric field externally applied by means of a lateral p-i-n junction can be used to extract carriers from the silicon core (Liu et al. 2006), or the built-in field of the junction can be exploited (Fathpour et al. 2006). However, such a carrier sweep-out is only partially effective because the TPA-generated carriers may screen the junction field at high pump intensities (Dimitropoulos et al. 2005). Exponential tapering of the effective mode area would permit arbitrary total gain, but is limited by the minimum effective area practically achievable (Renner et al. 2005). Finally, bidirectional pumping has been investigated, where the pump power is injected into the silicon waveguide from both ends instead of only one waveguide end. This reduces the peak powers inside the laser waveguide and thus avoids excessive FCA (Krause et al. 2005).

5.3.7 Raman Isolators

The nonreciprocal Raman gain discussed in Section 5.2.4.4 can be used to realize an all-optical dynamically reconfigurable isolator (Krause et al. 2008; Krause and Brinkmeyer 2008). Figure 5.14 shows a diagram of such a component. It comprises a balanced Mach–Zehnder interferometer (MZI) whose arms can be Raman-pumped in opposite directions with the pump powers P_A and P_B through wavelength-selective couplers. The device acts as an optical isolator between ports 1 and 2, as we show in the following.

Before proceeding, we note that the characteristics of this device depend both on the Raman gains and on the pump-induced phase shifts experienced by the partial signals in the two MZI arms. However, the phase shifts can be eliminated using a (reciprocal) phase shifter as discussed in Krause et al. (2008a). Therefore, we here assume that the two MZI arms have equal optical path lengths, and we concentrate exclusively on the gain caused by the presence of the two pumps.

We start by considering a backward ("–") signal injected into port 2: it will only interfere destructively at port 1 (perfect isolation) when the backward gain in the upper arm, G_A^-, is equal to

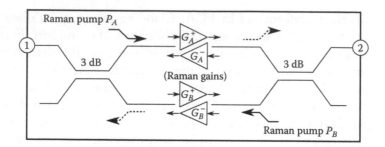

FIGURE 5.14 Schematic diagram of a Raman isolator. By choosing the Raman pump powers P_A and P_B adequately, the device can be configured so as to transmit only in the direction $1 \rightarrow 2$, while isolating in the $2 \rightarrow 1$ direction.

the backward gain in the lower arm, G_B^-. When the pump powers have been adjusted such that this is the case, we can now look at the forward signal $(1 \rightarrow 2)$: the forward ("+") gain in the upper arm, G_A^+, will be lower than the backward gain, because now the signal is copropagating with the pump (see Figure 5.14) and the copropagating Raman gain is weaker than the counterpropagating one, as shown in Section 5.2.4.4. On the other hand, the forward gain in the lower arm, G_B^+, will be larger than the backward gain because now the signal is counterpropagating with the pump. Therefore, destructive interference at port 2 in forward transmission is impossible, and the device isolates only in one direction. The two transmissions just discussed can be expressed as

$$T_{1 \rightarrow 2} = \frac{1}{4} \left| \sqrt{G_A^+} - \sqrt{G_B^+} \right|^2 \qquad T_{2 \rightarrow 1} = \frac{1}{4} \left| \sqrt{G_A^-} - \sqrt{G_B^-} \right|^2 \qquad (5.97)$$

To numerically evaluate these transmissions, we calculate the gains $G_{A,B}^{\pm}$ using the model of Section 5.2.7 assuming Raman-gain and TPA coefficients of 20 and 0.7 cm/GW, respectively. We assume unsatured amplifier operation and choose linear waveguide losses of $\alpha = 2.0\,\text{dB}/\text{cm}$, an effective free-carrier lifetime of $\tau_{\text{eff}} = 1$ ns, and a length of $L = 4$ cm.

The dependence of the transmission $T_{2 \rightarrow 1}$ on the two pump powers P_A and P_B is shown in Figure 5.15a. For pump-power combinations on the thick solid curve in Figure 5.15a, this transmission is zero; that is, we have the desired isolation. This curve is reprodced as a dotted curve in Figure 5.15b, which otherwise shows the transmission $T_{1 \rightarrow 2}$ From Figure 5.15b we can see that

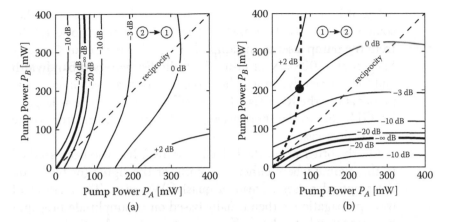

FIGURE 5.15 Simulated transmission characteristics (transmissions versus the two Raman pump powers) for the isolator proposed in Figure 5.14. (Reprinted, with permission, from Krause, M. and Brinkmeyer, E., *5th International Conference on Group IV Photonics (GFP)*, Paper ThP6, 2008, ©2008 IEEE.)

nonzero transmission in the $1 \to 2$ direction can be obtained while keeping isolation in the $2 \to 1$ direction. For example, at the pump powers $P_A = 65$ mW and $P_B = 205$ mW (marked with a dot in Figure 5.15b), the isolator becomes transparent in the $1 \to 2$ direction, while perfectly isolating in the opposite direction. On the other hand, as soon as the two pump powers are chosen equal (dashed lines in Figure 5.15), the device becomes reciprocal, and its transmission can be tuned between $-\infty$ and 0.5 dB. Finally, we can dynamically reconfigure the isolation direction by introducing a relative phase shift of π between the two MZI arms, for example, using a thermo-optic phase shifter.

5.3.8 Non-CW Operation

The results discussed so far in this chapter were concerned with the steady-state operation of silicon Raman amplifiers, that is, the amplification of monochromatic signals by a monochromatic pump. Here we summarize theoretical results from the literature that are relevant for nonmonochromatic amplifier operation.

On the one hand, for practical application of any amplifier its noise properties are of significant interest. This topic is treated, for example, by Dimitropoulos et al. (2008), who calculate the quantum noise imparted on a signal in an SRA in the presence of TPA and FCA for an ideal CW pump; their results suggest that the effective free-carrier lifetime τ_{eff} should be reduced to values below 0.1 ns in order to achieve useful noise figures. Another source of noise is the transfer of relative intensity noise (RIN) from the pump laser to the amplified signal, which is analyzed by Sang et al. (2008). Neglecting the finite Raman-gain bandwidth and free-carrier dynamics, the authors show that even in counter-propagation, the RIN-transfer cut-off frequency can be as large as 1.5 GHz due to the short waveguide lengths as compared to the optical-fiber case.

On the other hand, the behavior of SRA has also been studied in the regime where either the pump or the signal (or both) are short optical pulses instead of quasi-CW waves. The analysis of wave propagation is then usually based on an amplitude propagation model, where phase effects due to group-velocity dispersion and the finite Raman-gain bandwidth must be taken into account, which was unnecessary in the continuous-wave power-based model of Section 5.2.5. The necessary generalization of the model of Section 5.2 has been omitted here for brevity; compare Chen et al. (2006), for example. The corresponding free-carrier dynamics are usually modeled by a simple first-order differential equation for the carrier density. Numerical analyses of pulsed Raman scenarios have been performed on the basis of several amplitude-based models in Chen et al. (2006), Passaro and Leonardis (2006), and Leonardis and Passaro (2008), while Dissanayake et al. (2009) perform simulations of Maxwell's equations in a fully spatially discretized waveguide structure using a finite-difference time-domain algorithm into which a simplified SRS model has been incorporated. The precise evolution of pump and Stokes pulses propagating in the same silicon waveguide is complex and depends on many factors. For example, the pulses may be broadened, compressed, chirped, or otherwise distorted during propagation due to the various non-linear effects as well as group-velocity dispersion. Moreover, the Raman amplification process critically depends on the width of the pulses relative to the Raman response time of 10 ps: the signal may be distorted and the effective Raman gain reduced when its temporal variations are too fast. To gain more insight, Roy et al. (2009) reduce the full amplitude-based wave-propagation model using a variational formalism with a chirped-Gaussian ansatz function for the Stokes pulse. Neglecting the Raman-induced group-velocity change (Okawachi et al. 2006), relatively simple ordinary differential equations for the evolution of the pulse amplitude, width, phase, and chirp in the presence of a CW pump wave have been obtained.

5.4 Lasers

In this section, we discuss the basic characteristics of continuous-wave Raman lasers in silicon waveguides.

5.4.1 Basic Model

5.4.1.1 Operating Principle

Figure 5.16 shows a schematic of the silicon Raman laser we analyze in this section. It consists of a silicon waveguide of length L, into the left-hand side of which the pump laser at λ_p is coupled. When the latter is switched on, spontaneous Raman scattering inside the silicon will generate new light at the Stokes wavelength λ_s (which is offset from the pump by about 15.6 THz). Any Stokes light inside the waveguide will be amplified by SRS—it experiences Raman gain proportional to the pump power. At the ends of the waveguide, part of the Stokes light leaves the waveguide (forming the output beam), and part of it is reflected back into the waveguide. At high enough pump powers, this feedback and the amplification through SRS lead to an increasing buildup of optical power at the Stokes wavelength. Eventually, the laser reaches a steady state in which pump radiation at λ_p is continuously converted to Stokes radiation at λ_s.

The continuous-wave silicon Raman lasers reported in Rong et al. (2005), Rong et al. (2006), Rong et al. (2007), and Rong et

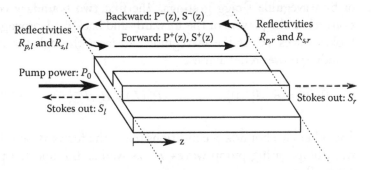

FIGURE 5.16 Schematic of a basic silicon Raman laser. Pump power is injected at the left-hand side into a silicon waveguide. The Stokes (lasing-wavelength) cavity is formed by reflectivities $R_{s,l}$ and $R_{s,r}$ at the ends of the waveguide, while reflectivities at the pump wavelength $R_{p,l}$ and $R_{p,r}$ can be used to influence the longitudinal pump-power distribution. (Reprinted, with permission, from Krause, M., Efficient Raman amplifiers and lasers in optical fibers and silicon waveguides: New concepts, Ph.D. thesis, Technische Universität Hamburg-Harburg, Hamburg, Germany. Cuvillier Verlag, Göttingen, 2007, ©2007 Cuvillier Verlag, Göttingen.)

al. (2008) have been observed to operate in a single longitudinal mode, that is, at a well-specified Stokes wavelength, in spite of the use of broadband cavity reflectors. The power-based model of Section 5.2.7 may therefore be used to describe the steady-state operation of these devices. The question remains which Raman-gain constant g to employ in (5.69) and (5.70), since the Raman gain depends on the spectral separation of the lasing line from the pump-laser wavelength, which in turn is determined by the longitudinal mode the laser chooses to run on. A typical laser cavity with a length of 1 cm has a free spectral range (FSR) of $c/(2Ln)$ ≈ 5 GHz, which is much smaller than the width of the Raman-gain spectrum of 100 GHz. The lasing Stokes wavelength will therefore deviate from the wavelength of maximum Raman gain by a negligible amount of at most 2.5 GHz, and we may simply assume the maximum Raman gain constant g_{max} of (5.42).

5.4.1.2 Boundary Conditions

As just discussed, the longitudinal evolution of the forward- (+) and backward-propagating (–) pump ("p") and Stokes ("s") powers $P_p^\pm$ and $P_s^\pm$ is described by the model summarized in Section 5.2.7. We only need to add suitable boundary conditions at the left-hand ("l") and right-hand ("r") ends of the waveguide, where the pump and Stokes wavelengths see power reflectivities of $R_{p,l}$, $R_{p,r}$, $R_{s,l}$, and $R_{s,r}$, respectively. The reflectors could be formed by polished waveguide end surfaces, by thin-film coatings on the end surfaces, or by waveguide Bragg gratings. The first two boundary conditions relate the powers of the forward- and backward-propagating Stokes waves $P_s^\pm$ at the waveguide end faces through the corresponding power reflectivities,

$$P_s^+(0) = R_{s,l} P_s^-(0), \qquad\qquad P_s^-(L) = R_{s,r} P_s^+(L) \qquad (5.98a,b)$$

The other two boundary conditions relate the forward- and backward-propagating pump waves $P_p^\pm$ as well as the injected pump power P_0 as

$$P_p^+(0) = T_p P_0 + R_{p,l} P_p^-(0), \qquad\qquad P_p^-(L) = R_{p,r} P_p^+(L), \qquad (5.99a,b)$$

where P_0 is the pump-laser power and T_p is the coupling efficiency into the waveguide. We assume lossless reflectors, that is, $T_p = 1 - R_{p,l}$. Equation (5.99a) assumes an incoherent superposition of the reflected backward-propagating pump wave and the externally injected pump power. This boundary condition is thus applicable whenever the pump-laser spectrum is so broad that it spans several free spectral ranges (FSRs) of the silicon-waveguide cavity. For example, the pump laser used in Claps et al. (2003) is a Raman fiber laser with a spectral width of several tens of GHz, while a silicon-waveguide cavity with a length of 1 cm has an FSR of only $c/(2Ln) \approx 5$ GHz.

On the other hand, it is also possible to formulate boundary conditions for coherent (narrow-linewidth) pumping of the laser cavity. Here, the electric-field amplitudes $\underline{E}$ add up instead of the powers as in (5.99a), that is, $E_p^+(0) = \sqrt{T_p}E_0 + \sqrt{R_{p,l}}E_p^-(0)$ for on-resonance pumping, that is, constructive interference of the externally injected pump and the reflected backward-propagating pump wave. In terms of powers $P \propto E^2$, the "coherent" boundary condition reads

$$\sqrt{P_p^+(0)} = \sqrt{T_p}\sqrt{P_0} + \sqrt{R_{p,l}}\sqrt{P_p^-(0)} \qquad (5.100)$$

which can be used instead of (5.99a) when coherent pumping is to be simulated. The other boundary conditions remain unchanged.

5.4.1.3 Numerical Simulation

The differential equations for the forward- and backward-propagating pump and Stokes waves as well as the boundary conditions form a well-posed two-point boundary-value problem (BVP) that can be solved using a variety of mathematical techniques. Prescribed is the pump power P_0, and what we look for is the longitudinal pump and Stokes power distribution inside the cavity.

We use the collocation method bvp4c available in Matlab, which requires an initial guess for the solution. Such a guess is difficult to find, especially when simulating cascaded lasers (Sections 5.4.3 and 5.4.4), where the model includes a large number of laser lines at different wavelengths. If the initial solution guess is too different from the desired exact solution, the collocation method may converge to an entirely different, mathematically possible but nonphysical solution (such as one with negative powers). We avoid this problem using parameter continuation: we start with the known solution for a pump power of $P_0 = 0$, where all powers along the entire waveguide are simply zero. We then repeatedly solve the BVP, each time using the existing solution as the solution guess for the next iteration where the pump power is slightly increased by some ΔP_0, until the solution for the desired pump power is reached. To avoid the nonphysical, "trivial" solutions of the BVP where the Stokes lines are exactly zero even though the laser is above the threshold, we include additional auxiliary terms in the equations for each line that could be interpreted as representing an increase of the line's power due to spontaneous Raman scattering. However, these terms are included only as a tool for finding the desired solution of the BVP; once a solution for the final pump power P_0 has been reached, it is used as the starting guess for one final calculation with the auxiliary terms removed. We then have the desired solution of the original BVP. The time needed for the entire solution process depends on the strength of the auxiliary terms, on the tolerances of the collocation solver, and on the size of the pump-power increments ΔP_0. For the latter, an automatic step-size control algorithm is helpful since in the neighborhood of the lasing threshold the BVP solution changes strongly, and the ΔP_0 has to be chosen particularly small here, while for larger pump powers ΔP_0 may be larger.

The numerical solution process can be based on the longitudinally varying powers of the pump and Stokes waves as the unknown functions in the BVP for both incoherent and coherent pumping, since the appropriate boundary conditions in both cases [(5.99a) and (5.100)] can be formulated in terms of the powers. However, we have found that the numerical stability is much improved for coherent pumping when the BVP is formulated in terms of the square root of the powers as the unknown function (the desired powers are then obtained by squaring the solution).

In another approach to simulating SRLs, the steady-state model of Section 5.2.7 could be generalized to a time-dependent model by replacing $d/dz \rightarrow \partial/\partial z + (1/\upsilon_{\text{group}}) \cdot \partial/\partial t$ on the left-hand side of (5.46). We then have a hyperbolic initial-value problem that can be solved starting from an initial time with an initial arbitrary power distribution until the temporal steady state is reached. For example, when neglecting the difference in the group velocities of the participating waves, the method of integration along characteristics could be used (de Sterke et al. 1991; Cierullies et al. 2005). A method resembling the latter has been proposed for modeling SRLs by Vermeulen et al. (2006).

5.4.2 Basic Properties of Silicon Raman Lasers

Now we discuss the basic threshold and input-output character-istics of SRLs (Krause et al. 2004). We initially concentrate on lasers with noncoated waveguide end-faces and assume that all reflectivities are due to the silicon–air interface, $R_{p,l} = R_{p,r} = R_{s,l} = R_{s,r} = 30\%$. We base our simulations on the waveguide structure of Rong et al. (2005). At the wavelengths chosen here (pump and Stokes wavelengths of $\lambda_p = 1427$ nm and $\lambda_s = 1542$ nm, respec-tively), the effective areas occurring in the model of Section 5.2.7 are all approximately equal to $A = 1.6\ \mu m^2$, while the confinement factors η are close to unity. We choose conservative values for the bulk-silicon Raman-gain and TPA constants in order to avoid too optimistic results: $g = 20$ cm/GW, $\alpha_p = \alpha_s = 1.0$ dB/cm, $\beta_{pp} = 0.7$ cm/GW. According to the scaling law of Section 5.4.3.1, the coefficients for pump-Stokes TPA and Stokes TPA are $\beta_{sp} = 0.57$ cm/GW and $\beta_{ss} = 0.47$ cm/GW.

5.4.2.1 Lasing Threshold

At the lasing threshold, the Stokes powers are much smaller than the pump powers, $P_s^{\pm} \ll P_p^{\pm}$, and we can accordingly simplify the full model of Section 5.2.7 by neglecting the depletion of the pump waves due to SRS and nondegenerate TPA. Furthermore the steady-state carrier density (5.71) is caused exclusively by pump-wave TPA. Using (5.68), we can thus write

$$\bar{N} = \frac{\beta_{pp}\tau_{\text{eff}}}{2h\nu_p A^2}(P_p^{+2} + P_p^{-2} + 4P_p^+ P_p^-) \qquad (5.101)$$

The laser is at threshold when the Stokes round-trip net gain equals the losses due to outcoupling at the left-hand and right-hand end faces with reflectivities $R_{s,l}$ and $R_{s,r}$. By integrating the local gain over a roundtrip, we can thus obtain the oscillation condition

$$R_{s,l}R_{s,r}\exp\left[2\int_0^L -\alpha_s - \eta_s\bar{\varphi}\lambda_s^2\bar{N}(z) + \left(\frac{g-2\beta_{sp}}{A}\right)[P_p^+(z) + P_p^-(z)]dz\right] = 1$$

$$(5.102)$$

In order to find the threshold pump power P_{th}, we numerically calculate the longitudinal pump-power distribution from (5.69) and (5.99) for varying pump powers P_0 until $P_p^+(z)$ and $P_p^-(z)$ fulfill (5.102).

We consider first the case of a laser in which both TPA and FCA are absent; that is, we artificially set $\beta_{pp} = \beta_{sp}$ and $\tau_{eff} = 0$ The threshold pump power of such a laser is shown in Figure 5.17 as a function of the waveguide length L, for three different loss coefficients $\alpha = \alpha_p = \alpha_s$ (solid curves). According to these results, it should be possible to make an SRL lase using a pump laser with a power on the order of 1 W, provided the effects of TPA and FCA are negligible. In that case, the conversion efficiencies of the SRL are comparable to those obtainable from Raman fiber lasers (Headley and Agrawal 2005). The dashed curves in Figure 5.17 represent the simulated threshold power of the SRL when the TPA coefficients have realistic nonzero values, yet FCA is still absent. We can observe a slight increase of the threshold pump powers due to TPA.

The effect of FCA, however, can be much more dramatic than that of TPA, which is illustrated in Figure 5.18. The linear waveguide losses are now fixed at $\alpha_p = \alpha_s = 1.0$ dB/cm. The dashed line reproduces the TPA-less threshold power from Figure 5.17 for comparison, whereas the solid lines show the threshold powers in the presence of TPA and several different free-carrier lifetimes τ_{eff}. As expected, a larger τ_{eff} results in an increased threshold.

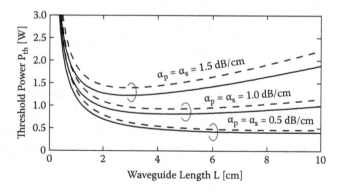

FIGURE 5.17 Threshold pump power of silicon Raman lasers versus waveguide length L. End-face reflectivities are 30%, and free-carrier absorption is assumed to be negligible (τ_{eff}). Solid curves: No two-photon absorption ($\beta = 0$). Dashed curves: With two-photon absorption, $\beta = 0.7$ cm/GW. (Reprinted, with permission, from Krause, M., Efficient Raman amplifiers and lasers in optical fibers and silicon waveguides: New concepts, Ph.D. thesis, Technische Universität Hamburg-Harburg, Hamburg, Germany. Cuvillier Verlag, Göttingen, 2007, ©2007 Cuvillier Verlag, Göttingen.)

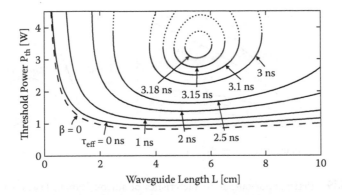

FIGURE 5.18 Threshold pump power of silicon Raman lasers versus waveguide length L for several effective carrier lifetimes τ_{eff} and $\alpha_p = \alpha_s = 1.0$ dB/cm. The solid and dotted curves show, for a given τ_{eff}, the lasing and shutdown thresholds, respectively. Dashed curve (included for comparison): threshold in the absence of TPA and FCA. (Reprinted, with permission, from Krause, M., Efficient Raman amplifiers and lasers in optical fibers and silicon waveguides: New concepts, Ph.D. thesis, Technische Universität Hamburg-Harburg, Hamburg, Germany. Cuvillier Verlag, Göttingen, 2007, ©2007 Cuvillier Verlag, Göttingen.)

Also, there is a limited usable range of waveguide lengths outside of which the laser has no threshold at all. Outside this range (e.g., for L = 8 cm and τ_{eff} > 3 ns), the waveguide will never start lasing, no matter how large the pump power is. The usable range becomes increasingly smaller with increasing τ_{eff}, until at $\tau_{\text{eff}} \approx$ 3.2 ns it vanishes completely. In other words, there is a maximum effective carrier lifetime that can be tolerated for lasing. This limitation is due to the increase of the overall cavity losses due to FCA, which is quadratic in the pump power [compare condition (5.91) for SRAs]. If TPA and FCA were absent, the overall cavity losses would remain constant with respect to the pump power, and for any given waveguide there would be a pump-power level above which the component will start lasing; that threshold pump power can be given explicitly (AuYeung and Yariv 1979).

5.4.2.2 Input-Output Characteristics

We now turn to the input-output characteristics of SRLs. These are shown in Figure 5.19 for various effective carrier lifetimes τ_{eff}. These characteristics were calculated from the full model described in Section 5.2.7, and we defined the output power of the laser as $P_{\text{out}} = P_s^+(L)(1 - R_{s,r})$. Directly above the threshold, increasing the pump power also increases the output power. However, there exists a roll-over point, that is, a critical pump power beyond

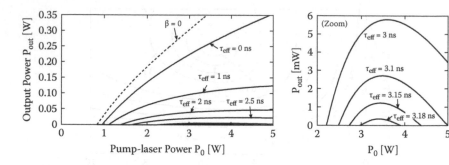

FIGURE 5.19 (left:) Input-output characteristics of silicon Raman lasers with $L = 55$ mm and several values of the effective carrier lifetime τ_{eff}. The dashed curve corresponds to absence of TPA and FCA. (right:) Zoom into the characteristics corresponding to large τ_{eff}. (Reprinted, with permission, from Krause, M., Efficient Raman amplifiers and lasers in optical fibers and silicon waveguides: New concepts, Ph.D. Thesis, Technische Universität Hamburg-Harburg, Hamburg, Germany. Cuvillier Verlag, Göttingen, 2007, ©2007 Cuvillier Verlag, Göttingen.)

which a further increase of the pump power actually results in a decrease of the output power. Finally, at the "shutdown threshold," laser operation shuts down completely (Krause et al. 2004). As in the case of silicon Raman amplifiers (see Figure 5.6), this is due to the increased nonlinear losses for both the pump and the Stokes waves due to TPA and FCA at large pump powers. Figure 5.19 shows that for increasing τ_{eff}, the maximum conversion efficiency of the lasers dramatically decreases and the lasing and shutdown thresholds come closer to each other. The existence of this roll-over point as well as its dependence on the free-carrier lifetime as shown in Figure 5.19 has been observed experimentally by Rong et al. (2005) and Rong et al. (2007).

Note that the shutdown threshold has already been obtained numerically in Section 5.4.2.1: for every given waveguide length L, there is either no threshold at all or two threshold pump powers can be found in the presence of FCA. While the solid curves in Figure 5.18 show the lower of the two threshold powers (i.e., the lasing threshold), the dotted curves show the upper threshold (i.e., the shutdown threshold). The laser can only operate between these two pump-power levels, with a maximum output power somewhere in between.

5.4.3 Cascaded Fabry–Perot Lasers

It has been shown theoretically (Krause et al. 2006) and experimentally (Rong et al. 2008) that cascaded silicon Raman lasers, in which the pump light undergoes multiple Stokes shifts in a single silicon waveguide, can efficiently convert near-infrared to mid-IR radiation. The use of silicon waveguides for this purpose is particularly interesting, since the TPA vanishes for wavelengths above 2.2 μm, such that FCA becomes insignificant if all wavelengths are large enough, and very efficient lasing should become possible. If the pump wavelength lies significantly below 2.2 μm, however, the effects of TPA and FCA become increasingly significant, and the question arises: What is the shortest pump wavelength that can be used for a given waveguide?

The basic structure of a cascaded SRL is shown in Figure 5.20. An external pump laser at wavelength λ_{pump} is injected into the silicon waveguide, and Stokes light downshifted in frequency by 15.6 THz is generated by SRS. This first-order Stokes light circulates inside an intermediate cavity formed by highly reflecting elements such as Bragg gratings or dielectric coatings and acts itself as the pump source for the second-order Stokes line generated in the same waveguide. This cascading process continues until the desired wavelength is reached. In this section, we aim at designing lasers with output at 3 μm.

5.4.3.1 Model

We base the following simulations on the rib-waveguide structure of Liu et al. (2006), assuming that the pump and all Stokes lines propagate in the fundamental quasi-TE mode of the waveguide. This could be enforced in practice by making use of waveguide birefringence such that only the quasi-TE-mode reflection spectrum of the Bragg gratings lies inside the Raman-gain spectrum.

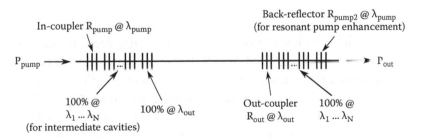

FIGURE 5.20 Schematic of a cascaded Fabry–Perot silicon Raman laser. The two reflectors R_{pump} and R_{pump2} at the pump wavelength λ_{pump} can serve to resonantly enhance the externally applied pump power P_{pump}.

The longitudinal evolution of the pump wave and the various Stokes lines is governed by differential equations that are a straightforward extension of the model of Section 5.2.7 to an arbitrary number of Stokes lines. We assume that the waveguide is dispersive enough such that processes that are not in general automatically phase-matched (such as FWM and CARS between the various laser lines) play no significant role and may be neglected. In that case, the evolution of the forward- (+) and backward-propagating (−) laser lines i can be described in terms of powers $P_i^{\pm}$ as in the noncascaded lasers of Section 5.4.2:

$$\pm \frac{1}{P_i^{\pm}} \frac{dP_i^{\pm}}{dz} = -\alpha_i - \frac{\beta_{ii}}{A_{ii}^{\text{TPA}}}(P_i^{\pm} + 2P_i^{\mp}) - \sum_{j \neq i} \frac{\beta_{ij}}{A_{ij}^{\text{TPA}}}(2P_j^{+} + 2P_j^{-}) - \eta_i \bar{\varphi} \lambda_i^2 \bar{N}$$

$$+ \sum_{j<i} \left[\frac{g_{ij}}{A_{i+j+}^{\text{Raman}}} P_j^{\pm} + \frac{g_{ij}}{A_{i+j-}^{\text{Raman}}} P_j^{\mp} \right] - \sum_{j>i} \left[\frac{\lambda_j}{\lambda_i} \frac{g_{ji}}{A_{i+j+}^{\text{Raman}}} P_j^{\pm} + \frac{\lambda_j}{\lambda_i} \frac{g_{ji}}{A_{i+j-}^{\text{Raman}}} P_j^{\mp} \right]$$

$$(5.103)$$

where $i = 1$ corresponds to the pump waves, and the remaining lines $i = 2 \dots n$ are ordered by increasing wavelength λ_i, with λ_n denoting the final (output) wavelength. The Raman gain constants g_{ij} are for the gain exerted by line j on line i and are nonzero only for $i = j + 1$. From (5.103), the steady-state carrier density follows as (compare Section 5.2.6.3)

$$\bar{N} = \frac{M}{2} \sum_{i=1}^{n} \frac{1}{h\nu_i} \left[\frac{\beta_{ii}}{A_{ii}^{\text{TPA}}} (P_i^{+2} + 4P_i^{+}P_i^{-} + P_i^{-2}) + 2(P_i^{+} + P_i^{-}) \sum_{j \neq i} \frac{\beta_{ij}}{A_{ij}^{\text{TPA}}} (P_j^{+} + P_j^{-}) \right]$$

$$(5.104)$$

As the wavelengths occurring in the analyzed cascaded lasers vary in a wide range (1.5 … 3.0 μm, i.e., by a factor of two), it is essential that we take into account the wavelength dependence of the mode fields [and thus of the effective areas A^{TPA} (5.38) and A^{Raman} (5.44)] and of the bulk TPA and Raman-gain coefficients. The mode fields are calculated with a full-vectorial mode solver (Krause et al. 2006), and the integrals in (5.38) and (5.44) are performed numerically for all wavelength combinations. For the degenerate-TPA coefficients b_{ii}, we use the model of Dinu (2003),

$$\beta_{ii} = \beta_{\deg}(\nu_i) = C \cdot \frac{(2h\nu_i / E_{ig} - 1)^4}{(2h\nu_i / E_{ig})^7} \qquad (5.105)$$

where $E_{ig} \approx 1.12$ eV is the indirect-band-gap energy of silicon, and C is a constant that we calibrate such that the degenerate-TPA coefficient at a wavelength of 1427 nm has one of the higher values reported, 0.7 cm/GW (Claps et al. 2004), in order not to underestimate the effects of nonlinear absorption. The degenerate TPA coefficient (5.105) is plotted versus wavelength as the thick solid curve in Figure 5.21. It vanishes for $\lambda > 2.2\,\mu m$, where the photon energy is less than half the indirect band gap of silicon. As for nondegenerate TPA, it is suggested in Dinu et al. (2003) to approximate the coefficient β_{ij} (where $\nu_i \neq \nu_j$) by the degenerate-TPA coefficient of (5.105) evaluated at the mean frequency. In our modeling, we therefore use $\beta_{ij} = (\nu_i / \nu_j)^{1/2} \beta_{\deg}[(\nu_i + \nu_j)/2]$, which is consistent with the requirement that $\beta_{ij} = (\nu_i / \nu_j)\beta_{ji}$ (Shen 1984). First considerations of three-photon absorption have found that its effect on Raman amplification is probably negligible (Raghunathan et al. 2006; Vermeulen et al. 2007; Pearl et al. 2008).

As for the spectral dependence of the Raman-gain coefficient, we assume that the Raman-tensor components (Grimsditch and Cardona 1980) and the Raman linewidth of silicon are constant in the wavelength range of interest. The Raman-gain constant then

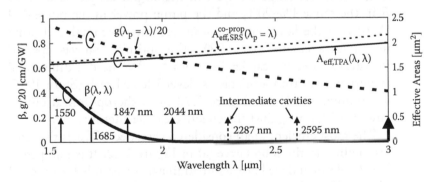

FIGURE 5.21 Spectral dependence of the bulk coefficients (thick) and effective areas (thin) for degenerate TPA (solid) and codirectional SRS (dotted). The arrows indicate wavelengths offset by 15.6 THz from the desired output wavelength of 3.0 μm. (Reprinted, with permission, from Krause, M., Fraheim, R., Renner,H., and Brinkmeyer, E. Cascaded silicon Raman lasers as midinfrared sources, *Electron. Lett.* 42(21): 1224–1226, 2006, ©2006, The Instituion of Engineering and Technology.)

scales essentially inversely with the Stokes wavelength (Boyd 2003). The thick dotted curve in Figure 5.21 shows this dependence; we calibrated this curve such that $g(\lambda_p = 1427\,\text{nm}) = 20\,\text{cm}/\text{GW}$ as in Jalali et al. (2005). As for the wavelength dependence of the linear waveguide losses, however, realistic models for such a large wavelength range are not available in the literature to the best of our knowledge; therefore, we will simply consider various different wavelength-independent loss coefficients in our simulations. Potentially large losses caused by OH impurities in the silica cladding beyond 2 μm could be circumvented by undercutting of the silicon waveguide (Jalali et al. 2005). For the numerical solution of the boundary-value problem describing the laser, we use the methods described in Section 5.4.1.3. The boundary conditions are (5.98) for the various Stokes lines, and (5.99) for the pump line (i.e., here we assume no pump resonator, $R_{\text{pump}} = R_{\text{pump2}} = 0$).

5.4.3.2 Conversion Efficiency versus Lifetime and Pump Wavelength

Since our goal is to obtain laser output at 3.0 μm, we must choose a shorter pump wavelength that is offset from the desired output wavelength by an integral multiple of the silicon Raman shift of 15.6 THz. The solid upright arrows in Figure 5.21 represent four possible choices that we analyze in the following text. We need several intermediate cavities (dashed upright arrows) in all of these four arrangements. For each laser, we have optimized its length L in the range 0.5–30 cm, the output-coupler reflectivity R in the range 1%–90%, and the pump power in the range 0–4 W so that the laser output power becomes maximal. All other reflectivities remain fixed at 99%. This optimization was repeated for several effective carrier lifetimes τ_{eff}. We used a differential evolution algorithm available as devec3 for Matlab (Storn 1997) to optimize the lasers. Figure 5.22 plots the optimized output power at 3.0 μm against the effective carrier lifetime.

The most efficient of all considered cascaded SRLs is that corresponding to the solid curve (a) in Figure 5.22. It is pumped at $\lambda_p = 2044$ nm and we have assumed linear waveguide losses of 0.3 dB/cm. Even for a comparatively large effective carrier lifetime of $\tau_{\text{eff}} = 10$ ns, the SRL output power exceeds 1 W at a pump power of 4 W and thus has a conversion efficiency comparable to that of fiber-based cascaded Raman lasers (Headley and Agrawal 2005). This is due to the fact that the TPA coefficient at the pump wavelength of 2044 nm is so small (see Figure 5.21) that only few

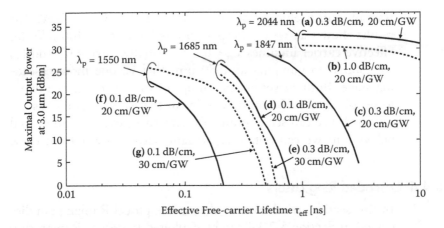

FIGURE 5.22 Maximal output power of cascaded silicon Raman lasers emitting at 3.0μm versus effective carrier lifetime τ_{eff} for four different pump wavelengths. Length L and output reflectivity R have been optimized.(Reprinted, with permission, from Krause, M. et al. *Electron. Lett.* **42**(21): 1224–1226, 2006. ©2006, The Institution of Engineering and Technology.)

carriers are generated by TPA—thus, FCA becomes significant only for very long effective carrier lifetimes $\tau_{eff} >$ 10 ns. For comparison, curve (b) in Figure 5.22 shows the optimization result for rather large linear losses of 1.0 dB/cm, where the SRL output powers still exceed 0.5 W.

When going toward shorter pump wavelengths, we expect that the efficiency of the SRLs will decrease, since the TPA coefficient increases toward shorter wavelengths (see thick solid curve in Figure 5.21). Consequently, more carriers are generated, leading to stronger FCA. Curve (c) in Figure 5.22 shows the maximal output power as a function of the effective carrier lifetime for a laser pumped at 1847 nm, that is, one more cascade than for curve (a) and otherwise unchanged parameters. The output power exceeds 0.25 W for effective carrier lifetimes $\tau_{eff} <$ 1 ns. However, it is not possible to obtain lasing at all for the larger value of τ_{eff} = 10 ns.

Finally, when the pump needs to be spectrally located at even shorter wavelengths, we found that waveguides with subnanosecond carrier lifetimes τ_{eff} are necessary. Curves (d) and (e) in Figure 5.22 correspond to pumping at 1685 nm. To achieve lasing, we assumed slightly more favorable simulation parameters than for curve (c); for (d), we assumed linear losses of only 0.1 dB/cm, and for (e), we assumed a more optimistic value for the

Raman-gain coefficient of $g(\lambda_p = 1427 \text{ nm}) = 30 \text{ cm/GW}$. The results show that we get output powers larger than 100 mW for effective carrier lifetimes $\tau_{eff} < 0.3$ ns. Finally, curves (f) and (g) correspond to pumping at 1550 nm, that is, one more cascading stage. Both curves are for an optimistic choice for the linear losses of $\alpha = 0.1$ dB/cm; they differ in the Raman-gain constant assumed. We can see that only for effective carrier lifetimes well below 1 ns we can obtain significant output power from these SRLs.

5.4.4 Cascaded Ring Lasers

In this section, we show how the nonreciprocal Raman gain discussed in Section 5.2.4.4 can be exploited to design Raman ring lasers that are more efficient than their Fabry–Perot counterparts of Section 5.4.3.

To this end, we have to scale down the waveguide dimensions and go from the micron-sized rib waveguides used so far (Rong et al. 2005; Rong et al. 2006; Rong et al. 2007; Rong et al. 2008) toward submicron photonic wires. On the one hand, it is expected that due to the smaller effective mode areas, the pump power requirements of the SRLs can be lowered (Rong et al. 2005). On the other hand, the Raman gain becomes nonreciprocal: as shown in Section 5.2.4.4, a Stokes wave propagating in a direction opposite to the pump wave can experience a local gain that is 50% larger than the gain experienced by a copropagating Stokes wave. In an RR-SRL based on small photonic wires, this nonreciprocity could be strong enough to force the laser to operate unidirectionally (Krause et al. 2009b). Note that in RR-SRLs based on the larger rib waveguides, lasing has been reported to be bidirectional (Rong et al. 2006). A practical advantage of unidirectional lasing is that the entire output power exits the laser at one port. In addition, it makes RR-SRLs more efficient than Fabry–Perot (FP) SRLs, since in the latter the laser light always experiences the less efficient codirectional amplification at least during one half of the round trip, as we show in the following (Krause et al. 2009a).

In a ring-resonator SRL, shown in Figure 5.23, the laser resonator is formed by a ring that is coupled to an access waveguide. The coupler is designed such that it couples over certain power fractions at the pump and output wavelengths while keeping the intermediate wavelengths inside the ring. In order to optimize various noncascaded and cascaded RR- and FP-SRLs, we use the model of Section 5.4.3.1. The various effective areas and confinement

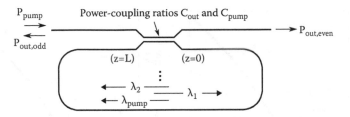

FIGURE 5.23 Schematic of a cascaded ring-resonator silicon Raman laser. (Reprinted, with permission, from Krause et al. (2009a), ©2009, The Institution of Engineering and Technology.)

factors occurring in that model are numerically calculated for the photonic-wire waveguide of Section 5.2.4.4, which has a strong Raman nonreciprocity. For both FP- and RR-SRLs (Figures 5.20 and 5.23, respectively), we assume resonant coupling of the pump laser at $\lambda_{\text{pump}} = 1455\,\text{nm}$ into the SRL, thereby allowing for resonant pump-power enhancement (see Section 5.4.1.2). Thus, the boundary conditions for an RR-SRL are $P_i^{\pm}(0) = P_i^{\pm}(L)$ for the intermediate lines $i = 2 \dots n - 1$, $P_n^{\pm}(0) = (1 - C_{\text{out}})^{\pm 1} P_n^{\pm}(L)$ for the last (output) Stokes wavelength ($i = n$), and

$$\sqrt{P_1^+(0)} = \sqrt{C_{\text{pump}}}\sqrt{P_{\text{pump}}} + \sqrt{(1 - C_{\text{pump}})}\sqrt{P_1^+(L)}$$

for the pump line $i = 1$; see Figure 5.23. For FP-SRLs, we use (5.98) for the various Stokes lines, while for the pump lines we use (5.99b) and (5.100) with $R_{\text{pump2}} = 100\%$. When simulating the RR-SRLs, we assume unidirectional propagation of each Stokes order as discussed at the beginning of this section. Because the contradirectional Raman gain is larger than the codirectional one, successive Stokes orders are assumed to propagate in alternating directions. To maximize the laser output power, we optimized the cavity-coupling ratios for the pump and output wavelengths (R_{pump}, R_{out}, C_{pump}, and C_{out} in Figures 5.20 and 5.23), the cavity length L, and the pump power (up to 100 mW) using a differential evolution algorithm (Storn 1997).

The resulting characteristics of the optimized *noncascaded* SRLs are shown in Figure 5.24a for fixed linear waveguide losses of $\alpha = 1.0\,\text{dB}/\text{cm}$. The FP-SRL (dotted curve) emits up to 17 mW at a pump power of 100 mW, while the RR-SRL is, as expected, significantly more efficient, emitting up to 24 mW (solid curve). Optimized characteristics for first- and second-order cascaded

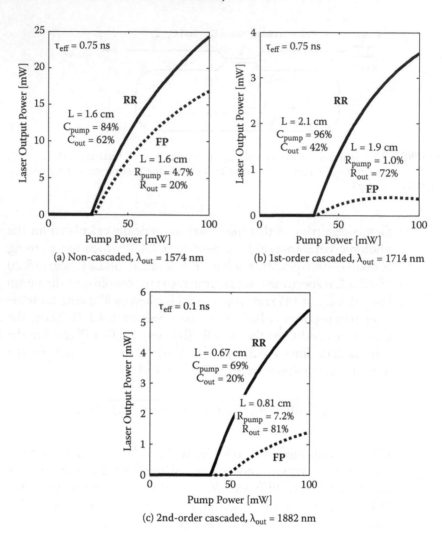

FIGURE 5.24 Characteristics of optimized SRLs for α = 1.0 dB/cm. (Reprinted, with permission, from Krause et al. (2009a), ©2009, The Institution of Engineering and Technology.)

SRLs are shown in Figures 5.24b and 5.24c, respectively. They show an even clearer superiority for the RR-SRL. While for the noncascaded and first-order cascaded lasers, we assumed τ_{eff} = 750 ps, for second-order cascading we had to assume a lower value of τ_{eff} = 100 ps (Tanabe et al. 2007) in order to achieve lasing at all. Longer carrier lifetimes could be tolerated when the pump is moved to larger wavelengths where TPA is less effective (compare the results of Section 5.4.3.2).

References

Agrawal, G. P., *Nonlinear Fiber Optics*, 3rd ed., Academic Press, San Diego, CA. 2001.

AuYeung, J. and Yariv, A., Theory of cw Raman oscillation in optical fibers, *J. Opt. Soc. Am.* **69**(6): 803–807, 1979.

Boyd, R. W., *Nonlinear Optics*, 2nd ed., Academic Press, San Diego, CA. 2003.

Boyraz, Ö and Jalali, B., Demonstration of a silicon Raman laser, *Opt. Express* **12**(21): 5269–5273, 2004.

Bristow, A. D., Rotenberg, N., and van Driel, H. M., Two-photon absorption and Kerr coefficients of silicon for 850-2200 nm, *Appl. Phys. Lett.* **90**: 191104, 2007.

Butcher, P. N. and Cotter, D., *The Elements of Nonlinear Optics*, Cambridge: Cambridge University Press, 1990.

Chen, X., Panoiu, N. C., and Osgood, R. M. Jr., Theory of Raman-mediated pulsed amplification in silicon-wire waveguides, *IEEE J. Quantum Electron.* **42**(2): 160–170, 2006.

Cierullies, S., Krause, M., Renner, H., and Brinkmeyer, E., Experimental and numerical study of the switching dynamics of Raman fiber lasers, *Appl. Phys.* B **80**(2): 177–183, 2005.

Claps, R., Dimitropoulos, D., Han, Y., and Jalali, B., Observation of Raman emission in silicon waveguides at 1.54 μm, *Opt. Express* **10**(22): 1305–1313, 2002.

Claps, R., Dimitropoulos, D., and Jalali, B., Stimulated Raman scattering in silicon waveguides, *Electron. Lett.* **38**(22): 1352–1354, 2002.

Claps, R., Dimitropoulos, D., Raghunathan, V., Han, Y., and Jalali, B., Observation of stimulated Raman amplification in silicon waveguides, *Opt. Express* **11**(15): 1731–1739, 2003.

Claps, R., Raghunathan, V., Dimitropoulos, D., and Jalali, B., Influence of nonlinear absorption on Raman amplification in silicon waveguides, *Opt. Express* **12**(12): 2774–2780, 2004.

de Leonardis, F. and Passaro, V. M. N., Ultrafast Raman pulses in SOI optical waveguides for nonlinear signal processing, *IEEE J. Sel. Top. Quantum Electron.* **14**(3): 739–751, 2008.

de Sterke, C. M., Jackson, K. R., and Robert, B. D., Nonlinear coupled-mode equations on a finite interval: a numerical procedure, *J. Opt. Soc. Am.* B **8**(2): 403–412, 1991.

Dimitropoulos, D., Fathpour, S., and Jalali, B., Limitations of active carrier removal in silicon Raman amplifiers and lasers, *Appl. Phys. Lett.* **87**: 261108, 2005.

Dimitropoulos, D., Houshmand, B., Claps, R., and Jalali, B., Coupled-mode theory of the Raman effect in silicon-on-insulator waveguides, *Opt. Lett.* **28**(20): 1954–1956, 2003.

Dimitropoulos, D., Jhaveri, R., Claps, R., Woo, J. C. S., and Jalali, B., Lifetime of photogenerated carriers in silicon-on-insulator rib waveguides, *Appl. Phys. Lett.* **86**: 071115, 2005.

Dimitropoulos, D., Solli, D. R., Claps, R., Boyraz, Ö., and Jalali, B., Noise figure of silicon Raman amplifiers, *J. Lightwave Technol.* **26**(7): 847–852, 2008.

Dinu, M., Dispersion of phonon-assisted nonresonant third-order non-linearities, *IEEE J. Quantum Electron.* **39**(11): 1498–1503, 2003.

Dinu, M., Quochi, F., and Garcia, H., Third-order nonlinearities in silicon at telecom wavelengths, *Appl. Phys. Lett.* **82**(18): 2954–2956, 2003.

Dissanayake, C. M., Rukhlenko, I. D., Premaratne, M., and Agrawal, G. P., Raman-mediated nonlinear interactions in silicon waveguides: copropagating and counterpropagating pulses, *IEEE Photon. Technol. Lett.* **21**(19): 1372–1374, 2009.

Doylend, J. K., Cohen, O., Lee, M. R., Raday, O., Xu, S., Sih, V., Rong, H., and Paniccia, M., Tunable ring resonators for silicon Raman laser and amplifier applications, in C. M. Greiner and C. A. Waechter (Eds), *Integrated Optics: Devices, Materials, and Technologies XII, Proc. SPIE*, vol. 6896, 2008.

Dvorak, M. D., Schroeder, W. A., Andersen, D. R., Smirl, A. L., and Wherrett, B. S., Measurement of the anisotropy of two-photon absorption coefficients in zincblende semiconductors, *IEEE J. Quantum Electron.* **30**(2): 256–268, 1994.

Fathpour, S., Tsia, K. K., and Jalali, B., Energy harvesting in silicon Raman amplifiers, *Appl. Phys. Lett.* **89**: 061109, 2006.

Först, M., Niehusmann, J., Plötzing, T., Bolten, J., Wahlbrink, T., Moormann, C. and Kurz, H., High-speed all-optical switching in ion-implanted silicon-on-insulator microring resonators, *Opt. Lett.* **32**(14): 2046–2048, 2007.

Grimsditch, M. and Cardona, M., Absolute cross-section for Raman scattering by phonons in silicon, *Phys. Stat. Sol. (B)* **102**: 155–161, 1980.

Headley, C. and Agrawal, G. P. (Eds), *Raman Amplification in Fiber Optical Communication Systems*, Elsevier, Burlington, MA. 2005.

Hellwarth, R. W., Third-order optical susceptibilities of liquids and solids, *Prog. Quant. Electr.* **5**: 1–68, 1977.

Jalali, B., Can silicon change photonics?, *Phys. Stat. Sol. (A)* **205**(2): 213–224, 2008.

Jalali, B., Boyraz, Ö., Dimitropoulos, D., Raghunathan, V., Claps, R., and Koonath, P., Silicon Raman amplifiers, lasers, and their applications, *Conference on Group IV Photonics (GFP)*, Paper ThA1, 2005.

Jones, R., Rong, H., Liu, A., Fang, A. W., Paniccia, M. J., Hak, D., and Cohen, O., Net continuous wave optical gain in a low loss silicon-on-insulator waveguide by stimulated Raman scattering, *Opt. Express* **13**(2): 519–525, 2005.

Kagawa, T. and Ooami, S., Polarization dependence of two-photon absorption in Si Avalanche photodiodes, *Jpn. J. Appl. Phys.* **46**(2): 664–668, 2007.

Krause, M., Efficient Raman amplifiers and lasers in optical fibers and silicon waveguides: new concepts, Ph.D. thesis, Technische Universität Hamburg-Harburg, Hamburg, Germany. Cuvillier Verlag, Göttingen, 2007, available online: urn:nbn:de:gbv:830-tubdok-5769.

Krause, M. and Brinkmeyer, E., Silicon-waveguide duplexers using nonreciprocal Raman gain, *5th International Conference on Group IV Photonics (GFP)*, Paper ThP6, 2008.

Krause, M., Draheim, R., Renner, H., and Brinkmeyer, E., Cascaded silicon Raman lasers as mid-infrared sources, *Electron. Lett.* **42**(21): 1224–1226, 2006.

Krause, M., Müller, J., Pagel, T., Renner, H., and Brinkmeyer, E., Nonreciprocal Raman scattering in silicon waveguides, *7th International Conference on Group IV Photonics (GFP)*, Paper ThA3, 2010.

Krause, M. and Renner, H., Optimized pumping schemes for mid-infrared silicon Raman amplifiers, *Optical Fiber Communication Conference*, San Diego, CA. Paper OWM2, 2008.

Krause, M., Renner, H., and Brinkmeyer, E., Analysis of Raman lasing characteristics in silicon-on-insulator waveguides, *Opt. Express* **12**(23): 5703–5710, 2004.

Krause, M., Renner, H., and Brinkmeyer, E., Mitigation of free-carrier absorption and efficiency increase in silicon waveguide lasers by bidirectional pumping, *2nd International Conference on Group IV Photonics*, Paper P15, 2005.

Krause, M., Renner, H., and Brinkmeyer, E., Polarization-dependent curvature loss in silicon rib waveguides, *IEEE J. Sel. Top. Quantum Electron.* **12**(6): 1359–1362, 2006. (Part II, Special Issue on Silicon Photonics.)

Krause, M., Renner, H., and Brinkmeyer, E., Optical isolation in silicon waveguides based on nonreciprocal Raman amplification, *Electron. Lett.* **44**(11): 691–693, 2008a.

Krause, M., Renner, H., Fathpour, S., Jalali, B., and Brinkmeyer, E., Gain enhancement in cladding-pumped silicon Raman amplifiers, *IEEE J. Quantum Electron.* **44**(7): 692–704, 2008b.

Krause, M., Renner, H., and Brinkmeyer, E. Raman lasers in silicon photonic wires: unidirectional ring lasing versus Fabry-Perot lasing, *Electron. Lett.* **45**(1): 42–43, 2009a.

Krause, M., Renner, H., and Brinkmeyer, E., Strong enhancement of Raman-induced nonreciprocity in silicon waveguides by alignment with the crystallographic axes, *Appl. Phys. Lett.* **95**: 261111, 2010a.

Krause, M., Renner, H., and Brinkmeyer, E., Silicon Raman amplifiers with ring-resonator-enhanced pump power, *IEEE J. Sel. Top. Quantum Electron.* **16**(1): 216–225, 2010.

Krause, M., Renner, H., Brinkmeyer, E., Fathpour, S., Dimitropoulos, D., Raghunathan, V., and Jalali, B., Efficient Raman amplification in cladding-pumped silicon waveguides, *3rd International Conference on Group IV Photonics (GFP)*, Paper P6, 2006.

Kuo, Y.-H., Rong, H., and Paniccia, M. High bandwidth silicon ring resonator Raman amplifier, *3rd International Conference on Group IV Photonics (GFP)*, Paper FB2, 2006.

Liang, T. K. and Tsang, H. K., Role of free carriers from two-photon absorption in Raman amplification in silicon-on-insulator waveguides, *Appl. Phys. Lett.* **84**(15): 2745–2747, 2004.

Liu, A., Rong, H., Jones, R., Cohen, O., Hak, D., and Paniccia, M., Optical amplification and lasing by stimulated Raman scattering in silicon waveguides, *J. Lightwave Technol.* **24**(3): 1440–1455, 2006.

Liu, Y. and Tsang, H. K., Nonlinear absorption and Raman gain in helium-ion-implanted silicon waveguides, *Opt. Lett.* **31**(11): 1714–1716, 2006.

McKelvey, J. P. *Solid State and Semiconductor Physics*, Harper & Row, New York, NY 1966.

Okawachi, Y., Foster, M. A., Sharping, J. E., Gaeta, A. L., Xu, Q., and Lipson, M., All-optical slow light on a photonic chip, *Opt. Express* **14**(6): 2317–2322, 2006.

Passaro, V. M. N. and de Leonardis, F., Space-time modeling of Raman pulses in silicon-on-insulator optical waveguides, *J. Lightwave Technol.* **24**(7): 2920–2931, 2006.

Pavesi, L. and Lockwood, D. J. (Eds), *Silicon Photonics*, Springer-Verlag, Berlin, Heidelberg, 2004.

Pearl, S., Rotenberg, N., and van Driel, H. M., Three photon absorption in silicon for 2300-3300 nm, *Appl. Phys. Lett.* **93**: 131102, 2008.

Raghunathan, V., Borlaug, D., Rice, R. R., and Jalali, B., Demonstration of a mid-infrared silicon Raman amplifier, *Opt. Express* **15**(22): 14355–14362, 2007.

Raghunathan, V., Shori, R., Stafsudd, O. M., and Jalali, B., Nonlinear absorption in silicon and the prospects of mid-infrared silicon Raman lasers, *Phys. Stat. Sol. (A)* **203**(5): R38–R40, 2006.

Renner, H., Silicon Raman amplifiers in the limit of zero free-carrier lifetime, *4th International Conference on Group IV Photonics (GFP)*, Paper WP30, 2007.

Renner, H., Upper limit for the amplifiable stokes power in saturated silicon waveguide Raman amplifiers, *7th International Conference on Group IV Photonics (GFP)*, Paper P1.15, 2010

Renner, H. and Krause, M., Maximal total gain of non-tapered silicon-on-insulator Raman amplifiers, *Optical Amplifiers and Their Applications (OAA) Topical Meeting*, Paper OMD2, 2006.

Renner, H., Krause, M., and Brinkmeyer, E., Maximal gain and optimal taper design for Raman amplifiers in silicon-on-insulator waveguides, *Integrated Photonics Research and Applications Topical Meeting (IPRA)*, Paper JWA3, 2005.

Rong, H., Jones, R., Liu, A., Cohen, O., Hak, D., Fang, A., and Paniccia, M., A continuous-wave Raman silicon laser, *Nature* **433**: 725–727, 2005.

Rong, H., Kuo, Y.-H., Xu, S., Liu, A., Jones, R., Paniccia, M., Cohen, O., and Raday, O., Monolithic integrated Raman silicon laser, *Opt. Express* **14**(15): 6705–6712, 2006.

Rong, H., Liu, A., Nicolaescu, R., Paniccia, M., Cohen, O., and Hak, D., Raman gain and nonlinear optical absorption measurement in a low-loss silicon waveguide, *Appl. Phys. Lett.* **85**(12): 2196–2198, 2004.

Rong, H., Xu, S., Cohen, O., Raday, O., Lee, M., Sih, V., and Paniccia, M., A cascaded silicon Raman laser, *Nature Photonics* **2**: 170–174, 2008.

Rong, H., Xu, S., Kuo, Y.-H., Sih, V., Cohen, O., Raday, O., and Paniccia, M., Low-threshold continuous-wave Raman silicon laser, *Nat. Photonics* **1**: 232–237, 2007.

Roy, S., Bhadra, S. K., and Agrawal, G. P., Raman amplification of optical pulses in silicon waveguides: effects of finite gain bandwidth, pulse width, and chirp, *J. Opt. Soc. Am. B* **26**(1): 17–25, 2009.

Rukhlenko, I. D., Premaratne, M., Dissanayake, C., and Agrawal, G. P., Continuous-wave Raman amplification in silicon waveguides: beyond the undepleted pump approximation, *Opt. Lett.* **34**(4): 536–538, 2009.

Salem, R. and Murphy, T. E., Polarization-insensitive cross correlation using two-photon absorption in a silicon photodiode, *Opt. Lett.* **29**(13): 1524–1526, 2004.

Sang, X., Dimitropoulos, D., Jalali, B., and Boyraz, Ö., Influence of pump-to-signal RIN transfer on noise figure in silicon Raman amplifiers, *IEEE Photon. Technol. Lett.* **20**(24): 2021–2023, 2008.

Seeger, K., *Semiconductor Physics*, 5th ed., Springer, 1991.

Shen, Y.-R., *The Principles of Nonlinear Optics*, Wiley, New York, NY. 1984.

Shen, Y. R. and Bloembergen, N., Theory of stimulated Brillouin and Raman scattering, *Phys. Rev. A* **137**(6): 1787–1805, 1965.

Siegman, A. E., *Lasers*, University Science Books, Mill Valley, CA. 1986.

Sipe, J. E., de Sterke, C. M., and Eggleton, B. J., Rigorous derivation of coupled mode equations for short, high-intensity grating-coupled, co-propagating pulses, *J. Mod. Opt.* **49**(9): 1437–1452, 2002.

Snyder, A. W. and Love, J. D., *Optical Waveguide Theory*, London: Chapman and Hall, 1983.

Soref, R. A. and Bennett, B. R., Kramers-Kronig analysis of electro-optical switching in silicon, in Mark A. Mentzer (Ed), *Integrated Optical Circuit Engineering IV, Proc. SPIE*, vol. 704, pp. 32–37, 1986.

Soref, R. A. and Bennett, B. R., Electrooptical effects in silicon, *IEEE J. Quantum Electron.* **QE-23**(1): 123–129, 1987.

Soref, R. A., Emelett, S. J., and Buchwald, W. R. Silicon waveguided components for the long-wave infrared region, *J. Opt. A: Pure Appl. Opt.* **8**: 840–848, 2006.

Sparacin, D. K., Spector, S. J., and Kimerling, L. C., Silicon waveguide sidewall smoothing by wet chemical oxidation, *J. Lightwave Technol.* **23**(8): 2455–2461, 2005.

Storn, R., devec3—Differential Evolution for MATLAB, 1997, Available online: http://www.icsi. berkeley.edu/~storn/code.html.

Tanabe, T., Nishiguchi, K., Shinya, A., Kuramochi, E., Inokawa, H., Notomi, M., Yamada, K., Tsuchizawa, T., Watanabe, T., Fukuda, H., Shinojima, H., and Itabashi, S., Fast all-optical switching using ion-implanted silicon photonic crystal nanocavities, *Appl. Phys. Lett.* **90**: 031115, 2007.

Tsang, H. K. and Liu, Y., Nonlinear optical properties of silicon waveguides, *Semicond. Sci. Technol.* **23**: 064007, 2008.

Tzolov, V. P., Fontaine, M., Godbout, N., and Lacroix, S., Nonlinear self-phase-modulation effects: a vectorial first-order perturbation approach, *Opt. Lett.* **20**(5): 456–458, 1995.

Vermeulen, N., Debaes, C., Fotiadi, A. A., Panajotov, K., and Thienpont, H., Stokes-anti-stokes iterative resonator method for modeling Raman lasers, *IEEE J. Quantum Electron.* **42**(11): 1144–1156, 2006.

Vermeulen, N., Debaes, C., and Thienpont, H., Modeling mid-infrared continuous-wave silicon-based Raman lasers, in Peter E. Powers (Ed), *Nonlinear Frequency Generation and Conversion: Materials, Devices, and Applications VI, Proc. SPIE*, Vol. 6455, p. 64550U, 2007.

Yamada, H., Shirane, M., Chu, T., Yokoyama, H., Ishida, S., and Arakawa, Y., Nonlinear-optic silicon-nanowire waveguides, *Jpn. J. Appl. Phys.* **44**: 6541–6545, 2005.

Zhang, J., Lin, Q., Piredda, G., Boyd, R. W., Agrawal, G. P., and Fauchet, P. M., Anisotropic nonlinear response of silicon in the near-infrared region, *Appl. Phys. Lett.* **91**: 071113, 2007.

Zhao, H., Temperature dependence of ambipolar diffusion in silicon on insulator, *Appl. Phys. Lett.* **92**: 112104, 2008.

CHAPTER 6

Silicon Photonics for Biosensing Applications

Jenifer L. Lawrie and Sharon M. Weiss

Contents

6.1 Introduction

Although reports of biosensing devices have only occurred over the past 50 years, humans have essentially been performing biological and chemical detection for millennia. Antibodies in our immune system identify and neutralize foreign invaders; enzymes in our taste buds distinguish between sour and sweet; and olfactory receptor neurons can recognize thousands of different scents. With improved technology and understanding of biological organisms in the last century, researchers have looked toward many of these highly efficient and highly specialized biomechanisms as inspiration for new sensing devices. Evolution has produced biorecognition molecules of unmatched specificity. As a result, the vast majority of biosensing devices utilize a biorecognition system either taken directly from a living organism, or designed to mimic their function.

When it comes to building biosensing devices, however, a signal transducer is required in order to take the reaction of the bioreceptor and convert it to a measurable signal. There are a number of possible signal transduction methods available, generally falling under the categories of optical, electrochemical, or mass-related transduction. Of these categories, optical transduction is by far the most widely utilized, given the variety of spectrochemical properties available to be recorded for any given material. Optical transducers reported in the literature include absorption, reflection, fluorescence, Raman, and surface-enhanced Raman scattering. It is in the area of optical signal transduction that the unique optical properties of silicon play a crucial role in the development of new and innovative biosensing devices.

Biosensors and lab-on-a-chip devices may be classified by either their bioreceptor unit or their transducer. Depending on the desired application, there is a large variety of possible bioreceptors and transducer materials, and each transduction method may be utilized with a number of different bioreceptor-target systems. In order to highlight the role of silicon photonics in biosensing applications, the sensing devices discussed herein will be classified by their transducer microstructure, and by the accompanying spectroscopy technique used to detect the analyte of interest.

This review of silicon photonic applications in biosensing will begin with a review of the variety of bioreceptor molecules frequently employed in sensor devices (Section 6.2) and the surface chemistry considerations in attaching receptor to transducer

(Section 6.3). A discussion of the major classes of silicon photonic biosensing devices classified by the specific signal transduction method utilized will follow (Sections 6.4–6.7).

6.2 Bioreceptors

6.2.1 Antibody/Antigen

The antibody–antigen biorecognition system is often viewed as a gold standard in biosensing. Antibodies are complex proteins that may be composed of hundreds of amino acid subunits, and corresponding antigen targets typically must exceed 5000 Da in order to prompt an immunogenic response. In such a system, the biological molecules known as antibodies act as a receptor site toward a large protein target molecule, in what is often described as a "lock and key" fit. The "lock and key" refers to the antibody's specific conformation, which allows for binding of the target antigen based on specific characteristics such as size, geometric conformation, and active chemical functional groups. Figure 6.1 illustrates a generic antibody/antigen biosensor, in which the antibody molecules immobilized on the surface are capable of selectively binding a target antigen.

6.2.2 Enzymes

With a few exceptions, enzymes are generally a specific type of protein—a biological macromolecule composed of amino acids. Some enzymes are composed only of amino acids, while others include cofactors (often transition metal ions) or coenzymes (complex organometallic molecules). Often, enzymes are selected for use in biosensors for their specific binding capability or catalytic activity. Catalytic activity, in particular, can impart unique properties to a biosensing device, as enzyme-coupled reactors can be designed to modify the bioreceptor mechanism. For example, ligand binding at the receptor may change the activity of an enzyme. Additionally, biosensors using enzyme bioreceptors can potentially show very high sensitivity toward a particular target, since the biorecognition event is not a one-to-one binding, but rather an enzymatic activity. For example, consider a glucose-sensing fiber-optic device that has the enzyme glucose oxidase immobilized on the surface. Instead of directly sensing glucose, a detector measures the presence of oxygen in solution; the level of oxygen changes when the enzymatic reaction between glucose and glucose oxidase occurs (Rosenzweig and Kopelman 1996).

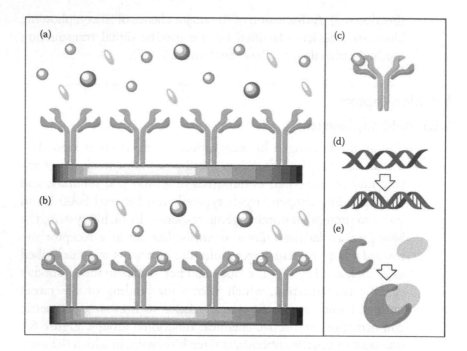

FIGURE 6.1 On the left, selective capture of target biomolecules by bioreceptors attached to a surface is illustrated: (a) A sensor surface functionalized with antibody molecules (i.e., bioreceptors) is exposed to a solution containing target and mismatched antigens. (b) Only the target antigens are bound to the immobilized antibodies. On the right, examples of common bioreceptor and target pairs are shown: (c) Antibody and antigen pair. (d) DNA as a receptor for a complementary DNA oligo. (e) An enzyme and target protein.

6.2.3 Nucleic Acids

Nucleic acids are a common choice for biorecognition molecules, where the hybridization event involving deoxyribonucleic acid (DNA) or ribonucleic acid (RNA) is detected. For DNA sensors, the hybridization reaction between the biorecognition element and target molecule results from the pairing up of complementary bases on corresponding oligos, with cytosine (C) bases in the oligo binding guanine (G) on the complement, and adenine (A) bases binding thymine (T) on the complement molecule. Nucleic acids have also been of considerable interest for biochip technologies that use nucleic acids as bioreceptors for detecting molecules other than complementary DNA strands. The considerable flexibility in defining a nucleic acid sequence and length allows for the generation of a large library of DNA sequences, which increases the likelihood of finding a suitable bioreceptor for a particular target of interest.

6.2.4 Cellular Structures

Cellular structure bioreceptors encompass a broad category that could include anything from using nonenzymatic transport proteins for target molecule binding to bioluminescent proteins for toxin detection in bacteria. When using cellular structures as bioreceptors, an entire cell or microorganism may be included in the sensor, or just a particular cell component. A few examples of such biorecognition systems are the protein cytochrome c' used for nitric oxide detection (Barker et al. 1998) and lipopeptides for virus detection (Boncheva et al. 1996).

6.3 Surface Chemistry and Passivation for Biosensing

Of critical importance in the development of silicon biosensing platforms and devices is the silicon surface preparation, allowing for appropriate passivation of the material toward aqueous solutions and functionalization for recognition of the biomolecule target. The focus of this section is on porous silicon surface functionalization, but much of the discussion is applicable to other silicon-based sensors described in this chapter. Passivation of the surface to provide a stable material for detection in a variety of biological solutions is especially important for porous silicon due to its biocompatibility, which causes degradation in physiologic fluids (Canham 1995). Detection of a number of biomolecules of interest, particularly in lab-on-a-chip flow cells, occurs in physiologic fluids. Surface degradation could lead to dramatic changes in effective optical thickness, which would significantly affect sensor performance and reliability (Bocking et al. 2008). This effect becomes amplified when the surface-to-volume ratio increases.

Immediately after fabrication, porous silicon architectures have silicon–hydride-terminated pore walls. The silicon–hydride bond is highly susceptible to oxidation under ambient conditions, so immediate stabilization of the material is required. Given the disparity between the refractive indices of silicon oxide (n = 1.45) and silicon (n = 3.5), fixing the silicon-to-oxygen ratio in the sensing layers of the material is critical. There are two general methods of surface passivation currently utilized in silicon photonics. The first method involves controlled growth of an oxide layer on the porous silicon architecture, and the second method involves passivation of the surface via formation of silicon–carbon bonds.

The more common method for porous silicon passivation is the growth of an oxide layer. The popularity of this method is more likely a reflection of the method's relative ease and repeatability, and not necessarily the quality of the resulting surface. An oxide layer on the surface of the silicon matrix can be produced thermally or by ozone. The thickness of the oxide layer grown thermally depends on the temperature and time for oxidation, as well as the oxygen content of the furnace. Growth of a sufficiently thick surface oxide layer prevents further changes in the effective optical thickness due to ambient conditions, but may not be sufficient to protect the surface from dissolution in aqueous solutions. After oxidation, silane chemistry is frequently utilized to form a protective alkyl silane monolayer on the silicon surface, further limiting attack on the silicon matrix by water molecules. Using alkyl silanes, the silicon sensor architecture can be tuned to sense a range of targets by selecting an appropriate distal functional group on the silane monomer, providing appropriate attachment groups for many of the bioreceptors described in Section 6.2. The major advantage of silane attachment is that it can be performed under ambient conditions; however, there is a strong tendency for silanes to form multilayers instead of monolayers, and the polarity of the silicon–oxygen bond formed upon silanization is highly susceptible to hydrolysis (Steinem et al. 2004).

For many applications, oxidization and silanization provide an appropriately passivated and functionalized surface for sensing. An alternative surface preparation method exists in hydrosilylation. The procedure involves exposing the hydride-terminated silicon to an alkene or alkyne, forming a very stable alkyl monolayer on the surface. Because the resulting silicon–carbon bond is nonpolar, this method forms a far more stable monolayer than with silanization. Additionally, since there is no possibility for cross-linking, this method does not result in multilayer formation (Buriak 2002). Surface passivation by hydrosilylation does have a major limitation in that it must occur under inert atmosphere, as formation of silicon oxides on the surface of the matrix will significantly compromise the quality of the monolayer formed. Hydrosilylation has been demonstrated as a highly effective surface preparation technique for biosensing, and has been used to attach a number of bioreceptors to various porous silicon architectures (Kilian et al. 2009). Figure 6.2

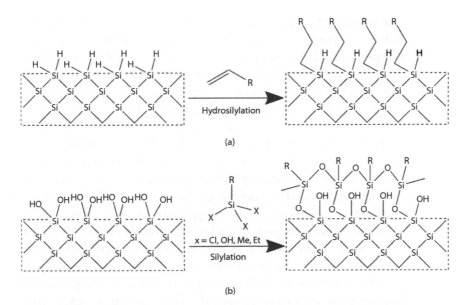

FIGURE 6.2 Sample reactions for porous silicon surface passivation. (a) Hydrosilylation reactions, involving either alkene or alkyne reagents, form silicon–carbon bonds. (b) Silylation reactions attach silane molecule to oxidized porous silicon via linkage to surface oxygen.

illustrates the difference in surface attachment between silanized and hydrosilylized silicon surfaces. Note that both surface functionalization methods are generally only a starting point for bioreceptor attachment, and further surface chemistry is often required for sensing.

6.4 Optical Reflectance Transducers in Porous Silicon

Although porous silicon was first reported in 1956 (Uhlir 1956), interest in the material was rather limited until its room temperature photoluminescence was observed (Canham 1990). Since then, there has been considerable interest in the integration of porous silicon components into optoelectronic devices, and more recently biochemical devices, based on the tunability of the physical and optical properties of the material. The pore diameters, which can range from nanometer to micron, as well as the porosity and thickness of the porous layer, can be tuned depending on the formation conditions, typically in an electrolyte containing hydrofluoric acid. When the pore diameter is much smaller than the wavelength of light, porous silicon can be treated as an effective medium for which the refractive index is a weighted average of the refractive indices of separate components of the composite matrix (Theiss 1997). Adjusting the porosity of the silicon allows for a wide range of refractive indices to be achieved. Depending on the conditions, a refractive index range of approximately 1.15 to 3.30 can be produced by appropriate tuning of the fabrication conditions. Multilayer thin-film structures, such as Bragg mirrors, rugate filters, microcavities, and waveguides, can be formed by simply varying the porosity through electrochemical etching. Since the etching preferentially proceeds at the pore tips where the electric field is concentrated and charge carriers are present, under most conditions the previously formed porous silicon layers are not etched further as the formation of subsequent layers proceeds. As a result, sharp interfaces between porous silicon layers of different porosity (different refractive index) are produced. Scanning electron microscopy (SEM) images of typical porous silicon films are shown in Figures 6.3a and 6.3b.

Over the past several years, there has been an increasing level of interest generated in the biological applications of porous silicon, such as drug delivery, environmental and chemical sensing, and filters for molecular separations. One of the most significant advantages of porous silicon is the enormous internal surface area, which can range up to a few hundred square meters per

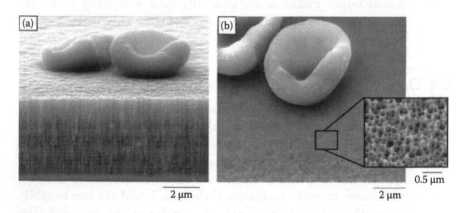

FIGURE 6.3 (a) Cross-sectional SEM view of a porous silicon microcavity. (b) Top view SEM of a porous silicon microcavity, including an inset of the surface pores with average diameter 88 nm. Note the erythrocytes on the surface, which are filtered out of the porous matrix and cross-linked to the surface via gluteraldehyde fixation, and demonstrating the size exclusion properties of the sensor even within complex media. (After Bonanno, L. M. and Delouise, L. A. 2007. *Biosens. Bioelectron.*, 23, 444–448. With permission.)

cubiccentimeter. Since many more biological molecules can attach to porous silicon compared to planar materials, its structure provides a significant advantage for capturing and detecting gaseous, chemical, and biological species. In the following sections, several different porous silicon optical structures are discussed for their application as biosensors.

6.4.1 Single-Layer Thin Films

The most basic transduction method utilizing porous silicon is based on the change in effective optical thickness of porous silicon films upon a binding event between receptor and target molecules in a single-layer thin film. This sensing mechanism takes advantage of the difference between the phase of light reflected at the top and bottom interfaces of a porous silicon layer. A reflectance spectrum for such a porous silicon thin film shows a characteristic Fabry–Perot interference pattern; the fringe spacing is dependent on the optical thickness of the porous silicon layer. Infiltration of bioreceptors into the pores, and the subsequent biorecognition reaction with a target molecule, results in a change in the optical thickness of the thin film due to the increase in the effective refractive index as the biological target replaces air in the pores.

A porous silicon interferometric biosensing device was first introduced by Sailor and colleagues in 1997, a breakthrough that sparked significant interest in porous silicon as a platform for optical biosensing (Lin et al. 1997). A porous silicon thin film was fabricated via electrochemical etch, oxidized, and functionalized with a chemical linker. Either DNA oligonucleotides or proteins were attached as bioreceptors in the films. To monitor the DNA hybridization or protein interaction, the reflectance spectrum was taken before and after exposure to the target molecule of interest, and the Fabry–Perot fringes were monitored. In the initial work by Sailor and colleagues, a blue shift (e.g., shift to shorter wavelengths) in the interference pattern—instead of the expected red shift (e.g., shift to longer wavelengths) corresponding to an increase in optical thickness—was observed. This blue shift of the DNA was later attributed to the corrosion of the porous silicon layer upon exposure to the analyte solution (Steinem 2004). This unanticipated dissolution of the film, however, resulted in a sensitivity that remains unmatched in similar single-layer porous silicon sensors using optical reflectance measurements. Detection limits in the subpicomolar range were observed (Lin 1997).

Recently, it has been shown that labeling DNA oligonucleotides with charge-carrying metal complexes can increase the porous silicon corrosion rate, thus improving the sensitivity of the detector. With a labeled system in a single-layer interferometer sensor, detection limits in the submicromolar range can be achieved for a number of bioreceptor and target pairs (Voelker et al. 2008).

The dissolution of the porous silicon thin film, and thus the blue shift of the Fabry-Perot spectrum, can be prevented by selecting appropriate surface passivation chemistry. Essentially, the observation of a blue shift indicates that the surface was not effectively passivated against chemical attack by the analyte solution. By adjusting the oxidation or silanization conditions appropriately, the surface dissolution in analyte is prevented, allowing for unambiguous detection of the target molecules through a red shift in the Fabry–Perot optical signature. This red shift in the optical reflectance spectrum has been observed for affinity interactions (Janshoff et al. 1998), protein–protein interactions (Dancil et al. 1999), DNA hybridization (De Stefano et al. 2006), and small molecule binding (Tinsley-Bown et al. 2005) in single-layer porous silicon. These systems, however, are considerably less sensitive than those relying on porous silicon dissolution. Typical detection limits of red-shifted sensors are in the micromolar range (Jane et al. 2009).

In recent years, there has been further development of Fabry–Perot optical materials for biosensing, such as two-layer interferometric biosensors with different pore sizes in each layer. A major limitation of the single-layer devices is the drift observed in the effective optical thickness of the thin film due to complex media being introduced and inferring with the sensor. Drift can be difficult to account for in real-time measurements, and has a significant impact on sensor performance. The two-layer architecture provides a top porous layer for the biorecognition event to occur within, and adds a bottom porous layer to act as the reference layer. The porosity of the bottom layer is sufficiently small to exclude biomolecules based on size, ensuring that the target biorecognition event occurs only in the top porous layer. When reflectance data is taken for these two-layer systems, the three superimposed interference patterns that result (i.e., due to top layer, bottom layer, combined layers) allow for a built-in control to compensate for signal drift by separately providing data on the binding, adsorption, and partitioning in the layers. Size-dependent filtration and molecular detection was demonstrated using bovine serum albumin (BSA; 66 kDa) and sucrose (0.34

kDa). The porosities of the layers were tuned such that the larger BSA molecule could only penetrate the top porous layer, but the smaller sucrose molecule could access both the top and bottom pores (Pacholski et al. 2005).

6.4.2 Bragg Mirrors

Bragg mirrors are a class of one-dimensional photonic crystal that have been produced in porous silicon for biosensing. Fabrication builds upon the methods discussed in the previous section. By periodically changing the applied current density during the electrochemical etch using a step function, alternating layers of high and low porosities of porous silicon are obtained. The condition of $\lambda/4$ for the optical thickness is typically applied to the Bragg mirrors, and gives a near 100% reflectance band over the spectral region of interest. When biomolecules infiltrate into the porous layers, there is a red shift in the wavelength of the stop band, allowing one to monitor changes in the reflectance spectrum of the Bragg mirror to detect the presence of biomolecules.

Some limited work in biosensing with Bragg mirrors has been done, including a set of theoretical (Moretti et al. 2007) and experimental analyses comparing the sensitivity of Bragg mirrors to single-layer porous silicon layers (Anderson et al. 2003). When a sucrose solution was introduced to the different porous silicon architectures, experiments suggested that Bragg mirror biosensors may be more sensitive than single-layer biosensing structures (Anderson et al. 2003).

6.4.3 Microcavities

Introduction of a defect layer in the middle of a Bragg mirror results in a porous silicon architecture known as a microcavity. The defect layer is simply a layer that has a different optical thickness than either of the repeated two layers that form the Bragg mirror. A resonant microcavity is often formed within a Bragg mirror with an optical thickness that is an integral multiple of $\lambda/2$. Breaking up the periodicity of the Bragg mirror with the resonant cavity results in a sharp resonance feature in the reflectance spectrum, located within the Bragg stop band. This resonance occurs as light becomes trapped and resonates in the defect in the microcavity (Reece et al. 2004). The resonance can be quite sharp and narrow, which suggests that a microcavity may be a highly sensitive system capable of detecting very small shifts in the optical spectrum. As a result, microcavities have received considerably more attention than Bragg mirrors for biosensing applications. Subnanometer linewidth cavity resonances are possible in microcavities with optimized parameters (Lerondel et al. 2004).

As with the Bragg mirror, in order for the microcavity to be effective in biosensing applications, the pore size must be sufficiently large to allow for infiltration of the target biomolecules into the pores. In particular, for the microcavity, the biomolecules must pass into the defect layer deep within the Bragg mirror layers if any signal from the sensor is to be seen. This limits the possible applications of Bragg mirrors and microcavities in macromolecule detection. In order to expand the list of potential target molecules, techniques have been developed to expand the pores found in microcavity architectures. One such technique employs postanodization etching of porous silicon in an alkaline solution (DeLouise et al. 2005), which has been shown to enlarge the pores and allow infiltration of protein into the defect layer (Stefano 2006). Pore widening does, however, result in a broadening of the cavity resonance. For macromolecules, such as proteins, optimizing and etching a porous silicon microcavity with larger pores by utilizing n-type silicon has been shown to be an effective way to preserve reasonable optical quality of microcavity sensors with pore diameters of approximately 100 nm (Ouyang et al. 2005; Ouyang et al. 2007).

Recent work on the optical detection limits achievable in porous silicon microcavities through a study using glutathione-S-transferase as the bioreceptor molecule provided insight into the relationship between microcavity response and structure. There was no dependence on microcavity thickness or quality factor for the sensitivity of the device. However, the sensitivity was highly dependent on the initial resonance wavelength of the sensor (DeLouise et al. 2005). Similar microcavity structures have also shown promise in label-free detection of specific antigens in serum and whole blood samples, indicating that the inherent size exclusion properties of these sensors could prove effective in direct point-of-care analysis of complex biological samples (Bonanno 2007).

6.4.4 Rugate Filters

As opposed to the stepwise variations in porosity to achieve porous silicon microcavities, rugate filters are produced by sinusoidal variation of the current density applied during the electrochemical etch. The resulting reflectance spectrum of a rugate filter is characterized by a relatively narrow high-reflectance stop-band with suppressed side lobes. Similar to the architectures previously described, small changes to the effective optical thickness of the porous matrix will result in a shift of the characteristic reflectance stop-band. Thus, the optical reflectance characteristics again allow the porous silicon to act as a signal transducer upon bio-recognition and binding of the target molecule. One significant advantage of the rugate filter over Bragg mirrors or microcavity sensors is the low contrast between the high- and low-porosity layers. Generally, the porosity variation is less than 1%, meaning the difference in diffusion rates for the biomolecules varies little from one layer to another.

Porous silicon rugate filters have received considerable interest in the literature for their application in detecting enzymes. Sailor and colleagues utilized a rugate filter architecture in their "smart dust" microparticles (Sailor et al. 2005). The porous layers in their work were functionalized via methylation, which prevented the infiltration of large biomolecules into the pores. The protein zein was subsequently adsorbed onto the upper surface of the porous silicon. Exposure to the enzyme pepsin led to digestion of the zein, breaking the protein into smaller peptide fragments that could easily filter into the rugate filter's pores. As these digestion products infiltrated into the pores, a red shift in the reflectance spectrum was observed. This device showed a color change visible to the naked eye when the protease concentration was higher than 14 µM. The detection limit of this "smart dust" toward protease was 7 µM (Orosco et al. 2006). Gao and colleagues were able to replicate this system using gelatin instead of zein, and detected the presence of the enzyme gelatinase, as shown in Figure 6.4 (Gao et al. 2008). Rugate filters with the ability to detect protease via digestion products could have applications in biological laboratories, possibly in the development of petri dishes with a built-in system to monitor the health of cell cultures by the presence of protease.

An alternative system for protease detection in rugate filters was developed by Gooding and colleagues (Kilian et al. 2007). Instead of adsorption on the top surface of the porous matrix, short peptides were covalently immobilized in the internal surfaces of the porous silicon matrix. Subsequent exposure to the target protease resulted in a blue shift of the stop-band, as the peptides within the matrix are digested and the resulting small fragments are washed out of the sensor. Because a single protease may digest many peptides within the sensor, a low detection limit

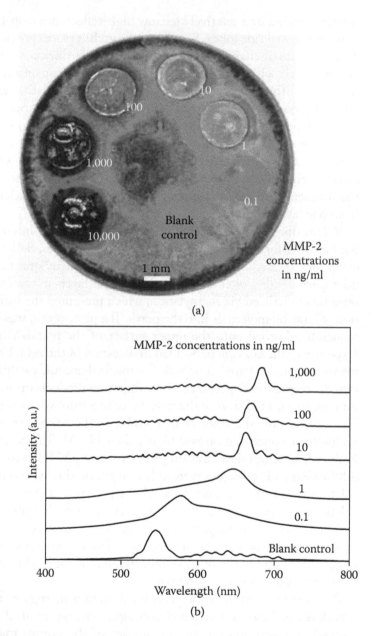

(a)

(b)

FIGURE 6.4 (a) Optical image and (b) intensity spectra for the detection of an important gelatinase (MMP-2), often associated with metastatic potential in tumors. Using a porous silicon transducer resulted in a decrease in the detection limit by two orders of magnitude when compared with the standard clinical detection methods toward this enzyme. (After Gao, L. Z. et al. 2008. *Anal. Chem.*, 80, 1468–1473. With permission.)

of 7.2 pM of protease was achieved. Similar systems have been developed for detection of protein loading, cellular processes, and other small biomolecules.

6.4.5 Waveguides

Porous silicon waveguides for biosensing are a relatively recent advance (Saarinen et al. 2005), although porous silicon waveguides were used for solvent detection as early as 1998 (Arrand et al. 1998). A significant advantage of the waveguide structure is that light is guided in the plane of the thin film, facilitating the integration of such a sensor, along with source and detector components, into lab-on-a-chip devices. Additionally, the active sensing layer in a porous silicon waveguide is the top porous layer, eliminating the infiltration difficulties that can plague Bragg mirrors, rugate filters, and microcavities, where biomolecules must filter through many layers of both high and low porosity. As a result, waveguide sensors display electric field localization, sharp resonance peaks, and a thin sensing layer, qualities that facilitate a fast response and a high sensitivity detection of molecules.

A prototypical porous silicon waveguide consists of a waveguide layer of low-porosity (high-index) porous silicon in which light is guided via total internal reflection, bounded by a cladding layer of high-porosity porous silicon (low-index) and air as shown in Figure 6.5. Porous silicon waveguides have also been

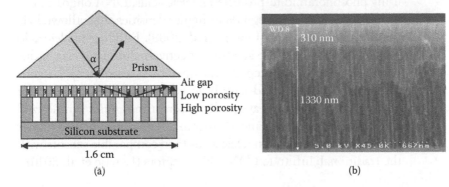

(a)   (b)

FIGURE 6.5 (a) Porous silicon waveguide consisting of a low-porosity waveguiding layer and a high-porosity cladding layer. The prism couples light at a specific angle α across the air gap and into the waveguide through an evanescent wave. When biomolecules infiltrate into the waveguide, the effective refractive index, and thus the angle at which light is coupled into the waveguide, change. (b) Cross-sectional SEM of a porous silicon waveguide with a 310 nm low-porosity layer and a 1330 nm high-porosity layer. (After Rong, G. et al. 2008b. *Biosens. Bioelectron.*, 23, 1572–1576. With permission.)

demonstrated using a single layer of porous silicon with a low-index polymer cladding (Rong et al. 2008a; Jiao and Weiss 2010). A prism or grating (Wei et al. 2008) can be used to couple light into the waveguide from the top. Under the proper conditions (e.g., angle of incidence, wavelength of light), a characteristic dip or peak in the reflectance spectrum appears. Varying the effective porous silicon optical thickness by biorecognition events in the waveguide results in a shift in the reflectance resonance, and allows for quantitative analysis of the target biomolecules present.

Weiss and colleagues demonstrated the use of porous silicon waveguides for detection of target DNA oligos of 8-, 16-, and 24-mer in length, with the lowest detection limit reported in the nanomolar range (Rong et al. 2008a; Rong et al. 2008b; Jiao and Weiss 2010). Bioreceptor and target molecule infiltration for a given waveguide porosity was shown to be dependent on the DNA oligo length (Rong and Weiss 2009). Additionally, the pore size and DNA bioreceptor density were shown to play a significant role in the detection sensitivity of porous silicon waveguide sensors (Lawrie et al. 2010).

In an effort to increase the sensitivity of porous silicon waveguide sensors, *in situ* synthesis of DNA oligonucleotide bioreceptors has been utilized, in contrast to the traditional method of infiltrating and attaching bioreceptor molecules on the internal surfaces of the pores. Applying a commercial, solid-phase synthesis method using phosphoramidite-protected nucleic acids, DNA oligos can be immobilized in the waveguide in higher densities than allowed via the traditional method (Lawrie et al. 2009). Building nucleic acid probe sequences one base at a time effectively eliminates size exclusion as a factor in bioreceptor attachment. As a result, DNA conformation, flexibility, and length do not play a role in bioreceptor density. Porous silicon waveguide sensors using *in situ* synthesized DNA bioreceptors of 8- and 16-mer in 20 nm pores showed higher sensitivity toward target nucleic acids than comparable sensors with the traditional, infiltrated DNA bioreceptors (Lawrie et al. 2010).

6.5 Optical Reflectance Transducers in Other Silicon Nanostructures

6.5.1 Ring Resonators

Owing to their low loss and comparatively fast switching times, ring resonators have traditionally found applications in optical switches, modulators, and lasers. In 2001, two theoretical papers proposed the use of ring resonators in both labeled (Blair and Chen 2001) and label-free (Boyd and Heebner 2001) biosensing applications. The strong electric field enhancement within the ring as a result of the high quality factor and low mode volume in the resonator make this system a good candidate for detection of biomolecules in low analyte concentrations.

Ring resonator devices typically consist of laterally or horizontally coupled rings, disks, toroids, or cascading resonators, which are separated by nanoscale gaps from a straight bus waveguide. The bus waveguide allows for light to evanescently couple into the ring when a phase match occurs between the incident light and the whispering gallery mode of the resonator. Once in the ring, light is trapped in that cavity by total internal reflection, leading to sharp resonances in the measured optical spectrum. Not only can a shift of the resonance be used to quantify the biorecognition of target molecules, but changes in peak intensity and quality factor can be used as well. One significant disadvantage of ring resonators that should be taken into consideration for sensing applications is the susceptibility of the resonance to changes in ambient temperature, especially for very high quality factor structures (Weiss et al. 2003).

There have been a few prominent examples of silicon ring and disk resonators for biosensing. In their work with silica microtoroids on silicon wafers, Armani and Vahala were able to distinguish between D_2O and H_2O at the ppm level by monitoring the change in Q-factor of the ring resonator (Armani and Vahala 2006). The change in Q-factor of the resonator resulted from the difference in absorption affinities between the D_2O and H_2O, with H_2O being preferentially absorbed onto the resonator, resulting in a larger change to the intrinsic quality factor when compared to D_2O. More recently, the same group demonstrated single molecule detection of interleukin-2 using an ultrahigh Q-factor microtoroid (Armani et al. 2007). Other groups have demonstrated silicon-on-insulator and silica-based ring resonators for the monitoring of cell growth (Wang et al. 2009) and detection of DNA, bacteria, and proteins (Yalcin et al. 2006; De Vos et al. 2007; De Vos et al. 2009), including a cancer biomarker in undiluted serum by monitoring shifts of the resonance wavelength (Washburn et al. 2009).

6.5.2 Slot Waveguides

The traditional configuration for a waveguide includes a thin film of high refractive index in which light is guided, bounded by films of a lower refractive index. Total internal reflection at the film boundaries enables confinement of light in the waveguide. An alternative waveguide architecture is the slot waveguide, for which a narrow slot of low-index material is bound by large strips of high-index material. This results in a significant electric field discontinuity at the material interfaces, allowing for light to be concentrated within the low-index slot. The low-index slot is typically very narrow, on the order of 100 nm thickness, and the waveguide mode in the slot is near lossless (Almeida et al. 2004). These properties result in order-of-magnitude increases in sensitivity of the slot waveguide compared with a traditional waveguide, making the structure of particular interest in biosensing applications where very low concentrations of analyte are present. Sensitivity can be even further enhanced by using multiple slot waveguides instead of a single slot waveguide (Sun et al. 2007).

Given the relatively recent interest in slot waveguides, only a few experimental demonstrations of slot waveguides for biomolecule detection have been reported (Barrios 2009). In one report, BSA and anti-BSA binding events in slot waveguide microring resonators were monitored via the transmission spectrum. When a binding event occurs in the slot, the resulting change in refractive index of that region shifts the resonant wavelengths in the spectrum. This label-free molecular biosensor showed detection limits in the ng/mm^2 level toward the BSA protein (Barrios et al. 2008). For slot waveguide biosensors with slot widths below 100 nm, many of the same size-dependent infiltration and immobilization challenges that arise in porous silicon sensors become an issue.

6.5.3 2D Silicon Photonic Crystals

In contrast with the out-of-plane one-dimensional porous silicon photonic crystal systems described in Section 6.4, two-dimensional, planar photonic crystals have a distinct sensing advantage in that light propagation in such systems occurs in-plane, making it a convenient architecture for integration onto single-chip biosensing devices. These 2D structures can be fabricated using standard lithography and semiconductor fabrication processes on high refractive index contrast materials, and often consist of a hexagonal array of air holes in silicon (Krauss and De La Rue 1999).

Similar to the porous silicon structures previously described, a photonic crystal microcavity results when the periodicity of the dielectric function is broken by introducing a defect in the etched pattern and strong field localization. Early work demonstrated a sensitivity of $\Delta n = 0.002$ using a silicon photonic crystal microcavity with a small, single hole defect that was exposed to various liquids (Chow et al. 2004). More recently, silicon photonic crystal waveguide microcavities, and drop filters have been demonstrated for solvent (Dorfner et al. 2008) and protein (Buswell et al. 2008) detection. In most cases, the addition of analyte was detected by monitoring shifts of features in the transmission spectrum. It has been predicted that photonic crystal microcavities with single small hole and missing three hole (L3) defects are capable of detecting as little as 2.5 fg (Lee and Fauchet 2007a) and 4.5 fg of analyte (Dorfner et al. 2009), respectively, using a photonic crystal total area much smaller than 100 μm^2. Moreover, it has been shown that photonic crystals are highly sensitive to changes in surface coverage with protein, indicating concentration-dependent measurements are possible with these systems (Skivesen et al. 2007).

One method proposed for improving the Q-factor and lowering the detection limit of 2D photonic crystal sensors involves the use of multihole defect regions (Kang and Weiss 2008). Replacing a single hole with a cluster of three or more smaller holes with an overall defect "footprint" of similar size can significantly increase the available surface area for target molecule binding, as well as increase the Q-factor of the cavity. These multihole defects can provide up to 25 times greater overall surface area for each defect site given reasonable fabrication tolerances. Surface area in biomolecule detection is critically important, given that the highest sensitivity occurs when there is the greatest overlap of field with molecules. Multihole defects enable strong interaction between the resonant mode and attached molecules on the inner walls of the defect holes. A photonic crystal waveguide sensor with small hole defects was shown to have a 40% improved bulk refractive index response to solvents (Buswell et al. 2008), and, very recently, a multihole L3 photonic crystal defect (three lattice holes were replaced with three smaller defect holes) was demonstrated for improved solvent detection (Kang et al. 2010). Figure 6.6 illustrates several photonic crystal defect configurations.

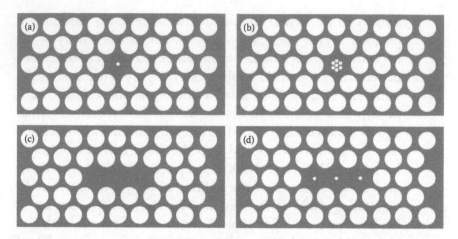

FIGURE 6.6 Schematics of photonic crystal microcavities with (a) single hole defect, (b) multihole defect, (c) L3 defect, and (d) multihole L3 defect.

6.6 Intensity Measurements with Mach–Zehnder Interferometers

Mach–Zehnder interferometers (MZIs) have proved to be among the most sensitive devices for label-free detection of biomolecules that can be integrated into lab-on-a-chip devices. A typical MZI biosensor is composed of an input waveguide that is split into two waveguide branches: one containing a sensing window and one serving as a reference. The sensing waveguide arm is exposed to the target biomolecules, while the reference arm remains isolated. The arms are then recombined into a single waveguide, and light from the sensing and reference arms interferes based on the change in optical path length. The output signal has sinusoidal variation, and the phase shift and intensity change provide information about the amount of molecular binding in the sensing region.

The built-in reference arm of the MZI sensor has some notable advantages over systems previously discussed. In particular, the simultaneous measurements in a reference arm in close proximity to the sensing arm (typically ~100 μm) leads to measurements that are nearly independent of temperature, humidity, and other environmental effects, since both arms of the sensor are nearly identical in their environment. This effectively reduces the problem of signal drift that plagues some of the reflectance-based systems. As a result, the system is more sensitive to small changes in optical thickness (Prieto et al. 2003).

In 2006, a completely integrated MZI lab-on-a-chip device was demonstrated as a detector for a 58-mer DNA oligo, capable of detecting as little as 300 pM of the target oligo in buffer solution. With better optimization of the device, detection limits as low as 10 pM were predicted (Sepulveda et al. 2006). In addition to nucleic acid detection, fully integrated MZI interferometers have been demonstrated using antibody bioreceptors (Brosinger et al. 1997). The MZI configuration has also been utilized with silicon wire waveguides that are designed to create a strong evanescent field at the waveguide surface for biomolecule interaction (Densmore et al. 2008). Surface coverage of less than 0.3 pg/mm^2 for antibody–antigen reactions can be detected using this configuration (Densmore et al. 2009).

6.7 Photoluminescence Transducers

Some of the earliest sensing work using porous silicon took advantage of the photoluminescence of the material, in particular, room-temperature photoluminescence, which was first observed for porous silicon in the early 1990s. While photoluminescent tags are commonly used in labeled biomolecule detection, the examples that follow are all label-free systems, where the biorecognition event induces a change in the photoluminescent properties of the material. Porous silicon biosensors using photoluminescent transduction have been developed for organic solvents, explosives, and hazardous wastes.

There are limited reports of detection of biomolecules due to porous silicon photoluminescence, likely due to the greater measurement error compared with the methods described. The photoluminescence of porous silicon is known to be highly sensitive to a number of molecular species, via mechanisms such as electron transfer and interfacial charging, while exposed to biological media, which makes it difficult to account for background effects and drift in the photoluminescence peak (Sailor and Wu 2009). Nevertheless, biosensors utilizing photoluminescence as the transduction method have been reported based on both photoluminescence quenching and spectral shifts. Immunocomplex formation between an antibody and antigen within a porous silicon matrix has been observed to quench the material's photoluminescence (Starodub et al. 1996), and a similar quenching mechanism was observed for DNA hybridization in porous silicon (Di Francia et al. 2005). In both reports, the photoluminescence quenching was attributed to a nonradiative recombination mechanism. Resonance shifts in the photoluminescence spectra of porous silicon microcavities when functionalized and subsequently exposed to target bacteria (Chan et al. 2001) and DNA oligos (Chan et al. 2000) have also been reported.

6.8 Conclusions

A number of biosensing mechanisms and fully integrated devices utilizing silicon photonics have been described. Table 6.1 provides a comparison of bioreceptor, transducer structure, and signal output type for a number of the systems described in this chapter. For each example, an experimentally derived detection range or detection limit is provided. As the table shows, there are a number of bioreceptor–transducer combination options available when designing a silicon photonic device. Selecting the appropriate detection system requires careful consideration of the desired signal output and also the concentration of the analyte of interest.

Acknowledgment

Work in the authors' laboratory related to silicon-based biosensors has been supported in part by the Army Research Office and National Science Foundation.

TABLE 6.1 Summary Data for Silicon Photonic Biosensors Described within the Chapter

Bioreceptor	Transducer	Signal	Detection range	Detection limit	Reference
Antibody	MZI	Intensity change (phase shift)	10 µg/mL		Bronsinger 1997
Antibody	MZI	Intensity change (phase shift)		0.25 pg/mm²	Densmore et al. 2009
Antibody	PSi microcavity	Reflectance shift	2–10 mg/mL		Bonanno 2007
Antibody	PSi microcavity	Reflectance shift	0.5–2.5 mg/mL		Ouyang 2005
Antibody	PSi microcavity	Photoluminescence quenching	10–1000 µg/mL	10 µg/mL	Starodub et al. 1996
Antibody	PSi single layer	Reflectance shift	2.5 mg/mL		Dancil 1999
Antibody	PSi single layer	Reflectance shift	10 µM	1 ng/mm²	Janshoff 1998
Antibody	Ring resonator	Transmission shift	0.1 aM to 1 µM	5 aM	Armani et al. 2007
Antibody	Ring resonator	Transmission shift	1–199 ng/mL	25 ng/mL	Washburn et al. 2009
Antibody	Slot waveguide	Transmission shift	2–75 µg/mL	16 pg/mm²	Barrios et al. 2008
Biomimetic	PSi rugate filter	Reflectance shift	0.01–1 µM		Kilian 2007
Cell structure	PSi microcavity	Photoluminescence shift		1.7 µg	Chan et al. 2001
Cell structure	Ring resonator	Transmission shift	0.005–0.5 mg/L		Wang 2009
Coenzyme	2D photonic crystal	Transmission shift	10 mM		Buswell et al. 2008
Coenzyme	PSi microcavity	Reflectance shift	0–2 mg/mL	0.3 ng/mm²	Ouyang 2005
Coenzyme	Ring resonator	Transmission shift	10 ng/mL to 10 µg/mL	10 ng/mL	De Vos et al. 2007
DNA	PSi double layer	Reflectance shift	1 pM to 10 µM	55 fg/mm²	Steinem 2004
DNA	PSi microcavity	Photoluminescence shift		1 µM	Chan et al. 2000

—continued

TABLE 6.1 (Continued) Summary Data for Silicon Photonic Biosensors Described within the Chapter

Bioreceptor	Transducer	Signal	Detection range	Detection limit	Reference
DNA	PSi microcavity	Photoluminescence qunching	10 μM		DiFrancia 2005
DNA	PSi single layer	Reflectance shift	10 nM to 1 fM	9 fg/mm²	Lin 1997
DNA	PSi single layer	Reflectance shift	1 μM		Voelker et al. 2008
DNA	PSi waveguide	Reflectance shift	50 μM		Rong et al. 2008a
DNA	PSi waveguide	Reflectance shift	1–10 μM	42 nM	Rong et al. 2008b
DNA/virus	PSi microcavity	Photoluminescence shift		194.2 fM	Chan et al. 2000
Enzyme	PSi microcavity	Reflectance shift	1–40 μM	50 pg/mm²	DeLouise et al. 2005
Enzyme	PSi microcavity	Reflectance shift	4–15 μM		Ouyang 2007
Protein	2D photonic crystal	Transmission shift	10 pM to 0.1 mM	500 pg/mm²	Dorfner et al. 2009
Protein	PSi rugate filter	Reflectance shift	7–14 μM		Orosco 2006

References

Almeida, V. R., Xu, Q., Barrios, C. A., and Lipson, M. 2004. Guiding and confining light in void nanostructures. *Opt. Lett.*, 29, 1209–1211.

Anderson, M. A., Tinsley-Bown, A., Allcock, P., Perkins, E. A., Snow, P., Hollings, M., Smith, R. G., Reeves, C., Squirrell, D. J., Nicklin, S., and Cox, T. I. 2003. Sensitivity of the optical properties of porous silicon layers to the refractive index of liquid in the pores. *Phys. Status Solidi A—Appl. Res.*, 197, 528–533.

Armani, A. M., Kulkarni, R. P., Fraser, S. E., Flagan, R. C., and Vahala, K. J. 2007. Label-free, single-molecule detection with optical microcavities. *Science*, 317, 783–787.

Armani, A. M. and Vahala, K. J. 2006. Heavy water detection using ultra-high-Q microcavities. *Opt. Lett.*, 31, 1896–1898.

Arrand, H. F., Benson, T. M., Loni, A., Arens-Fischer, R., Kruger, M., Thonissen, M., Luth, H., and Kershaw, S. 1998. Novel liquid sensor based on porous silicon optical waveguides. *IEEE Photon. Technol. Lett.*, 10, 1467–1469.

Barker, S. L. R., Kopelman, R., Meyer, T. E., and Cusanovich, M. A. 1998. Fiber-optic nitric oxide-selective biosensors and nanosensors. *Anal. Chem.*, 70, 971–976.

Barrios, C. A. 2009. Optical slot-waveguide based biochemical sensors. *Sensors*, 9, 4751–4765.

Barrios, C. A., Banuls, M. J., Gonzalez-Pedro, V., Gylfason, K. B., Sanchez, B., Griol, A., Maquieira, A., Sohlstrom, H., Holgado, M., and Casquel, R. 2008. Label-free optical biosensing with slot-waveguides. *Opt. Lett.*, 33, 708–710.

Blair, S. and Chen, Y. 2001. Resonant-enhanced evanescent-wave fluorescence biosensing with cylindrical optical cavities. *Appl. Opt.*, 40, 570–582.

Bocking, T., Kilian, K. A., Gaus, K., and Gooding, J. J. 2008. Modifying porous silicon with self-assembled monolayers for biomedical applications: The influence of surface coverage on stability and biomolecule coupling. *Adv. Functional Mater.*, 18, 3827–3833.

Bonanno, L. M. and Delouise, L. A. 2007. Whole blood optical biosensor. *Biosens. Bioelectron.*, 23, 444–448.

Boncheva, M., Duschl, C., Beck, W., Jung, G., and Vogel, H. 1996. Formation and characterization of lipopeptide layers at interfaces for the molecular recognition of antibodies. *Langmuir*, 12, 5636–5642.

Boyd, R. W. and Heebner, J. E. 2001. Sensitive disk resonator photonic biosensor. *Appl. Opt.*, 40, 5742–5747.

Brosinger, F., Freimuth, H., Lacher, M., Ehrfeld, W., Gedig, E., Katerkamp, A., Spener, F., and Cammann, K. 1997. A label-free affinity sensor with compensation of unspecific protein interaction by a highly sensitive integrated optical Mach–Zehnder interferometer on silicon. *Sens. Actuat. B—Chem.,* 44, 350–355.

Buriak, J. M. 2002. Organometallic chemistry on silicon and germanium surfaces. *Chem. Rev.,* 102, 1271–1308.

Buswell, S. C., Wright, V. A., Buriak, J. M., Van, V., and Evoy, S. 2008. Specific detection of proteins using photonic crystal waveguides. *Opt. Express,* 16, 15949–15957.

Canham, L. T. 1990. Silicon quantum wire array fabrication by electrochemical and chemical dissolution of wafers. *Appl. Phys. Lett.,* 57, 1046–1048.

Canham, L. T. 1995. Bioactive silicon structure fabrication through nanoetching techniques. *Advanced Materials,* 7, 1033–1037.

Chan, S., Fauchet, P. M., Li, Y., Rothberg, L. J., and Miller, B. L. 2000. Porous silicon microcavities for biosensing applications. *Phys. Status Solidi A—Appl. Res.,* 182, 541–546.

Chan, S., Horner, S. R., Fauchet, P. M., and Miller, B. L. 2001. Identification of gram negative bacteria using nanoscale silicon microcavities. *J. Am. Chem. Soc.,* 123, 11797–11798.

Chow, E., Grot, A., Mirkarimi, L. W., Sigalas, M., and Girolami, G. 2004. Ultracompact biochemical sensor built with two-dimensional photonic crystal microcavity. *Opt. Lett.,* 29, 1093–1095.

Dancil, K. P. S., Greiner, D. P., and Sailor, M. J. 1999. A porous silicon optical biosensor: Detection of reversible binding of IgG to a protein A-modified surface. *J. Am. Chem. Soc.,* 121, 7925–7930.

De Stefano, L., Rotiroti, L., Rendina, I., Moretti, L., Scognamiglio, V., Rossi, M., and D'auria, S. 2006. Porous silicon-based optical microsensor for the detection of L-glutamine. *Biosens. Bioelectron.,* 21, 1664–1667.

De Vos, K., Bartolozzi, I., Schacht, E., Bienstman, P., and Baets, R. 2007. Silicon-on-Insulator microring resonator for sensitive and label-free biosensing. *Opt. Express,* 15, 7610–7615.

De Vos, K., Girones, J., Popelka, S., Schacht, E., Baets, R., and Beinstman, P. 2009. SOI optical microring resonators with poly(ethylene glycol) polymer brush for label-free biosensor applications. *Biosens. Bioelectron.,* 24, 2528–2533.

Delouise, L. A., Kou, P. M., and Miller, B. L. 2005. Cross-correlation of optical microcavity biosensor response with immobilized enzyme activity. Insights into biosensor sensitivity. *Anal. Chem.,* 77, 3222–3230.

Densmore, A., Vachon, M., Xu, D. X., Janz, S., Ma, R., Li, Y. H., Lopinski, G., Delage, A., Lapointe, J., Luebbert, C. C., Liu, Q. Y., Cheben, P., and Schmid, J. H. 2009. Silicon photonic wire biosensor array for multiplexed real-time and label-free molecular detection. *Opt. Lett.,* 34, 3598–3600.

Densmore, A., Xu, D. X., Janz, S., Waldron, P., Mischki, T., Lopinski, G., Delage, A., Lapointe, J., Cheben, P., Lamontagne, B., and Schmid, J. H. 2008. Spiral-path high-sensitivity silicon photonic wire molecular sensor with temperature-independent response. *Opt. Lett.*, 33, 596–598.

Di Francia, G., La Ferrara, V., Manzo, S., and Chiavarini, S. 2005. Towards a label-free optical porous silicon DNA sensor. *Biosens. Bioelectron.*, 21, 661–665.

Dorfner, D., Zabel, T., Hurlimann, T., Hauke, N., Frandsen, L., Rant, U., Abstreiter, G., and Finley, J. 2009. Photonic crystal nanostructures for optical biosensing applications. *Biosens. Bioelectron.*, 24, 3688–3692.

Dorfner, D. F., Hurlimann, T., Zabel, T., Frandsen, L. H., Abstreiter, G., and Finley, J. J. 2008. Silicon photonic crystal nanostructures for refractive index sensing. *Appl. Phys. Lett.*, 93, 181103-1–181103-3.

Gao, L. Z., Mbonu, N., Cao, L. L., and Gao, D. 2008. Label-free colorimetric detection of gelatinases on nanoporous silicon photonic films. *Anal. Chem.*, 80, 1468–1473.

Jane, A., Dronov, R., Hodges, A., and Voelcker, N. H. 2009. Porous silicon biosensors on the advance. *Trend. Biotechnol.*, 27, 230–239.

Janshoff, A., Dancil, K. P. S., Steinem, C., Greiner, D. P., Lin, V. S. Y., Gurtner, C., Motesharei, K., Sailor, M. J., and Ghadiri, M. R. 1998. Macroporous p-type silicon Fabry-Perot layers. Fabrication, characterization, and applications in biosensing. *J. Am. Chem. Soc.*, 120, 12108–12116.

Jiao, Y. and Weiss, S. M. 2010. Design parameters and sensitivity analysis of polymer-cladded porous silicon waveguides for small molecules detection. *Biosens. Bioelectron.*, 25, 1535–1538.

Kang, C. and Weiss, S. M. 2008. Photonic crystal with multiple-hole defect for sensor applications. *Opt. Express*, 16, 18188–18193.

Kang, C., Phare, C., Weiss, S. M., Vlasov, Y. A., and Assefa, S. 2010. Photonic crystal defects with increased surface area for improved refractive index sensing. *Conf. on Lasers and Electro-Optics*, May.

Kilian, K. A., Bocking, T., Gaus, K., King-Lacroix, J., Gal, M., and Gooding, J. J. 2007. Hybrid lipid bilayers in nanostructured silicon: a biomimetic mesoporous scaffold for optical detection of cholera toxin. *Chem. Commun.*, 1936–1938.

Kilian, K. A., Boecking, T., and Gooding, J. J. 2009. The importance of surface chemistry in mesoporous materials: Lessons from porous silicon biosensors. *Chem. Commun.*, 630–640.

Krauss, T. F. and De La Rue, R. M. 1999. Photonic crystals in the optical regime—past, present and future. *Progr. Quantum Electron.*, 23, 51–96.

Lawrie, J. L., Xu, Z., Rong, G. G., Laibinis, P. E., and Weiss, S. M. 2009. Synthesis of DNA oligonucleotides in mesoporous silicon. *Phys. Stat. Solidi A—Appl. Mater. Sci.*, 206, 1339–1342.

Lawrie, J. L., Jiao, Y., and Weiss, S. M. 2010. Size-dependent infiltration and optical detection of nucleic acids in nanoscale pores. *IEEE Trans. Nanotechnol.*, Submitted Dec. 22, 2009.

Lee, M. and Fauchet, P. M. 2007a. Two-dimensional silicon photonic crystal based biosensing platform for protein detection. *Opt. Express*, 15, 4530–4535.

Lerodel, G., Reece, P., Bruyant, A., and Gal, M. 2004. Strong light confinement in microporous photonic silicon structures. *Mater. Res. Soc. Symp. Proc.*, 797, W1.7.1–W1.7.6.

Lin, V. S. Y., Motesharei, K., Dancil, K. P. S., Sailor, M. J., and Ghadiri, M. R. 1997. A porous silicon-based optical interferometric biosensor. *Science*, 278, 840–843.

Moretti, L., Rea, I., De Stefano, L., and Rendina, I. 2007. Periodic versus aperiodic: Enhancing the sensitivity of porous silicon based optical sensors. *Appl. Phys. Lett.*, 90, 191112-1–191112-3.

Orosco, M. M., Pacholski, C., Miskelly, G. M., and Sailor, M. J. 2006. Protein-coated porous-silicon photonic crystals for amplified optical detection of protease activity. *Advanced Materials*, 18, 1393–1396.

Ouyang, H., Christophersen, M., Viard, R., Miller, B. L., and Fauchet, P. M. 2005. Macroporous silicon microcavities for macromolecule detection. *Adv. Functional Mater.*, 15, 1851–1859.

Ouyang, H., Delouise, L. A., Miller, B. L., and Fauchet, P. M. 2007. Label-free quantitative detection of protein using macroporous silicon photonic bandgap biosensors. *Anal. Chem.*, 79, 1502–1506.

Pacholski, C., Sartor, M., Sailor, M. J., Cunin, F., and Miskelly, G. M. 2005. Biosensing using porous silicon double-layer interferometers: Reflective interferometric Fourier transform spectroscopy. *J. Am. Chem. Soc.*, 127, 11636–11645.

Prieto, F., Sepulveda, B., Calle, A., Llobera, A., Dominguez, C., Abad, A., Montoya, A., and Lechuga, L. M. 2003. An integrated optical interferometric nanodevice based on silicon technology for biosensor applications. *Nanotechnology*, 14, 907–912.

Reece, P. J., Gal, M., Tan, H. H., and Jagadish, C. 2004. Optical properties of erbium-implanted porous silicon microcavities. *Appl. Phys. Lett.*, 85, 3363–3365.

Rong, G., Ryckman, J. D., Mernaugh, R. L., and Weiss, S. M. 2008a. Label-free porous silicon membrane waveguide for DNA sensing. *Appl. Phys. Lett.*, 93, 161109-1–161109-3.

Rong, G., Najmaie, A., Sipe, J. E., and Weiss, S. M. 2008b. Nanoscale porous silicon waveguide for label-free DNA sensing. *Biosens. Bioelectron.*, 23, 1572–1576.

Rong, G. and Weiss, S. M. 2009. Biomolecule size-dependent sensitivity of porous silicon sensors. *Phys. Stat. Solidi A—Appl. Mater. Sci.*, 206, 1365—1368.

Rosenzweig, Z. and Kopelman, R. 1996. Analytical properties and sensor size effects of a micrometer-sized optical fiber glucose biosensor. *Anal. Chem.*, 68, 1408–1413.

Saarinen, J. J., Weiss, S. M., Fauchet, P. M., and Sipe, J. E. 2005. Optical sensor based on resonant porous silicon structures. *Opt. Express*, 13, 3754–3764.

Sailor, M. J. and Link, J. R. 2005. "Smart dust": Nanostructured devices in a grain of sand. *Chem. Commun.*, 1375–1383.

Sailor, M. J. and Wu, E. C. 2009. Photoluminescence-based sensing with porous silicon films, microparticles, and nanoparticles. *Adv. Functional Mater.*, 19, 3195–3208.

Sepulveda, B., Del Rio, J. S., Moreno, M., Blanco, F. J., Mayora, K., Dominguez, C., and Lechuga, L. M. 2006. Optical biosensor microsystems based on the integration of highly sensitive Mach-Zehnder interferometer devices. *J. Opt. A—Pure Appl. Opt.*, 8, S561–S566.

Skivesen, N., Tetu, A., Kristensen, M., Kjems, J., Frandsen, L. H., and Borel, P. I. 2007. Photonic-crystal waveguide biosensor. *Opt. Express*, 15, 3169–3176.

Starodub, N. F., Fedorenko, L. L., Starodub, V. M., Dikij, S. P., and Svechnikov, S. V. 1996. Use of the silicon crystals photoluminescence to control immunocomplex formation. *Sens. Actuat. B—Chem.*, 35, 44–47.

Steinem, C., Janshoff, A., Lin, V. S. Y., Voelcker, N. H., and Ghadiri, M. R. 2004. DNA hybridization-enhanced porous silicon corrosion: Mechanistic investigators and prospect for optical interferometric biosensing. *Tetrahedron*, 60, 11259–11267.

Sun, R., Dong, P., Feng, N. N., Hong, C. Y., Michel, J., Lipson, M., and Kimerling, L. 2007. Horizontal single and multiple slot waveguides: optical transmission at $\lambda = 1550$ nm. *Opt. Express*, 15, 17967–17972.

Theiss, W. 1997. Optical properties of porous silicon. *Surf. Sci. Rep.*, 29, 95–192.

Tinsley-Bown, A., Smith, R. G., Hayward, S., Anderson, M. H., Koker, L., Green, A., Torrens, R., Wilkinson, A. S., Perkins, E. A., Squirrell, D. J., Nicklin, S., Hutchinson, A., Simons, A. J., and Cox, T. I. 2005. Immunoassays in a porous silicon interferometric biosensor combined with sensitive signal processing. *Phys. Stat. Solidi A—Appl. Mater. Sci.*, 202, 1347–1356.

Uhlir, A. 1956. Electrolytic shaping of germanium and silicon. *Bell Syst. Technical J.*, 35, 333–347.

Voelcker, N. H., Alfonso, I., and Ghadiri, M. R. 2008. Catalyzed oxidative corrosion of porous silicon used as an optical transducer for ligand-receptor interactions. *Chembiochem*, 9, 1776–1786.

Wang, S., Ramachandran, A., and Ja, S. J. 2009. Integrated microring resonator biosensors for monitoring cell growth and detection of toxic chemicals in water. *Biosens. Bioelectron.*, 24, 3061–3066.

Washburn, A. L., Gunn, L. C., and Bailey, R. C. 2009. Label-free quantitation of a cancer biomarker in complex media using silicon photonic microring resonators. *Anal. Chem.*, 81, 9499–9506.

Wei, X., Kang, C., Liscidini, M., Rong, G., Retterer, S. T., Patrini, M., Sipe, J. E., and Weiss, S. M. 2008. Grating couplers on porous silicon planar waveguides for sensing applications. *J. Appl. Phys.*, 104, 123113-1–123113-5.

Weiss, S. M., Molinari, M., and Fauchet, P. M. 2003. Temperature stability for silicon-based photonic band-gap structures. *Appl. Phys. Lett.*, 83, 1980–1982.

Yalcin, A., Popat, K. C., Aldridge, J. C., Desai, T. A., Hryniewicz, J., Chbouki, N., Little, B. E., King, O., Van, V., Chu, S., Gill, D., Anthes-Washburn, M., and Unlu, M. S. 2006. Optical sensing of biomolecules using microring resonators. *IEEE J. Sel. Topics Quantum Electron.*, 12, 148–155.

CHAPTER 7

Mid-Wavelength Infrared Silicon Photonics for High-Power and Biomedical Applications

Varun Raghunathan, Sasan Fathpour, and Bahram Jalali

Contents

7.1 Introduction

Mid-wave infrared (MWIR)/long-wave infrared (LWIR) sources operating at 3–5 μm and 8–12 μm wavelengths, respectively, have been the topic of active research for over two decades. Historically, the need for sources operating in this range has been primarily driven by military applications such as wind light detection and ranging (LIDAR), remote chemical and biological sensing, and IR countermeasures (IRCM). Over the past decade, such sources have also found use in a wide array of applications ranging from purely scientific uses, such as ring down and Fourier transform infrared (FTIR) spectroscopy, to clinical uses where tissue ablation is achieved by targeting the resonant absorption peaks in water, the amide bonds in collagen, and other tissue chromophors in the MWIR region. Additionally, industrial uses such as hydrocarbon detection from vehicles, oil fields, and industrial smoke stacks are of interest.

Optical parametric oscillators (OPOs) have become popular for generating tunable MWIR radiation for many of the previously mentioned applications [1]. The maturing diode-pumped solid-state laser technology combined with continued improvements of IR transmitting nonlinear crystals, especially with respect to reduced bulk absorption at the pump wavelength (typically between 1–3 μm) and the higher damage threshold, have contributed to the development of MWIR OPOs. Much effort has been spent on designing tandem OPO-based systems using a 1064 nm laser as the pump source [2]. While these systems use the most mature laser as a pump source (i.e., Nd:YAG laser), the tandem OPO configuration adds considerable complexity to the design and operation of the overall system and often results in significant penalty in terms of the effective electrical-to-optical conversion efficiency, when compared to a single-stage OPO system with the pump laser emitting between 2.1–3 μm. The advances in nonlinear material, IR coating, and OPO systems level design have resulted in operation over a broad range of pulse durations (femtosecond pulses and continuous wave [CW]), spectral region coverage, power scaling, and more importantly overall system robustness. However, such lasers are still a specialty item primarily due to the high cost and limited commercial availability of high-quality MWIR nonlinear crystals.

An alternative means of generating MWIR radiation, and one that mitigates some of these limitations, is the solid-state Raman laser (SSRL) [3]. Solid-state Raman lasers have also made remarkable advances in recent years. The development of SSRLs based on crystalline Raman materials pumped at the fundamental and frequency-doubled Nd-based and Ti:Sapphire lasers has resulted in generation of a broad range of wavelengths ranging from visible up to 1.6 μm, and more impressively the demonstration of CW Raman laser action at ~589 nm with power levels approaching few watts. Compared to the OPO, the SSRL utilizes a much simpler architecture for IR frequency shifting of the pump. SSRLs, although not as broadly tunable as OPOs, can be tuned via the pump wavelength.

The demonstration of nonlinear optical effects in silicon, especially Raman lasers and amplifiers, combined with excellent transmission of silicon in the MWIR, suggests that silicon nonlinear devices can be extended to the MWIR [4,5]. It is noteworthy that at near-IR wavelengths (~1550 nm telecom band), the main limitation of silicon nonlinear optical devices is presently the loss due to free carriers that are generated by two-photon absorption (TPA) [6,7]. Active carrier sweep-out using a p-n junction and short-pulse pumping has been proposed as a means to mitigate this problem [8,9]. TPA vanishes in the MWIR regime and three-photon absorption (3PA) and associated free-carrier effects are negligible, hence eliminating the main problem with silicon nonlinear devices [10]. This combined with (1) the unsurpassed quality of commercial silicon crystals, (2) the low cost and wide availability of the material, (3) extremely high optical damage threshold of 1–4 GW/cm^2 (depending on the crystal resistivity), and (4) excellent thermal conductivity of ~150 W/m-K renders silicon a very attractive nonlinear crystal.

This chapter is a review of some of the applications of silicon nonlinear optical devices operating in the MWIR. Section 7.2 discusses in detail the case for MWIR silicon devices. The measurement of low-linear absorption in the 1–7 μm wavelength region, along with negligible nonlinear absorption at wavelengths longer than the two-photon band edge (~2.2 μm), suggests that silicon is a suitable MWIR nonlinear crystal. Section 7.3 discusses certain design considerations pertaining to silicon nonlinear devices. The use of silicon-on-sapphire substrates for fabricating integrated silicon devices in the MWIR is discussed. Issues such as self-focusing and thermal-lensing effects, which limit the performance of solid-state lasers, are also studied. Section 7.4 discusses a wide slew of MWIR nonlinear optical devices. Our research work has mainly focused on Raman nonlinear optical devices such as Raman amplifiers and lasers; this is the main emphasis of the chapter. We review results of our recent observation of extreme value phenomenon (highly non-Gaussian) fluctuations during Raman amplification in silicon. It will be shown that the influence of the strongly nonlinear stimulated Raman scattering on input pump fluctuations causes amplified Stokes pulses to exhibit statistical distributions that resemble those observed in socioeconomic systems rather than Gaussian or other traditional distributions that one would expect in physical systems. Such unusual behavior has important implications for modeling of noise in Raman-amplified optical links. We have also highlighted other electronic-based nonlinear optical effects studied in silicon in the MWIR. The nonlinear optical effects presented in this chapter are from a device point of view. Details of the underlying physics can be found elsewhere [11,12].

7.2 The Case for MWIR Silicon Devices

First, the measurement of linear absorption in silicon using a standard FTIR apparatus is presented. Figure 7.1 shows the absorption coefficient of a standard silicon wafer in units of dB/cm as a function of wavelength in the range of 1–13 μm. The high losses at around 1 μm can be attributed to the indirect band-gap absorption corresponding to energy of 1.12 eV. The low-loss window following this absorption peak extends from a 1.2 to 6.5 μm wavelength range. This is very attractive for building optical devices in the near-IR, telecom as well as the mid-IR wavelengths. Beyond 7 μm, the increase in losses could be due to multiphoton absorption processes.

Next, measurement of nonlinear absorption in silicon as a function of pump intensity is discussed. Pulsed pump laser sources were coupled into a 1-in.-thick bulk silicon sample using a standard calcium fluoride (CaF$_2$) lens. At the output end, a slow

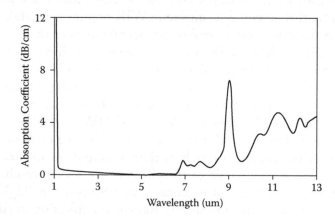

FIGURE 7.1 Linear absorption in silicon measured using an FTIR apparatus. (From Rashunathan Nonlinear Absorption in Silicon and the Prospects of Mid-infrared Silicon Raman Laser, Phys. Status Solidi—Rapid Research Letters Phys. Status Solidi (2006). With permission.)

photodetector was used to measure the energy of the pulse. The silicon sample was moved toward the focus of the lens to increase the intensity of the coupled optical beam (Z-scan technique). The following solid-state pump sources were used in this work: (1) CrTmHo-YAG crystal operating at 2.09 μm, free-running mode with a pulse width of 100 μs and energy of 1 J; (2) Er-doped YAG laser operating at 2.936 mm, Q-switched with a pulse width of 75 ns and energy of 25 mJ.

Figure 7.2 shows the transmission through the silicon sample at the pump wavelengths of 2.09 and 2.936 μm. The silicon sample was double-sided polished, and the reflection loss per facet was ~29%. Hence, the maximum transmission was ~53%.

At a 2.09 μm pump wavelength, which is close to the indirect band edge for the TPA process, the transmission reduces considerably with increasing pump intensity. This loss can be attributed to the TPA and associated free-carrier absorption (FCA) processes. As the pump photons are reduced in energy below half the band gap, the TPA process is expected to vanish. This is clearly observed in the transmission results corresponding to a 2.936 μm pump wavelength. The slight decrease in the transmission with increasing intensities could be attributed to the 3PA process. However, this process is expected to be weak. Thus, pumping beyond the TPA edge (i.e., wavelengths longer than 2.2 μm) is attractive for

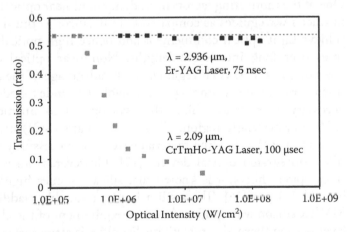

FIGURE 7.2 Optical transmission in silicon as a function of intensity. Two different pump sources at 2.09 μm and 2.936 μm were used in these experiments. The enhanced nonlinear losses at 2.09 μm due to TPA and FCA and the absence of these losses at 2.936 μm are clearly seen. (From V. Raghu. *Nonlinear Absorption in Silicon*, Wiley-VCH R38-40; Nathan et al. and Prospects of Midinfrared 2006, Silicon Raman Laser, Physica Status Solidi—Rapid Research Letters. With permission.)

building useful nonlinear optical devices with the TPA and associated free-carrier loss effects completely eliminated.

Evidently, the TPA effect that is the central issue with the near-IR nonlinear absorption effects in silicon is found to reduce significantly when pumped at energies less than half the band [10]. Two pump photons absorbed simultaneously lack the energy to exceed the energy gap in order to excite electron–hole pairs. The TPA coefficient has also been measured across a broad wavelength range from near- to mid-IR (1.2 to 2.4 μm) [13]. The possible nonlinear loss mechanism at these longer wavelengths is the 3PA process and the FCA associated with 3PA. This process is much weaker than TPA and is expected to be insignificant at typical pump intensities used to study nonlinear effects, before the onset of damage. The 3PA coefficient has been measured in silicon from 2300 to 3300 nm wavelengths using high-peak power pump subpicosecond pulses with the 3PA peaking at 2700 nm at ~0.035 cm³/GW² [14]. It should be noted that FCA scales with wavelength as λ^2 and can pose a serious problem in the MWIR. These carriers could either be due to background concentration or optical generation processes.

7.3 Design Considerations

One of the motivating factors in studying nonlinear optical effects in silicon waveguides as compared to bulk silicon is the ability to achieve tight optical confinement and hence high optical intensities over long interaction lengths. Nonlinear optical devices operating in the MWIR region can take advantage of the mature silicon waveguide technology. A suitable low-index cladding is necessary in order to achieve the tight optical confinement. At 1550 nm wavelength region, silicon-on-insulator (SOI) structures with silica as the cladding material have been successfully used in silicon-integrated optical devices [15]. However, from work on silica optical fibers, it is known that silica becomes highly lossy beyond 1.8 μm [16]. Thus, silica is not a suitable cladding for MWIR silicon waveguides. Another requirement of the cladding layer is good thermal conductivity. Sapphire is attractive from the point of view of low index (~1.6) and low loss in the MWIR and good thermal conductivity of 23 Wm⁻¹K⁻¹ (roughly 20× higher than silica). Figure 7.3 shows the absorption spectrum of sapphire exhibiting low-loss transmission at the MWIR wavelengths [17]. Thus, silicon-on-sapphire (SOS) structures could be potentially

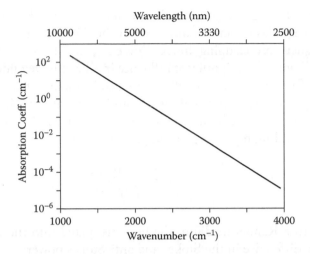

FIGURE 7.3 Absorption spectrum of sapphire versus wavenumber (lower abscissa) and wavelength (upper abscissa). (Adapted from Michael E. Thomas, Richard I. Joseph, and William, *J. Applied Optics* 27, 239–245 (1988).)

used to build integrated silicon MWIR Raman devices. Different waveguiding configurations including hollow air cladding have been discussed in [5].

The nonlinear optical effects studied at high optical intensities are often accompanied by the self-focusing effect that involves the spatial focusing or defocusing of the optical beam due to the nonlinear refractive index of the medium [18]. In addition to this, the heat dissipated into the nonlinear medium creates a thermal gradient that alters the refractive index and hence causes lensing effects [19]. These effects have to be included while designing nonlinear optical devices operating at high optical intensities. In a semiconductor Raman medium, the presence of free carriers generated due to TPA also alters the refractive index and hence causes a lensing effect. This effect is not relevant in silicon at MWIR wavelengths due to the presence of negligible carriers.

The self-focusing effect becomes significant when it becomes comparable to the diffraction effects of the Gaussian beam in bulk media [19]. However, when considering waveguide structures, the optical mode does not diverge because of the guiding achieved by the core-cladding index difference. Hence, a more relevant parameter that characterizes self-focusing is the index difference induced by the nonlinear refraction in comparison to the core-clad index difference. The index difference causing self-focusing is given as $\Delta n_{sf} = n_2 I$. At typical pump intensities of 100 MW/cm^2 and $n_2 =$

5×10^{-14} cm²/W [18], the change in core index is 5×10^{-6}. Silicon waveguides structures discussed in the following text have much higher core-cladding index difference, and hence, self-focusing will not have a significant influence in silicon waveguides.

The thermal lensing effect can also be quantitatively evaluated based on the earlier analysis. Assuming steady-state heat dissipation into the medium, the change in refractive index of the core of the guiding region is given approximately as

$$\Delta n = \frac{dn}{dT} \cdot \frac{P_{Heat} \cdot \tau}{\rho C_P V} \tag{7.1}$$

In a Raman medium, the heat dissipated into the medium is the difference in the Stokes and anti-Stokes power:

$$P_{Heat} = P_{Stokes}\left(\frac{\lambda_S}{\lambda_P} - 1\right) \text{ [3]. In silicon, } \frac{dn}{dT} = 1.86 \times 10^{-4} / K,$$

density is $\rho = 2.33$ gm/cm³, specific heat is $C_P = 0.7$ J/gm-K, and the thermal relaxation time is ~10–100 μs, depending on the cross-section over which heat is dissipated. The change in refractive index of the waveguiding region of area 100×100 μm and length of 1 cm is ~1.14 × 10⁻⁴ per watt of average heat dissipation. This is much smaller than the core-cladding index difference in silicon waveguides and hence does not cause a significant alteration in the mode profile. Thus, self-focusing and thermal lensing effects that have been found to be important in the design of Raman lasers in bulk materials [19] are not expected to have a significant effect on silicon waveguide Raman devices in the MWIR.

7.4 MWIR Nonlinear Silicon Devices

In this section, various MWIR nonlinear optical devices in silicon are discussed. Our research work has mainly focused on using Raman-based nonlinear optical devices in silicon. Experimental demonstration of MWIR silicon Raman amplifiers, the statistics of the Stokes amplification, simulation studies of a novel self-imaging Raman amplifier, and design of a cascade MWIR Raman laser are discussed. Following this, studies of other nonlinear effects such as the experimental demonstration of parametric amplifiers and wavelength converters using electronic nonlinearities design of a difference frequency generation device based on periodically poled silicon are also presented. Novel silicon- and germanium-based optical fibers are also discussed.

7.4.1 Raman Amplification

The experimental realization of Raman amplification in silicon in the MWIR wavelength region is discussed in the following text (see Figure 7.4). The pump and Stokes signal lasers used in the demonstration were at 2.88 and 3.39 μm, respectively. The pump laser is an optical parametric oscillator (OPO) and operated under pulsed condition with a pulse width of 5 ns and repetition rate of 10 Hz. The source for the Stokes signal is a helium–neon laser and operated under CW condition. The beam qualities of the pump and Stokes beams are $M^2 = 32$ and 2, respectively, as obtained from the manufacturer's specification. The M^2 parameter is a measure of the deviation of an optical beam from an ideal Gaussian beam in terms of its divergence properties [20]. In this case, the divergence of the pump is ~16 times more than the Stokes laser beam. At the input end, a dichoric mirror combiner is used to efficiently transmit the pump and reflect the Stokes beam and hence combine the two beams. A single plano-convex (PCX) lens of 75 mm focal length was used to focus both the pump and Stokes laser into the silicon sample. The focal spot radii of the pump and Stokes beams inside silicon were ~430 μm and 120 μm, respectively. This was determined using simulations based on a commercial optical system design software package (ZEMAX).

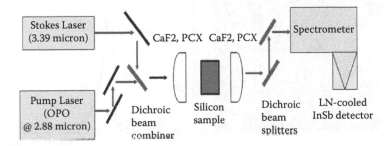

FIGURE 7.4 The experimental setup used to observe Raman amplification in silicon at MWIR wavelengths. The pulsed pump OPO (2.88 μm) and the CW Stokes HeNe laser (3.39 μm) are focused and coupled into 2.5-cm-long AR-coated silicon sample with suitable CaF_2 plano-convex (PCX) lenses. At the output end, dichroic filters are used to block the residual pump and look at the amplified Stokes using a fast InSb detector placed after a spectrometer. (From V. Raghu, Demonstration of a mid infrared, *OSA* 14355; Nathan et al. Silicon Raman Amplifier, 2007. With permission.)

The poor beam quality of the pump limited the focusing of the pump beam and hence the overlap of the pump and Stokes beams inside the silicon sample. The overlap between the pump and the Stokes beam inside the silicon sample was estimated to be ~14%. The high refractive index (~3.45) of silicon ensures a long Rayleigh length within the silicon sample and, hence, the overlap was approximately constant over the sample length.

The two beams were coupled into the silicon sample (both bulk silicon and slab/ridge waveguides). The two facets of the silicon samples were coated with a broadband antireflection coating to prevent incurring Fresnel reflection losses. At the output end, a single CaF$_2$ convex lens (focal length: 75 mm plano-convex for bulk silicon and 25 mm biconvex for waveguides) was used as the imaging lens. Two dichroic beam splitters were used to separate the strong residual pump from the weak amplified Stokes signal. A spectrometer was used to further filter the residual pump and observe the amplification at the Stokes wavelength. The time-resolved Stokes signal was detected using a cooled indium anti-monide (InSb) detector and observed using an oscilloscope.

Raman amplification was observed in multimode slab and ridge waveguide geometry [21,22]. The slab waveguides used were ~260 μm thick, and a typical mode profile obtained using an Electrophysics' pyroelectric camera at 3.39 μm is shown in Figure 7.5a. The time-resolved Raman gain measurement results

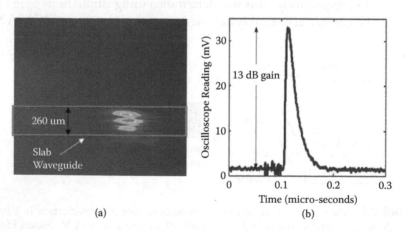

(a)   (b)

FIGURE 7.5 (a) The mode profile of a slab waveguide used in our experiments at Stokes wavelength of 3.39 μm is shown. Image was obtained using Electrophysics' pyroelectric camera. (b) The time-resolved Raman amplification measurement is shown. Maximum on–off Raman gain of 13 dB was achieved at 3.39 μm for a pump wavelength of 2.88 μm.

are shown in Figure 7.5b. Maximum on–off Raman amplification of ~13 dB has been achieved at 3.39 μm in this demonstration for pump energy level of ~3.5 mJ or peak pump intensity of ~220 MW/cm². This is the first demonstration of Raman amplification in silicon in the MWIR region to the best of our knowledge. Detailed measurements of Raman gain as a function of varying pump intensity and the Raman gain spectrum can be found in [21,22].

More recently, the statistics of the Raman amplification process in silicon in the MWIR was studied [23]. It was experimentally determined that the Raman amplification process of the weak input Stokes beam by a noisy pump source follows an L-shaped distribution. Figure 7.6a shows the distribution of pulse energy of 3000 pump pulses; it is found to follow a mean centric distribution, such as a Gaussian or a Rician distribution. Simultaneously, the Raman gain experienced by the Stokes beam is measured, as shown in Figure 7.6b. The observed distribution clearly follows an L-shaped extreme value behavior, highlighted by the high probability of the large outliers in the extended tail of the distribution [23]. In other words, the pump pulses with slightly larger power than the mean creates unproportionally larger gain through the nonlinear transfer function of the gain process. This is attributed to the inherently noisy pump source used in the experiments (Q-swtiched Nd-YAG laser-pumped OPO).

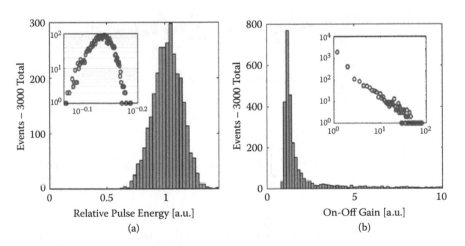

FIGURE 7.6 (a) Measured histogram of pump pulses. (b) Measured histogram of amplified Stokes pulses. Extreme value statistical behavior and outlier events are clearly evident in the tail of the L-shape distribution. The inset in both the curves shows the same data in log–log plot.

7.4.2 Multimode Raman Image Amplifier

Stimulated Raman interaction in multimode structures have been found to be inherently different when compared to that in single-mode waveguides due to the numerous possible ways in which the various pump and Stokes spatial modes can interact. Experimentally, interesting effects such as increased Raman gain in an amplifier due to interaction of higher-order pump and Stokes modes [24] and the so-called Raman beam clean-up effect [25] have been observed. Certain multimode waveguide structures also exhibit a Talbot self-imaging effect on account of constructive interference among the various waveguide modes every periodic length [26]. Talbot imaging has been reported in amplifying media consisting of solid-state medium with top or side pumping using diodes [27]. In this section, the design of a new multimode silicon waveguide Raman amplifier that consists of collinearly propagating MWIR pump and Stokes beams is discussed [28]. The waveguide amplifies and images the spatial profile of an input beam. The amplifier is characterized based on Raman gain and the image quality of the Talbot imaging process. Applications of this device as an image preamplifier are also discussed.

The evolution of amplified Stokes in a multimode waveguide can be described based on coupled-mode treatment [11,12] as follows:

$$\frac{dA_{S-mn}}{dz} = -\frac{\alpha_{S-mn}}{2}A_{S-mn} + \sum_{kl}\kappa_{mn-kl}\left|A_{P-kl}\right|^2 A_{S-mn} +$$

$$\sum_{\substack{k \neq m \, \& \\ l \neq n}} \kappa'_{mn-kl}A_{P-mn}A_{P-kl}{}^*A_{S-kl}e^{j\Delta\beta z} \qquad (7.2)$$

$$\kappa_{mn-kl} = \omega_S\varepsilon_o \int_0^b\int_0^a \phi_{P-kl}\phi_{P-kl}{}^*\left(\chi_{Raman}^{(3)}\right)\phi_{S-mn}\phi_{S-mn}{}^*.dxdy \quad (7.3)$$

$$\kappa'_{mn-kl} = \omega_S\varepsilon_o \int_0^b\int_0^a \phi_{P-kl}^*\phi_{P-mn}\left(\chi_{Raman}^{(3)}\right)\phi_{S-mn}{}^*\phi_{S-kl}.dxdy \quad (7.4)$$

$$\Delta\beta = \left(\beta_{P-mn} - \beta_{P-kl}\right) - \left(\beta_{S-mn} - \beta_{S-kl}\right) \qquad (7.5)$$

In these equations, subscripts S and P refer to the Stokes and pump, respectively, and the indices m and n refer to the mode indices in the X and Y directions, respectively. A refers to the mode coefficient of the propagation mode, and α refers to the linear propagation loss for the mode. There are two kinds of interactions considered in Equation 7.2: the first term in the right-hand side is the conventional Raman amplification with the overlap interaction $\kappa_{mn\text{-}kl}$, and the second term is the Raman Spatial Four-Wave Mixing (RS-FWM) term involving the interaction of the different spatial pump and Stokes modes with coupling coefficient $\kappa'_{mn\text{-}kl}$. $\chi^{(3)}_{Raman} = 1.6 \times 10^{-18}$ m^2/V^2 is the Raman susceptibility of silicon. The RS-FWM term is momentum mismatched by the factor $\Delta\beta$ (Equation 7.4). This analysis ignores the highly oscillatory RS-FWM terms as these terms average out to zero over the interaction length.

The preceding equations were solved numerically. The amplifier performance is discussed here. The pump and Stokes fields are considered in the MWIR region at 2.94 μm and 3.46 μm. This eliminates any nonlinear absorption in silicon [9]. Figure 7.7 shows the electric field amplitude profile (X-Z plane) as it evolves along the waveguide. The pump is considered to be quasi-CW with 1 kW peak power (achievable with typical Er-YAG

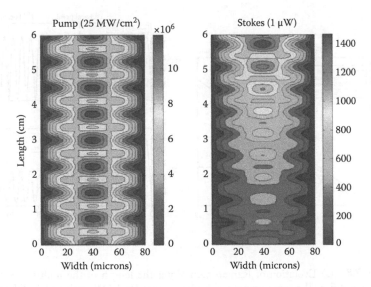

FIGURE 7.7 Contour profile of the electric field amplitude (X-Z profile) showing the self-imaging Raman amplifier with the evolution of the pump and Stokes along the length of the multimode silicon waveguide. (From V. Raghunathan et al. Self-imaging silicon Raman amplifier, *Opt. Express* 2007, 15, 4. With permission.)

Q-switched laser) and the Stokes is 1 μW average power. Both beams are launched as Gaussian beams centered at the waveguide. The amplification of the Stokes is clearly noticeable along with the self-imaging effect. The pump is not significantly depleted due to the weak input Stokes beam. Figure 7.8a shows the Raman gain achievable along the waveguide for varying waveguide propagation losses. The vertical dashed lines denote the location of the first and second focal points for Talbot imaging [26]. As seen in this figure, for the propagation losses considered here (0.1, 0.2, and 0.5 dB/cm), the gains at the first imaging lengths are 9, 8, and 5 dB, respectively. Figure 7.8b shows the beam quality of the Stokes beam along the waveguide. The well-known M^2 parameter is used to characterize the deviation of the Stokes image from an ideal Gaussian profile. For comparison, the top figure shows the Stokes image quality without the presence of a pump. The bottom figure in the presence of Raman amplification shows the Stokes image reproduction with a slight degradation of the image quality along the length of the waveguide.

There are two mechanisms that affect the desired self-imaging of the amplified Stokes beam. First, with increasing length and pump power, the Stokes modes that have highest overlap with the

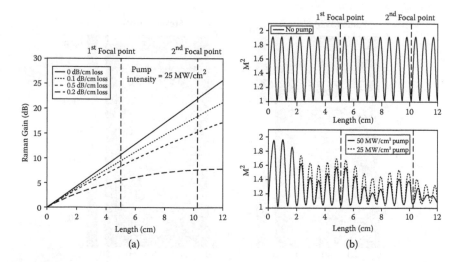

FIGURE 7.8 (a) Evolution of Raman gain along the length of the multimode silicon Raman amplifier. The input pump and Stokes power are 1 kW peak and 1 μW, respectively. (b) The beam quality (M^2) of the Stokes calculated along the waveguide. The top figure shows Stokes propagating through a passive waveguide (no pump launched), and bottom figure shows Stokes propagating through an active waveguide Raman amplifier with input pump intensities of 25 (red curve) and 50 MW/cm² (blue curve).

pump modes experience higher gains compared to the other Stokes modes. This leads to preferential amplification of certain Stokes modes and hence distortion of the image. Second, the RS-FWM effect changes the phases of Stokes-mode coefficients and further affects the image. Both lead to a trade-off between the achievable gain and the reproducibility and quality of the Stokes image at focal points. The use of multiple incoherent pump beams would further worsen the effect of the RS-FWM terms by adding additional phase shifts. From Figures 7.7a and 7.7b, it is found that gains close to 10 dB can be achieved in waveguide lengths of ~5 cm with minimal image distortion. The image distortion effects discussed here can be eliminated by employing certain restrictive mode excitation schemes. That is, by propagating the pump only in a single higher-order mode that the Stokes image does not support, it is possible to eliminate the two sources of image distortion discussed earlier. Such selective mode excitation has been employed in multimode optical fiber communication [29].

Such large-area MWIR waveguide amplifiers can be used as image preamplifiers for LADAR applications. There has been an increased interest in MWIR remote sensing due to the availability of efficient lasers and also the strong molecular vibrational resonances in this range. So far, bulk solid-state media have been used to implement near-IR image amplifiers [30]. The use of waveguides as described here would increases the length over which image quality can be maintained and also reduces the threshold pump power through confinement of the pump mode. The device also offers large field-of-view image amplification due to the high numerical aperture of the multimode silicon waveguide.

7.4.3 Design of Cascaded Raman Lasers

Another interesting feature of the Raman waveguide laser is the possibility of achieving cascaded Stokes emission in nested cavities with the Stokes beam reaching high intensities and hence acting as the pump for the successive Stokes emission. The absence of a nonlinear loss mechanism in silicon will allow higher-order Raman emission in a cascaded cavity, thus extending the addressable wavelength spectrum. Moreover, waveguiding is necessary to enable cascaded cavity operation, similar to the cascaded cavity fiber lasers [31]. Figures 7.9 and 7.10 show possible implementations of the integrated cascaded Raman laser cavities. Bragg mirrors can be implemented in silicon by deep etching of 1D photonic crystals on the waveguides. The high index of silicon makes it possible

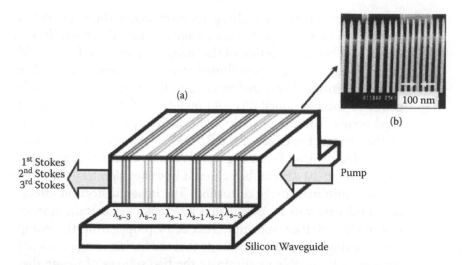

FIGURE 7.9 (a) An integrated cascaded Raman cavity in a silicon waveguide using Bragg mirror pairs to selectively reflect the various orders of the Stokes beam, (b) SEM photograph of chirped Bragg mirrors in silicon fabricated at UCLA using electron beam lithography and deep reactive ion etching. (From B. Jalali et al., *IEEE J. of Sel. Topics of Quant. Elec.* 12, 1618–1627 (2006). With permission.)

FIGURE 7.10 Schematic of the integrated cascaded cavity Raman laser using a microresonator structure. (From B. Jalali et al., *IEEE J. Sel. Topics of Quant. Elec.* 12, 1618–1627 (2006). With permission.)

to realize high reflection with only few periods. By implementing such Bragg mirrors for successive Stokes wavelengths, as shown in Figure 7.9, it will be possible to achieve cascaded Stokes lasing. The same can also be implemented in a microcavity configuration, as shown in Figure 7.10. In this case, the successive Stokes wavelengths have to be resonant in the cavity for sustained oscillation. This has been implemented in a silica micro-cavity in the near-IR [31]. Table 7.1 lists the successive wavelengths that can be achieved

TABLE 7.1 Cascaded Stokes Emission in a Silicon Raman Cavity

Pump Wavelength	1st-order Stokes	2nd-order Stokes	3rd-order Stokes
2.1 µm (CTH:YAG)	2.35 µm	2.68 µm	3.12 µm
2.69 µm (CTE:YAG)	3.13 µm	3.74 µm	4.64 µm
2.9 µm (Er:YAG)	3.42 µm	4.15 µm	5.30 µm
2 to 3.5 µm (Cr:CdSe)	2.23 to 4.2 µm	2.52 to 5.37 µm	2.9 to 7.45 µm

Note: Various pump sources that can be used are also listed.
Source: B. Jalali et al., *IEEE J. Sel. Topics Quant. Elec.* 12, 1618–1627 (2006). With permission.

in a cascaded silicon Raman laser. Some tunable laser pump sources that can operate at wavelengths above the TPA edge of silicon (λ > 2.2 µm) include the Cr/Tm/Er:YAG laser (cascade emission of ~6 laser lines emitted between 2.62–2.94 µm), Cr^{+2}, and other transition-metal-doped ZnSe/ZnS or CdSe laser (continuously tunable from ~2.1–2.7 µm (Cr:ZnSe) or continuously tunable from ~2.1–3.6 µm (Cr:CdSe)) [32].

Researchers at Intel Corp. have demonstrated the first-cascaded Raman laser in integrated silicon structure operating in the near-IR wavelength region [33]. In this demonstration, using a pump source at 1550 nm, Raman lasing has been observed at first and second Stokes orders at 1675 and 1848 nm with output power in excess 5 mW. One of the drawbacks of using near-IR pump sources is the associated nonlinear absorption due to two-photon and free-carrier effects that deplete the pump and generated Stokes, thus preventing further cascading of the generated Stokes radiation to higher orders. Starting with an MWIR pump source could mitigate this effect and be the potential route to multiple-order Stokes lasing. This could help achieve laser radiation at otherwise hard to reach wavelengths.

7.4.4 Other Parametric Nonlinear Effects

Realization of other parametric nonlinear effects has also been demonstrated or proposed recently. Though this chapter mainly focuses on our work on realization of Raman nonlinearities in the MWIR, we would like to briefly describe some of these other nonlinear effects studied. The discussion here highlights some of the prominent works and is not meant to be exhaustive.

The use of stress in silicon photonic structures to create second-order nonlinear effects has been demonstrated [34]. Second-order nonlinear effects are typically forbidden in bulk silicon due to the centrosymmetric crystal structure. The use of stress breaks this symmetry and creates second-order effects. At UCLA, this concept has been extended to design a novel MWIR nonlinear device in silicon, as discussed in detail in Chapter 3. The technique has been proposed to be used as a difference frequency generation device by pumping with standard lasers at 1.3 and 1.55 µm to generate MWIR radiation at 5.1 µm with a conversion efficiency as high as 50% using ~1.8 GW/cm^2 peak pump power [35].

Four-wave mixing, based on third-order nonlinear susceptibility, has been used to demonstrate wavelength conversion and parametric amplification. Turner et al. at Cornell University have demonstrated wavelength conversion from 1250 to 2056 nm using a pump at 1554 nm [36]. A broadband wavelength conversion of over 800 nm has been achieved with longest idler wavelength of ~2100 nm. Liu et al. at Columbia University have demonstrated MWIR parametric amplification and net gain from 4-mm-long silicon waveguide [37]. A parametric gain of ~28 dB has been demonstrated in the 2200–2300 nm wavelength range. These two demonstrations involve careful dispersion engineering of the waveguide to achieve phase matching.

Recently, silicon-based optical fibers have been proposed to be used as an active nonlinear medium [38]. The long interaction length is an advantage to build to the nonlinear effects considerably. However, propagation loss has to be optimized to prevent power depletion along the fiber length. About 30-m-long fibers were fabricated with 60–120 µm size core diameter. Propagation losses were measured at 4.3 dB/m at 2.936 µm wavelength. Germanium optical fibers have also been fabricated using similar techniques for nonlinear optics applications [39].

7.5 Biomedical Applications

Interaction of mid-IR wavelengths with biological tissues and organic materials has unique advantages that has led to various biophotonic applications. Mid-IR biophotonics offers the potential development of minimally invasive and safe medical diagnostic systems. Tissue ablation, tissue spectroscopy, bioimaging, and biosensors are some of the potential biomedical applications of mid-IR photonics. The silicon photonics devices and solutions provided in the previous sections of this chapter potentially lend themselves to these applications and could lead to compact and cost-effective mid-IR biophotonic systems. Some of the main biomedical applications of such systems are briefly reviewed in the following. Readers interested in more detailed information can refer to [40–45] and the references therein.

The MWIR range of 3.0 to 3.7 µm covers virtually all the dipole-active stretching vibrational modes of carbon–hydrogen (C-H) bonds in organic molecules. Interestingly, the absorption overtones of these mid-IR "C-H stretch" modes are typically much stronger than those in the near-IR. Hence, vibrational spectroscopy using MWIR sources can be used to observe and analyze the C-H stretch modes. In mid-IR spectroscopy, the absorption, transmission, or reflection of these dipole-active modes are measured. An alternative technique is using Raman spectroscopy [41]. One potential application is detection and analysis of organic molecules in human breath for medical diagnostics [42]. Another possibility is tissue spectroscopy. A major obstacle for mid-IR spectroscopy of tissues, however, is water absorption peaking at ~3 µm. This limits the penetration depth of MWIR light to 1 to 100 µm at wavelengths of 2.5–4 µm [43]. Consequently, mid-IR spectroscopy is limited to thin tissue layers. Nonetheless, techniques such as attenuated total reflection spectroscopy have been successful in reducing the impact of water absorption [41].

The high water absorption at MWIR discussed earlier becomes an advantage for precise photoablation applications. The low penetration depth into tissues ensures that ablation occurs superficially with minimal photoinduced damage of the regions surrounding the exposed area. Four basic mechanisms are considered to cause photoablation of tissues, namely, photochemical, photothermal, photomechanical, and photoelectrical processes [40]. Holmium yttrium-aluminum-garnet (Ho:YAG at 2.1 µm), Erbium YAG (at 2.94 µm), and even free-electron lasers (2.9–9.2 µm) have been used for surgical procedures as wide as ophthalmic microsurgery, corneal ablation, cortical bone ablation, tumor removal, skin resurfacing, endoscopy, and hair removal [41,44,45].

7.6 Conclusions

Silicon is considered as a potential candidate for building photonic devices in the technologically important mid-IR region of the wavelength spectrum. In the past, bulk silicon has been used as windows and attenuator in MWIR lasers. Recently, silicon has been evolving as active photonic material in the MWIR. This chapter is a summary of the wide array of applications of MWIR silicon nonlinear optical devices being studied. The low linear and nonlinear absorption losses in the MWIR region (2–7 μm region), combined with the good crystal quality, mature processing technology, especially for integrated optical devices, high optical damage threshold, and high thermal conductivity make silicon a good material system to build nonlinear optical devices in the MWIR. The use of silicon-on-sapphire technology to build integrated silicon devices and the potential deleterious effects of self-focusing and thermal lensing in high-power applications were also discussed. The various topics discussed in this chapter include the experimental demonstration of Raman amplification in silicon at 3.39 μm using pulsed 2.88 μm pump sources, the modeling of a novel self-imaging multimode silicon Raman amplifier, and the design of cascaded Raman lasers. We have also reviewed results of our recent observation of extreme value phenomenon (highly non-Gaussian) fluctuations during Raman amplification in silicon. The influence of the strongly nonlinear stimulated Raman scattering on input pump fluctuations causes amplified Stokes pulses to exhibit statistical distributions that resemble those observed in socioeconomic systems rather than Gaussian or other traditional distributions that one would expect in physical systems. Other parametric nonlinear effects such as design of MWIR difference frequency generation and the experimental demonstration of four-wave-mixing-based parametric amplification and wavelength conversion were discussed. The recent experimental studies of silicon and germanium optical fibers for MWIR applications were also discussed. The technologies discussed in this chapter can potentially expand the application space of silicon photonic technology beyond data/telecommunication and into biochemical sensing, laser medicine, LADAR, and numerous other MWIR applications.

References

1. N. Barnes et al., Diode-Pumped Ho, Tm:YLF laser pumping an AgGaSe2 parametric oscillator, *J. Opt. Soc. Am. B* 11, 2422–2426 (1994).
2. T. H. Allik et al., Tunable 7–12 µm optical parametric oscillator using a Cr,Er:YSGG laser to pump CdSe and $ZnGeP_2$ crystals, *Opt. Lett.* 22, 597–599 (1997).
3. H. M. Pask, The design and operation of solid-state Raman lasers, *Progr. Quant. Electron.* 27, 3–56 (2003).
4. B. Jalali, V. Raghunathan, R. Shori, S. Fathpour, D. Dimitropoulos, and O. Stafsudd, Prospects for silicon mid-IR Raman lasers, *IEEE J. of Sel. Topics of Quant. Elec.* 12, 1618–1627 (2006).
5. R. A. Soref, S. J. Emelett, and W. R. Buchwald, Silicon waveguided components for the long-wave infrared region, *J. Opt. A.* 8, 840–848 (2006).
6. T. K. Liang, and H. K. Tsang, Role of free carriers from two-photon absorption in Raman amplification in silicon-on-insulator waveguides, *Appl. Phys. Lett.* 84, 2745–2747 (2004).
7. R. Claps, V. Raghunathan, D. Dimitropoulos, and B. Jalali, Influence of nonlinear absorption on Raman amplification in silicon waveguides, *Opt. Express* 12, 2774–2780 (2004).
8. H. Rong, R. Jones, A. Liu, O. Cohen, D. Hak, A. Fang, and M. Pannicia, A continuous-wave Raman silicon laser, *Nature* 433, 725–728 (2005).
9. O. Boyraz and B. Jalali, Demonstration of a silicon Raman laser, *Opt. Express* 12, 5269–5273 (2004).
10. V. Raghunathan, R. Shori, O. M. Stafsudd, and B. Jalali, Nonlinear absorption in silicon and the prospects of mid-infrared silicon Raman laser, *Phys. Status Solidi.* (a) 203, R38–R40 (2006).
11. R. W. Boyd, *Nonlinear Optics*, 2nd edition, Academic Press, San Diego, CA. 2003.
12. A. Yariv, *Quantum Electronics*, 3rd edition, Wiley Publication, New York, NY. 1989.
13. J. Zhang, Q. Lin, G. Piredda, R. W. Boyd, G. P. Agrawal, and P. M. Fauchet, Anisotropic nonlinear response of silicon in the near infrared region, *Appl. Phys. Lett.* 91, 07113–07115 (2007).
14. S. Pearl, N. Rotenberg, and H. M. Van Driel, Three photon absorption in silicon for 2300–3300 nm, *Appl. Phys. Lett.* 93, 131102–131104 (2008).
15. L. Pavesi and D. J. Lockwood, Eds., *Silicon Photonics*, Springer, Germany, 2004.
16. G. P. Agrawal, *Nonlinear Fiber Optics*, 3rd edition, Academic Press, San Diego, CA. 2001.
17. Michael E. Thomas, Richard I. Joseph, and J. William, *Appl. Optics* 27, 239–245 (1988).

18. O. Boyraz, T. Indukuri, and B. Jalali, Self-phase modulation induced spectral broadening in silicon waveguides, *Opt. Express* 12, 829–834 (2004).

19. W. Koechner, *Solid-State Laser Engineering*, Springer series in optical sciences, 5th edition, New York, NY. 1999.

20. A. E. Siegman, How to (may be) measure laser beam quality, Tutorial OSA annual meeting (1997).

21. V. Raghunathan, D. Borlaug, R. Rice, and B. Jalali, Demonstration of a mid infrared silicon Raman amplifier, *Opt. Express* 15, 14355 (2007).

22. V. Raghunathan, D. Borlaug, R. Rice, and B. Jalali, Mid-infrared silicon Raman amplifier, *Adv. Solid-State Photonics Conf. Proc.*, Nara, Japan (2008).

23. D. Borlaug, S. Fathpour, and B. Jalali, Extreme value statistics in silicon photonics, *IEEE Photonics J.* 1, 33–39 (2009).

24. N. Bloembergen, Multimode effects in stimulated Raman emission, *Phys. Rev. Lett.* 13, 720–724 (1964).

25. S. H. Baek and W. B. Roh, Single-mode Raman fiber laser based on a multimode fiber, *Opt. Lett.* 29, 153–155 (2004).

26. L. B. Soldano and E. C. M. Pennings, Optical multi-mode interference devices based on self-imaging: Principles and applications, *IEEE J. Light. Tech.* 13, 615–627 (1995).

27. I. T. McKinnie, J. E. Koroshetz, W. S. Pelouch, D. D. Smith, J. R. Unternahrer, and S. W. Henderson, Self-imaging waveguide Nd:YAG laser with 58% slope efficiency, *Conference on Lasers and Electro-Optics (CLEO)*, CTuP2 (2002).

28. V. Raghunathan, H. Renner, R. Rice, and B. Jalali, Self-imaging silicon Raman amplifier, *Opt. Express* 3406–3408 (2007).

29. L. Raddatz, I. H. White, D. G. Cunningham, and M. C. Norwell, Influence of restricted mode excitation on bandwidth of multimode fiber links, *IEEE Photon. Technol. Lett.* 10, 534–536 (1998).

30. A. Brignon, G. Feugnet, J. P. Huignard, and J. P. Pocholle, Large-field-of-view, high-gain, compact diode-pumped Nd:YAG amplifier, *Opt. Lett.* 22, 1421–1423 (1997).

31. T. J. Kippenberg, S. M. Spillane, B. K. Min, and K. J. Vahala, Theoretical and Experimental study of stimulated and cascaded Raman scattering in ultrahigh Q optical microcavities, *IEEE J. Sel. Topics Quant. Electron.* 10, 1219–1228 (2004).

32. I. T. Sorokina, and K. L. Vodpyanov, Solid state mid-infrared laser sources, Springer-Verlag, Berlin-Heidleberg, Germany (2003).

33. H. Rong, S. Xu, O. Cohen, O. Raday, M. Lee, V. Sih, and M. Paniccia, A cascaded silicon Raman laser, *Nat. Photonics* 2, 170–174 (2008).

34. R. S. Jacobsen, K. N. Andersen, P. I. Borel, J. Fage-Pedersen, L. H. Frandsen, O. Hansen, M. Kristensen, A. V. Lavrinenko, G. Moulin, H. Ou, C. Peucheret, B. Zsigr, and A. Bjarklev, Strained silicon as a new electro optic material, *Nature* 441, 199–202 (2006).

35. N. K. Hon, K. K. Tsia, D. R. Solli, and B. Jalali, Periodically poled silicon, *Appl. Phys. Lett.* 94, 09116–09118 (2009).

36. A. C. Turner-Foster, M. A. Foster, R. Salem, A. L. Gaeta, and M. Lipson, Frequency conversion over two-thirds of an octave in silicon nanowaveguides, *Opt. Express* 18, 1904–1908 (2010).
37. X. Liu, R. M. Osgood, Jr., Y. A. Vlasov, and W. M. J. Green, Broadband mid-infrared parametric amplification, net off-chip gain, and cascaded four-wave mixing in silicon photonic wires, *Group IV Photonics Conference, Conference Proceedings* (2009).
38. J. Ballato, T. Hawkins, P. Foy, R. Stolen, B. Kokuoz, M. Ellison, C. McMillen, J. Reppert, A. M. Rao, M. Daw, S. R. Sharma, R. Shori, O. Stafsudd, R. R. Rice, and D. R. Powers, Silicon optical fiber, *Opt. Express* 16, 18675–18683 (2008).
39. J. Ballato, T. Hawkins, P. Foy, B. Yazgan-Kokuoz, R. Stolen, C. McMillen, N. K. Hon, B. Jalali, and R. Rice, Glass-clad single-crystal germanium optical fiber, *Opt. Express* 17, 8029–8035 (2009).
40. I. K. Ilev and R. W. Waynant, Mid-infrared biomedical applications, in *Mid-Infrared Semiconductor Optoelectronics*, A. Krier, Ed., Springer, Berlin/Heidelberg (2006).
41. W. Petrich, Mid-infrared and Raman spectroscopy for medical diagnostics, *Appl. Spectroscopy Rev.* 36(2&3), 181–237 (2001).
42. G. von Basum, H. Dahnke, D. Halmer, P. Hering, and M. Murtz, Online recording of ethane traces in human breath via infrared laser spectroscopy, *J. Appl. Physiol.* 95, 2583–2590 (2003).
43. D. Wieliczka, S. Weng, and M. Querry, Wedge shaped cell for highly absorbent liquids: Infrared optical constants of water, *Appl. Opt.* 28, 1714–1719 (1989).
44. G. M. Peavy, L. Reinisch, J. T. Payne, and V. Venugopalan, Comparison of cortical bone ablations by using infrared laser wavelengths 2.9 to 9.2 μm, *Lasers Surg. Med.* 26, 421–434 (1999).
45. C. A. Chaney, Y. Yang and N. M. Fried, Hybrid germanium/silica optical fibers for endoscopic delivery of Erbium:YAG laser radiation, *Lasers Surg. Med.* 34, 5–11 (2004).

CHAPTER **8**

Novel III-V on Silicon Growth Techniques

Diana L. Huffaker and Jun Tatebayashi

Contents

8.1 Introduction

Silicon-based complementary metal oxide semiconductor (CMOS) and the III-V-based optoelectronics industries have achieved considerable advancements over the past few decades according to the prediction by Gordon Moore, Chairman of Intel, that integrated circuits would double in performance every 18 months, which is called Moore's Law. This has been achieved largely through the CMOS and bipolar technology platforms. Innovations in the field of silicon technology now result from well-resourced and coordinated global research and development programs. The combination of these superior electrical devices with III-V-based optical semiconductor devices resulting in high-speed and high-power devices with mature silicon technology would be highly beneficial for the semiconductor industry since it would result in many electronic and optoelectronic applications including on-chip photonic devices integrated with CMOS. Recent developments in CMOS-integrated optoelectronics make III-V lasers on Si a highly desirable and researched device.

There are mainly two predominant schemes for the integration of III-V materials with Si: hybrid integration and monolithic integration. The hybrid integration usually involves growing the III-V device separately and then achieving subsequent integration through other methods, which include conventional wafer bonding (Wada and Kamijoh 1994; 1996), novel methods such as recess mounting of devices, and newer variations in wafer bonding that incorporate an intermediate layer such as polymers or spin on glass to bond the III-Vs to Si (Lin et al. 2002). On the other hand, a monolithic growth of III-V materials on Si offers intriguing features such as an efficient use of the integrating platform and reduced processing complexity compared to growth on GaAs or GaSb substrates (Deppe 1998). The monolithic approach utilizing GaAs/AlGaAs can produce room-temperature (RT) edge-emitting laser diodes and even vertical-cavity surface-emitting lasers (VCSELs) (Deppe et al. 1990; Egawa et al. 1994) on Si(100) substrates reported by many research groups. A wide range of the lasing wavelengths of ~0.7–0.8 (Van der Ziel et al. 1987; Deppe et al. 1987; Razeghi et al. 1988; Sugo et al. 1990; Egawa et al. 1990; Groenert et al. 2003), ~1.0 (Chriqui et al. 2003; Mi et al. 2005; Kwon et al. 2006), 1.3 (Razeghi et al. 1988), and 1.55 μm (Sugo et al. 1990) has been demonstrated by using different active materials. Recent demonstrations of the monolithic III-V lasers on Si show the device performance that parallels output powers and threshold currents of lasers grown directly on GaAs substrates. Some of the prominent results are reported using SiGe metamorphic buffers (MBs) (Groenert et al. 2003; Kwon et al. 2006) and GaAs MBs on Si (Mi et al. 2006) that achieve dislocation bending through InAs quantum dot (QD)-based strain fields (Mi et al. 2006). These results are very encouraging and promising, whose details are discussed in another chapter in this book. In general, one potential challenge for such hybrid devices is that they may suffer from reliability issues associated with material incompatibilities such as lattice mismatch (Figure 8.1), thermal expansion coefficient, and process temperature, and may hinder stable and repeatable production processes based on monolithic integration (Hwang et al. 1995).

Recently, a novel growth technique has been demonstrated that involves 90° interfacial misfit (IMF) arrays formed during the growth of AlSb on Si (001) (Balakrishnan et al. 2005). These IMF arrays can relieve 99% of the entire strain caused by the 13% lattice mismatch via a self-assembled two-dimensional (2D) array of 90° dislocations without using a very thick MB. The IMF arrays have been reported in several systems including GaP/Si (Kawanami et al. 1982), GaAs (or AlGaAs)/Si (Wang 1984), InAs/GaAs (Trampert et al. 1995), InAs/GaP (Chang et al. 1996), GaSb/GaAs (Rocher 1991), and InP/GaAs (Jin-Phillipp

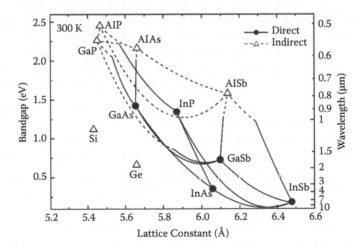

FIGURE 8.1 Lowest forbidden gap between conduction and valence bands as a function of lattice constants for III-V compound semiconductor materials (except for nitride) and Si/Ge at 300 K.

et al. 2001), besides AlSb/Si (Kim et al. 2006) over a range of lattice-mismatched conditions ranging from $\Delta a_o/a_o$= 0.4% (GaP/Si) to $\Delta a_o/a_o$= 13% (AlSb/Si). The IMF array forms at the III–V/Si interface and remains localized within that plane rather than propagating vertically into the material. Therefore, the IMF array immediately accommodates the hetero-interface strain by the bending and stretching of atomic bonds surrounding each 90° dislocation. It is noted that the 90° dislocation threads neither vertically nor laterally. A detailed explanation of the IMF formation via atomic self assembly and energy minimization has been reported (Jallipalli et al. 2007). The IMF growth mode on Si (001) results in low-defect density bulk epitaxy ($\sim$10^6/cm^2) that has enabled optically pumped VCSELs and superluminescent diodes (Balakrishnan et al. 2006a) In addition, the growth of AlSb on Si offers significantly better agreement of the substrate and epilayer thermal expansion coefficients compared to GaAs on Si (Kumar and Sastry 2001). At 300 K, AlSb has a thermal expansion coefficient of 2.55 × 10^{-6}/K, which is very close to that of Si (2.59 × 10^{-6}/K). In comparison, the thermal expansion coefficient of GaAs is 6.93 × 10^{-6}/K, resulting in a significant tensile strain during cooling down from the growth temperature to RT. However, there still remains a substantial issue of the formation of antiphase domains (APDs) that have inhibited the realization of laser diodes on Si (001) substrates.

The formation of APDs in the growth of AlSb on Si (001) is indeed an inherent issue with the growth of polar III-Vs on non-polar Si (Fan and Poate 1986; Choi et al. 1988). In the absence of step-free Si(001) substrates, the established method to achieve single domain III-Vs is to use miscut Si(001) substrates (Adomi et al. 1991; Andre et al. 2005; Sieg et al. 1998). Miscut Si (2.5° to 5°) substrates, typically characterized by a double atomic-step height (Barbier et al. 1991), facilitate registration of the III and V sublattices on the (001) plane, resulting in the suppression of APD formation. So far, high-quality III-V materials on Si have been produced using the APD annihilation or suppression combined with a strain-relief and defect-filtering mechanism, usually a thick buffer layer. These methods require a two-step growth process initiated at a rather low temperature to enable 60° and 90° dislocation formation followed by normal growth temperatures for metamorphic and bulk layer growth. Lattice-matched bulk GaAs epitaxy on miscut Ge has also been demonstrated to produce very low defect and low APD density (Tanoto et al. 2005).

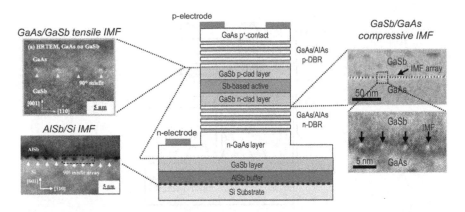

FIGURE 8.2 Future prospect of VCSELs on Si operating at the telecommunication wavelength utilizing three different kinds of IMF: GaSb/GaAs compressive IMF, GaAs/GaSb tensile IMF, and AlSb/Si IMF. (From Huang, S. H. et al. 2006. *Appl. Phys. Lett.* 88: 131911-1–3; Huang, S. H. et al. 2007. *Appl. Phys. Lett.* 90: 161902-1–3. With permission.)

There are several prospective applications of the IMF growth mode to the integration of CMOS and photonic devices, which includes the realization of VCSELs emitting at the telecommunications wavelength monolithically grown on Si substrates or the evanescent coupled lasers monolithically grown on silicon-on-insulator (SOI) substrates (as shown in Figure 8.2). Since the first invention of a VCSEL 30 years ago (Iga 2000), a lot of unique features of VCSELs have been proven, such as low power consumption, wafer-level testing, and small packaging capability. The commercialization of VCSELs has been growing up rapidly in recent years and they are now key devices in local area networks using multimode optical fibers. Long-wavelength VCSELs emitting at the telecommunication wavelengths of 1.3 μm or 1.55 μm are currently attracting growing interest for use in single-mode fiber metropolitan area and wide area network applications. Therefore, it would be advantageous if VCSELs would also be integrated with Si-based CMOS electronic devices with extremely small chip-size, low power consumption, and low cost. This structure is comprised of a GaSb-based active region emitting at the near-infrared (NIR) or even mid-wavelength-infrared (MWIR) regime cladded by GaAs/AlAs-based distributed Bragg reflectors (DBRs) monolithically grown on Si substrates. Such structures can be realized by using three different kinds of IMF arrays: GaSb/GaAs compressive IMF, GaAs/GaSb tensile IMF (Huang et al. 2007), and AlSb/Si IMF. In spite of a

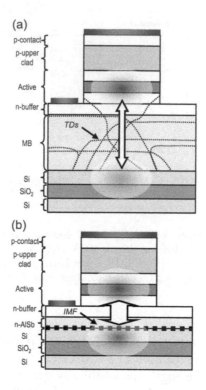

FIGURE 8.3 Schematic illustration of evanescent coupled lasers monolithically grown on SOI by using the (a) metamorphic buffer approach or (b) the IMF growth technique approach.

substantial problem of an increased turn-on voltage caused by a potential drop at the IMF array (Mehta et al. 2006), it would be advantageous to realize long-wavelength GaSb-based VCSELs with GaAs-based DBRs as they are well developed have higher reflectivity, and GaAs provides better metal contacts compared to the GaSb-based systems.

Evanescent coupled lasers monolithically grown on silicon-on-insulator (SOI) can also be realized with the IMF technology (Figure 8.3). So far, Fang et al. have demonstrated "hybrid" AlGaInAs quantum well (QW) lasers bonded to the silicon waveguide fabricated on SOI substrates (Fang et al. 2006). In these laser structures, an evanescent tail from the QW active region overlaps into the optical mode of the silicon waveguide region. Therefore, the IMF growth technique instead of growing very thick MB would enable the evanescent light from the active region to be coupled more efficiently to the silicon waveguide.

8.2 Growth and Interface Characterization of the Interfacial Misfit Array

8.2.1 Theory of Mismatched Growth

One of the main constraints on achieving new and unique devices with the present semiconductor technology is imposed by lattice mismatch. The foundation for future studies in mismatched crystalline growth with the misfit interface between an epitaxial film and its substrate has been established in the late 1940s by Frank and Van der Merwe. They show that a lattice mismatch that is smaller than ≅7% will be accommodated by a uniform elastic strain until a certain film thickness is reached (termed *critical thickness*). These studies were further advanced in the 1970s by Matthews and Blakeslee, who established that mismatched epitaxy results in coherent strain and not polycrystalline or amorphous incoherent growth (Matthews and Blakeslee 1974).

Lattice-matched epitaxy implies layer-by-layer deposition of material where the epilayer and the substrate have exactly the same crystallographic type and lattice constant. This results in a coherent strain-free growth. The growth of high-quality epilayer on either a substrate or another epilayer is limited by lattice mismatch between these layers, which can be described by the following equation:

$$f = \frac{\Delta a}{a} = \frac{a_e - a_s}{a_s} \qquad (8.1),$$

where f is a lattice mismatch, a_s is a lattice constant of the substrate, and a_e is a lattice constant of the epilayer. There is no strain associated with the lattice-mismatched growth mode, as shown in Figure 8.4. However, if the lattice constant of the epilayer does not match the lattice constant of the substrate, strained growth results. Depending on whether the epilayer is larger or smaller than the substrate, this strain is classified as compressive or tensile. In the case of compressive strain, the larger epilayer conforms to the smaller substrate so as to realize a one to one correspondence with the substrate. This results in the cell expanding along the direction perpendicular to the growth surface, as has been depicted in Figure 8.4b. Tensile strain has the opposite effect, as shown in Figure 8.4c. These instances of strain are called *pseudomorphic* because the epilayer

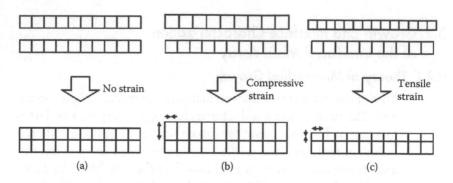

FIGURE 8.4 (a) Lattice-matched and (b), (c) lattice-mismatched growth modes.

takes on the morphology or the lattice constant of the underlying substrate, and the distortion of the material grown is termed *tetragonal distortion*.

As the growth of the strained material continues, the strain energy increases linearly. However, if this strain energy in the crystal becomes relatively large due to either a very large mismatch to begin with or a very thick epilayer, it is relieved through a network of dislocations. These dislocations are initially in the form of misfit dislocations, that is, sites where atoms have been skipped, and these misfit dislocation sites act as sources for threading dislocations, which as the name suggests, refers to defects that thread through the material breaking bonds as they propagate to one of the crystal surfaces.

The critical thickness arises as the result of competition between strain energy and chemical energy. Below the critical thickness, the minimum energy of the epilayer is achieved through strain. However, above the critical thickness, the minimum energy is achieved through dislocations. This critical thickness depends on the extent of the mismatch, the material parameters, and also on the nature of these dislocations forming at the interface when the critical thickness is exceeded. For example, when only misfit dislocations are considered forming at the point where the critical thickness is exceeded, the expression for critical thickness may be written as

$$h_c = \frac{a_0(1 - \frac{v_{PR}}{4})\left[\ln(\frac{h_c\sqrt{2}}{a_0}) + 1\right]}{2\sqrt{2}\pi|f|(1 + v_{PR})} \qquad (8.2),$$

where,

a$_o$ = Substrate lattice constant

f = Mismatch

v$_{PR}$ = Poisson's ratio (~1/3 for most semiconductors)

As growth proceeds beyond the critical thickness, *relaxation* occurs. Relaxation implies that the material that was once strained and contorted has now returned to its original shape. Depending on material properties, the crystal may take several microns from the onset of dislocation to completely relax. By doing so, the material has moved from a pseudomorphic phase to a metamorphic phase, where it has once again regained its native lattice form and lattice constant. At thicknesses between complete relaxation and complete strain, the following equation may be used:

Fraction of relaxation = 1 − fraction of strain still present

Also noteworthy is the fact that the process of relaxation is highly directional. For zinc blende structures, quite often the [110] direction relaxes faster compared to the [110] direction due to the inherent asymmetry in the lattice. While it may appear that the metamorphic approach could provide us with any lattice constant to grow on, thus eliminating the problem of limited substrates, the resulting dislocations render the material unusable for optoelectronic purposes. This is because these defects act as nonradiative recombination centers. Researchers have used several techniques to reduce defect densities, such as the use of strain layer superlattices to force the threading dislocations to terminate at the {110} growth planes rather than thread into the active region. However, this still requires the growth of very thick metamorphic buffers (a few microns), and there is no guarantee that the material that results will be defect free.

8.2.2 Burgers Vector and Types of Misfit Dislocations

The Burgers vector of a dislocation is a crystal vector, specified by Miller indices, that quantifies the difference between the distorted lattice around the dislocation and the perfect lattice. The direction of the vector depends on the plane of dislocation, which is usually on the closest-packed plane of the unit cell. The magnitude is usually represented by

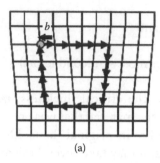 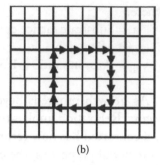

(a)   (b)

FIGURE 8.5 Schematic illustrations for determining the Burgers vector of an edge dislocation.

$$\|b\| = \frac{a}{2}\sqrt{h^2 + k^2 + l^2} \qquad (8.3)$$

where a is the lattice constant of the crystal, and h, k, and l are the components of Burgers vector, **b** = <h k l>.

In a 2D primitive square lattice, the Burgers vector of an edge dislocation can be determined using the following steps: Trace around the end of the dislocation plane to form a closed loop, as shown in Figure 8.5.

In a perfect lattice, trace out the same path, moving the same number of lattice vectors along each direction as before, as shown in Figure 8.5. This loop will not be complete, and the closure failure (green arrow) is the Burgers vector.

There are primarily two types of dislocations: edge dislocations and screw dislocations (Figure 8.6). Screw dislocations lie along the lines in the glide plane parallel to the direction of slip. Taylor dislocations or edge dislocations are formed by inserting an extra half-plane of atoms into a perfect crystal and is marked by an edge of an incomplete plane of atoms. An edge dislocation is often represented by the symbol ⊥, in which the horizontal line represents the glide plane and vertical line represents the tip of the inserted half-plane.

Edge dislocations are further categorized into various dislocations depending on the angle of the dislocation line with respect to the interface. In screw dislocations, the misfit dislocation is parallel to the interface and, hence, the angle is $0°$ between the dislocation line and the interface. Commonly observed edge dislocations are $30°$, $60°$, and $90°$ dislocations. The $30°$ dislocations form in systems with a small lattice mismatch. In highly lattice

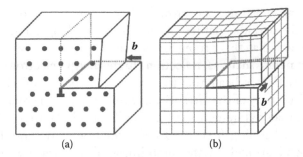

FIGURE 8.6 (a) Edge dislocations and (b) screw dislocations.

mismatched systems, 60° dislocations form to relieve strain after the critical thickness. In addition to direct formation, 90° dislocations can also be formed by the interaction between two 60° dislocations, where the dislocation line is perpendicular to the interface, which is parallel to the Burgers vector. The 60° dislocations are able to thread to the surface by forming threading dislocations. Threading dislocations can be measured either by etching the material using a corresponding etchant for the material such as potassium hydroxide (KOH) and a Normarski microscope for surface observations, or from the analysis of plan view transmission electron microscopy. The former can cover large areas and can be more efficient when the defect density is around 10^5–10^6 cm^{-2} and the latter is accurate in all cases; however, only a very small area can be analyzed. A Lomer dislocation is defined as a dislocation that lies on (001), is parallel to <110>, and is of the edge type.

8.2.3 An Alternative Mechanism for High-Quality Mismatched Growth

While the theory of strained growth states that a critical thickness has to be achieved prior to the onset of misfit dislocations, in certain materials systems such as GaSb on GaAs, a two-dimensional (2D) array of misfit dislocations is present at the interface of the GaSb on GaAs growth. This is a fundamentally different growth mode that results in low-defect or defect-free bulk material in which strain energy is solely relieved by laterally propagating (90°) misfit dislocations confined to the episubstrate interface.

The IMF formation does not proceed through the critical thickness route, but instead makes use of atomic arrangements on the substrate surface to spontaneously relax. However, if the growth conditions are not conducive to forming this atomic arrangement on the substrate, the growth then turns pseudomorphic. The highly periodic nature of the array, its long-range order, and the fact that this arrangement can be picked up by *in situ* measurements such as reflection high-energy electron diffraction (RHEED) indicate that the process may be one of self-assembly.

The study of the IMF dislocations can be done more effectively in a material system such as GaSb on GaAs than AlSb on Si. This is because, first, the IMF formation of the interfacial array depends strongly on the smoothness of the surface. A smooth surface can be achieved with GaAs through homoepitaxy, whereas Si homoepitaxy is not an option in the current configuration of the V80 molecular beam epitaxy (MBE) reactor used. Second, the lack of APDs in the GaSb on GaAs makes the transmission electron microscopy (TEM) study of the material and the interface easier and more accurate. Therefore, in this chapter, the theory of interfacial misfit dislocations using the GaSb/GaAs material system is developed and optimized, and the results are applied to the AlSb/Si material system.

8.2.4 Interfacial Misfit Dislocation Array-Based Nucleation of GaSb on GaAs

The growth of thick GaSb layers on GaAs was previously believed to start as islands form and then coalesce into a uniform layer. In these previous demonstrations, both 90° and 60° misfit dislocations were present (Qian et al. 1997). While the predominant strain relief mechanism was believed to be the 90° misfits, the minority 60° misfits were shown to cause threading dislocations in the GaSb. The source of the 60° misfits is still unclear but attributed to one or more of the following factors: island coalescence, growth temperature, and the degree of the mismatch. Island coalescence, in which the {111} planes of adjacent islands merge, has been shown to cause 60° misfits. Supporting data includes a strong correlation between island coalescence and the location of the 60° misfits. The growth temperature has been shown to be a strong factor in determining which type of misfit is produced, with GaSb grown at ~520°C favoring 90° misfits and 560°C favoring 60° misfits (Kim et al. 1998). However, 560°C may be too close to the melting temperature of GaSb to be a meaningful data point. Some researchers have stated that the lattice mismatch is of critical importance in the formation of 90° misfits (Mallard et al. 1989). Low-strain systems (<2%) have resulted in 60° misfits, moderate strain (3%–4%) in mixed 90° and

60° misfits, and high strain (>6%) in pure 90° misfits. This data is, however, based on experimental observations and is not backed by any theoretical analysis or calculations. It is noted that the formation of 90° seems to require balancing strain energy with adatom migration and is therefore a function of lattice mismatch, Sb overpressure, and temperature. Based on energy minimization calculations, the interfacial array of misfit dislocations is the result of a process of self-assembly at the substrate–epi interface. In the following sections, the nature of the GaSb–GaAs interface is studied in greater detail, followed by a more general theory for interfacial misfit dislocation array formation and, finally, a mathematical analysis based on molecular mechanics is performed to assess the strain distribution in interfacial misfit networks.

Under optimized growth conditions, a highly periodic array of 90° dislocations in the growth of GaSb on GaAs to yield a highly (95%) relaxed and defect-free GaSb buffer on GaAs can be realized. The growth of the GaSb bulk on GaAs is performed using the V80 MBE reactor, and the presence of the two-valved crackers for As and Sb sources allows us to exercise a great deal of control on the growth of the structures. The GaAs substrate is deoxidized at 600°C prior to the growth of 100 nm of GaAs at 560°C to obtain a smooth surface. Before the Sb growth is initiated, the As valve is closed, allowing As adatoms to desorb, leaving a Ga-rich surface. This process, confirmed by RHEED transition from an As-rich (2 × 4) to Ga-rich (4 × 2) surface, reduces As/Sb intermixing (Figure 8.7). At this point, an Sb flux is introduced on the (4 × 2) GaAs surface, resulting is an instant transformation to a (2 × 8) reconstructed surface. The Sb overpressure is stopped, and the substrate temperature is reduced to 510°C. The GaSb growth is now commenced at a nominal growth rate (0.2–0.8 μm/h) and the RHEED pattern assumes a (1 × 3), indicating that a thin film of GaSb has formed on the surface. The pattern is difficult to

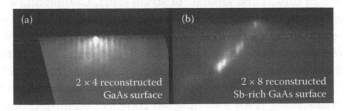

FIGURE 8.7 RHEED patterns of (a) 2 × 4 reconstructed GaAs and (b) Sb-rich 2 × 8 reconstructed GaAs. (From Balakrishnan, G. et al. 2005. *Appl. Phys. Lett.*, 86: 034105-1–3. With permission.)

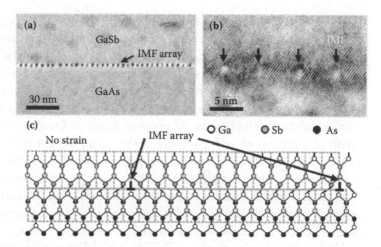

FIGURE 8.8 Cross-sectional TEM image of 120 nm of GaSb on GaAs. Part (a) shows a highly periodic array of misfit dislocations at the interface of the GaSb layer and the GaAs substrate. Part (b) shows a low-resolution image of the same interface. (c) Schematic illustration of the interface between GaAs and GaSb using IMF growth modes.

analyze during the first few GaSb monolayers (MLs). But, under optimized growth conditions, a clear 1×3 reconstruction pattern appears within the first 3–5 MLs, indicating a planar growth mode of relaxed bulk material. Unoptimized parameters yield a spotty RHEED pattern with continued deposition that indicates a defective growth mode.

Figures 8.8a and 8.8b show high- and low-resolution TEM bright-field images of a strain-relaxed, defect-free GaSb (120 nm) buffer on GaAs. The bright spots in both images correspond to misfit dislocation sites (Williams and Carter 1996). Figure 8.8b shows that the highly periodic misfit array is localized at the GaSb–GaAs interface and no misfit dislocations at any other location. Furthermore, no threading dislocations are visible. Careful examination of Figure 8.8a allows the identification of misfits and analysis of strain relief. Completing a Burgers circuit around one misfit dislocation indicates that the Burgers vector lies along the interface and identifies the misfit as 90° type. Measurement of the GaAs substrate and GaSb bulk lattice constants within several MLs of the interface yields $a_o = 0.398$ and $a_o = 0.430$ nm, respectively, which are equivalent to the published values along [110] and indicate complete strain relaxation. The misfit separation, measured to be ~56 Å, corresponds to exactly 13 GaSb lattice sites and 14 GaAs lattice sites. Thus, every 14th Ga atom has a pair of dangling bonds (one going into and out of the image plane) to accommodate the larger Sb atom in the next (001) plane.

Most of the strain energy generated by the GaSb/GaAs lattice mismatch is dissipated by the misfit array at the interface. In the following paragraph, the strain energy, E_ε, with the energy dissipated from a two-dimensional misfit array, E_d', is calculated and compared. This simulation is performed by calculating the strain energy associated with 1 ML of GaSb on GaAs and comparing it with the energy of a misfit dislocation array. Since the IMF arrays form within the first monolayer of growth, this should give us a very good estimate as to what percentage of the strain generated by the mismatch is relieved at the interface.

The strain energy E_ε is found using the following equation:

$$E_\varepsilon = \varepsilon^2 Bh = 0.0933955 \frac{J}{m^2} \qquad (8.4)$$

where
$$\varepsilon = \frac{a_s - a_f}{a_f},$$
a_f = in-plane lattice constant of the film,
a_s = in-plane lattice constant of the substrate, and
$$B = 2\,\mu_f(1+v)/(1-v).$$

In these equations, ε is the in-plane strain, B is a constant, $h = 0.38$ nm is the layer thickness (which in this case is taken to be 1 ML), $a_s = 0.565325$ nm is the in-plane lattice constant of the GaAs substrate, $a_f = 0.609593$ nm is the lattice constant of the relaxed GaSb film, $\mu = 2.4 \times 10^{10}$ N/m^2 is the GaSb shear modulus, and $v = 0.31$ is GaAs Poisson's ratio. Now,

$$f = |a_s - a_f|/a_s = 0.07831$$

is the GaSb/GaAs lattice mismatch, $b = a_f/\sqrt{2} = 0.43105$ nm is Burgers vector along the [110] direction in the GaAs substrate. Therefore, the dislocation energy dissipated by a 2D misfit array is

$$E_d' = \frac{E_d}{S} = 0.07952 \text{ J/m}^2 \qquad (8.5)$$

where $E_d \approx \mu_f b^2 / [4\pi(1-\nu)]$ is the energy per unit length of a single edge dislocation.

The misfit spacing, S, can be derived theoretically by

$$S = \frac{b}{f} = \frac{0.39975 \ nm}{0.07262} = 5.50468 \ nm \qquad (8.6)$$

which agrees very well with S = 5.60 nm measured from TEM images shown in Figure 8.8.

A comparison of values for E_ε = 0.0933955 $\times$ J/m^2 and E_d = 0.07952 J/m^2 of the strain energy generated by the GaSb/GaAs lattice mismatch at the growth temperature shows that ~0.014 J/m^2 is not relieved through misfit dislocations. The presence of low dislocation densities under optimized growth conditions for thicker GaSb layers (>100 nm) can be verified using KOH etch-pit measurements on the epilayer. The study involved growing nine samples with the GaSb layers on GaAs grown at temperatures of 480, 510, and 540°C, with GaSb thicknesses of 200, 1100, and 3100 nm for each growth temperature. These wafers were then etched in 20% KOH solution for 10 min, and roughly 100 nm of the epilayer was removed in each case. The etch-pit density results are shown in Figure 8.9 and clearly show that at the

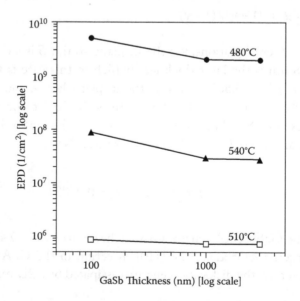

FIGURE 8.9 Defect densities at different thicknesses of GaSb on GaAs for different growth temperatures.

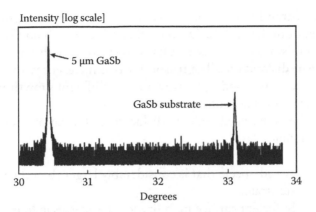

FIGURE 8.10 Triple axis (004) XRD scan of 5 μm of GaSb on GaAs showing an FWHM of ~20 arcseconds for both the substrate and the epilayer.

optimal growth temperature of ~510°C, the etch-pit density is the lowest. The 480°C sample had the highest decoration density of all the samples grown. Furthermore, the triple axis ω-2θ high-resolution x-ray diffraction (HR-XRD) studies of very thick buffer layers of GaSb on GaAs resulted in extremely low full-width at half-maximum (FWHM) values of ~20 arcseconds, shown in Figure 8.10. This value is the same as that of GaSb substrates.

8.2.5 Interfacial Array of Misfit Dislocations: A Self-Assembly Process

Thus far, the growth of high-quality low-defect density GaSb on GaAs through the formation of an interfacial array of misfit dislocations has been studied. It has been shown that the interfacial array of misfit dislocations formed at the GaSb and GaAs interface does not obey the usual pattern that mismatched growth follows. Instead of achieving a critical thickness, the material instead opts to relieve the entire mismatch at the interface and spontaneously relax through the formation of 90° misfit dislocations. The highly periodic nature of the misfit dislocations and their long-range order suggest that the mechanism for their formation might be self assembly, and this can further be verified through energy minimization calculations.

To understand how this interfacial array forms, two material system growths are looked into that have very similar mismatches and yet result in very different outcomes: InAs on GaAs and GaSb on GaAs. InAs on GaAs is a very popular material system. The 7.8% mismatch between the epilayer and the substrate results in a Stranski–Krastanow (SK) growth mode and highly strained quantum dots (QD). On the other hand, GaSb on GaAs has an 8% mismatch and yet results in relaxed growth with interfacial misfit dislocations. The reason for two materials with similar lattice constants opting to grow in two different growth modes can be explained as follows.

The interfacial misfit dislocation formation requires the following:

a. An ad-specie that is considerably larger than the underlying substrate.
b. Sufficient surface mobility for the ad-specie to move around and achieve optimal 2D-packing.
c. Sufficient surface residence time such that the ad-specie can form an interfacial array of dislocations.
d. Ability of the epilayer to wet the substrate so as to achieve a 2D layer of atoms.

$$\gamma_{substrate} > \gamma_{Epilayer} + E_{IMF} \qquad (8.7)$$

This means that to achieve a planar layer of misfit dislocations, the surface energy of the substrate must be greater than the surface energy of the atomic layer forming the misfit dislocation array and the energy of the misfit dislocations themselves.

The Sb atom meets all of the aforementioned conditions when deposited on GaAs. When exposed to a Ga-rich GaAs surface, at sufficient temperatures, the Sb atoms pack themselves on the GaAs surface, resulting in a 13:14 correspondence between the Sb atom and the underlying Ga atom. This effect can be seen in the RHEED pattern, as shown in Figure 8.7. When the 2D layer of packed Sb atoms is formed on the GaAs surface, it has a very special reconstructed surface—a 2 × 8. For the 2 × 8 RHEED pattern to be observed, the As coverage of the GaAs has to be replaced by an Sb coverage. To facilitate this, we grow GaAs at 560°C and at that high temperature both the Ga and the As fluxes are stopped. This causes the desorption of the As in the top monolayer of the GaAs, giving us a 2 × 4 to 4 × 2 transition. At this instant, if Sb is provided to the surface, a 2 × 8 reconstruction is immediately observed. If the same experiment is performed at a much lower temperature of ~520°C this exchange of As atoms with Sb atoms takes a much longer time to occur. This is due to the fact the As surface sticking coefficient is much improved at this temperature, requiring almost 30 min to transition from a 2×4 to a 2×8. At even lower temperatures of 460°C, this transition is never completed.

8.2.6 Modeling an Interfacial Array of Misfit Dislocations Using Molecular Mechanics

In this section, a mathematical model is described to elucidate the formation, atomic arrangement, and strain energy resulting from a periodic pure edge, 90° interfacial misfit dislocation array in highly mismatched III-V semiconductors. Using molecular mechanics methods, the strain energy is calculated at the atomic level by considering the stretch and bend of each bond in the system under consideration. The model predicts the misfit array periodicity and indicates a self-assembly process driven by energy minimization. Three highly mismatched systems—InAs/GaAs ($\Delta a_0/a_0$ ~7.2%), GaSb/GaAs (-$\Delta a_0/a_0$ ~7.8%), and AlSb/Si ($\Delta a_0/a_0$ ~13%)—are considered, and calculations are compared to experimental data.

The publications by People and Bean (1985) and Matthews et al. assume tetragonal distortion up to a thickness known as *critical layer thickness*. This critical layer thickness, h_c, can be calculated using one of these theories only for a moderate mismatch, $\Delta a_0/a_0 < 4\%$. Alternative strain relief mechanisms, such as interfacial misfit dislocations, cannot be described by the aforementioned theories since they do not consider the distribution of strain at the atomic level necessary to understand the nature or mechanics of interfacial self-assembled array formation.

Another genre of mathematical methods describes interfacial misfit dislocations, including the Frank and van der Merwe model, which uses the Frenkel and Kontorowa (FK) model of a truncated Fourier series (Van der Merwe 2002) to describe the interfacial misfit dislocation formation mechanics. These models describe the position of the atoms and the total strain energy of the system based on a mechanical energy minimization principle. However, they do not describe atom–atom interaction, bond energetics that drives the interfacial misfit dislocation formation, or implications of residual strain. Moreover, these previously reported models do not elucidate the self-assembled nature of periodic 90° interfacial misfit dislocation formations.

The method reported here is based on molecular modeling, or molecular mechanics, and treats the lattice as a collection of weights connected with springs, where the weights represent the nuclei and the springs represent the bonds. Strain energy is calculated by summing the individual distorted bond energies derived using the stretch and bend equations described later. Electrons are not considered explicitly, but rather it is assumed that the electrons will find their optimum distribution once the positions of the nuclei are known (Teufel 2003). Both the strain energy and atomic geometry at the lattice-mismatched hetero-interface are derived from isolated atomic parameters such as force constants and lattice spacing. Kuronen et al. have used molecular dynamics to study the lattice-mismatched systems characterized by gliding dislocations such as 30° and 60° misfits, but not interfacial (90°) misfit dislocations (Kuronen et al. 2001). In this chapter, 90° misfit arrays in three mismatched systems—AlSb on Si, GaSb on GaAs, and InAs on GaAs—are considered.

The ratio of lattice constants, a_f:a_s, is equal to the ratio of interfacial atoms that participate in the misfit dislocation, x_f:x_s. For compressively strained material, $a_f > a_s$ and $x_s = x_f + 1$. The ratio x_f:x_s determines the misfit periodicity and equals 13:14 in the case of GaSb/GaAs; that is, $x = 13$ lattice sites of GaSb and is equivalent in distance to 14 lattice sites in the GaAs substrate. Careful observation of the lattice shows that the misfit occurs at the 0th and again at the 14th lattice sites, and so on; the characteristic and undistorted zinc blende arrangement occurs exactly at the 7th atom from the misfit. A schematic of the interfacial atoms in Figure 8.8b illustrates atomic arrangement and bonding in the [100] plane around the interfacial misfit. The bonds appear undistorted between misfits, but begin bending and stretching until the some physical limit is surpassed to necessitate a skipped bond. The misfit forms when the bond length becomes so large that the energetics of the system cannot accommodate that bond. A lower energy bond is available with the nearest-neighbor interfacial atom. Therefore, the energy generated by bond distortion is the limiting factor in determining the misfit site. Thus, the misfit array formation is a self-assembled process driven by energy minimization. The total energy for each atomic bond is expressed as the sum of several force potentials. These force potentials are E_s, energy required for stretching the atom; E_θ, energy required for bending the atoms; $E_{s-\theta}$, energy required for stretch–bend interactions; E_{tor}, torsional strain; E_{vdw}, van der Waals interactions; and E_{Dp-Dp}, dipole–dipole interactions. The physical system considered in this work is sensitive primarily to two components, E_s and

E_θ, that is, $E_{total} = E_s + E_\theta$, and the magnitude of the rest of the components is very small in comparison and can be neglected. The energy required to stretch or compress an atomic bond is described using an equation similar to Hook's Law:

$$E_s = 143.82 \cdot \frac{K_s}{2}(l - l_0)^2 \qquad (8.8)$$

where K_s is the stretching force constant, l is the actual bond length and l_0 is the original bond length, and 143.82 is a constant used to convert the stretching energy to Kcal/mol. The same logic applies to bond bending as $E_\theta = 0.21914 \times K_\theta \times (\theta - \theta_0)^2$ and K_θ is the bending force constant, 0.21914 is a constant used to convert the bending energy to Kcal/mol, θ is the distorted bond angle, and θ_0 ($=90°$) is the original bond angle. It is clear from the foregoing equations that the distortion energy for bond stretching is directly proportional to the square of the deviation in bond length, $(l - l_0)$, and the distortion energy for bond bending is also directly proportional to the square of the deviation in bond angle, $(\theta - \theta_0)$. In our calculations, the total energy is in Kcal/mol.

Figures 8.11a and 8.11b illustrate the zinc blende atomic lattice in undistorted and distorted conditions. The figure identifies bond length, l; bond stretching, Δl; bond angle, θ; and bond bending, $\Delta\theta$. In the case of GaSb on GaAs, the radius of the Sb

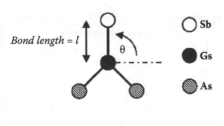

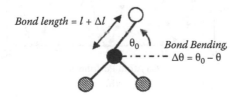

FIGURE 8.11 Schematic drawing of (a) undistorted and (b) distorted lattices showing $l, \Delta l, \theta, \Delta\theta$. (From Jallipalli, A. et al. 2007. *J. Cryst. Growth*, 303: 449–455. With permission.)

atom is large (1.38 Å) compared to the radius of the Ga (1.26 Å) atom. Therefore, when the GaSb epilayer is deposited on top of the GaAs substrate, the Sb atom must deviate slightly from the ideal zinc blende structure resulting in a new bond length $l + \Delta l$ and the new bond angle is $\theta + \Delta\theta$ in order to accommodate the atom. These new bond lengths and bond angles can be very easily calculated using simple geometry based on the lattice mismatch.

Figures 8.12a and 8.12b plot the calculated stretching and bending energies, respectively, as a function of the Ga atom position at the GaSb–GaAs interface. The periodicity of the misfit

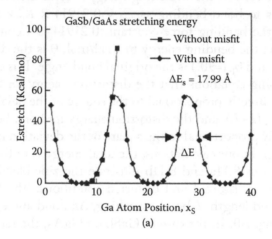

(a)

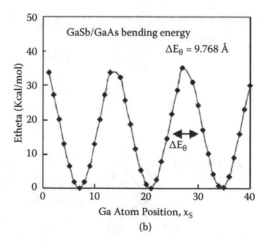

(b)

FIGURE 8.12 (a) Stretching and (b) bending energies plotted as a function of Ga atom position along the GaSb–GaAs interface in the [110] direction. (From Jallipalli, A. et al. 2007. *J. Cryst. Growth*, 303: 449–455. With permission.)

TABLE 8.1 Material and Calculated Parameters Associated with Misfit Arrays in the InAs/GaAs, GaSb/GaAs, and AlSb/Si Material Systems

Parameters	InAs/GaAs	GaSb/GaAs	AlSb/Si
Mismatch (%)	7.2	7.8	13
Lattice ratios (x_s:x_f)	14:15	13:14	8:9
Stretching force constant, K_s (N/m)	35.64	32.72	34.15
Bending force constant, K_θ (N/m)	6.51	5.95	6.22
Peak energy, E_{max} (Kcal/mol)	230	179.22	231.6
FWHM (number of atoms, Å)	6 Ga; 19.99	6.2 Ga; 20.79	3 Si; 9.22
Max. bond length deviation, Δl_{max}(Å)	1.71	1.54	1.68
Max. bond angle deviation, $\Delta\theta_{max}$ (o)	52.9	50.9	58.38

array is obvious from the plot as the first misfit is placed at x = 0, and the subsequent misfits form every 14 Ga lattice positions. Both energy components have a maximum value at the misfit dislocation position where E_s (max) ~56 Kcal/mol and E_θ (max) ~34 Kcal/mol. The maximum values for stretching and bending energy values indicate a maximum stretch and bend of bonds at the GaSb–GaAs interface. The maximum energy values can be calculated for the other material systems listed in Table 8.1. They are E_{max} for InAs–GaAs 115 Kcal/mol, and E_{max} for AlSb–Si 115.79 Kcal/mol. The plot also provides information about distribution of strain among the bonds as measured by full-width at half maximum (FWHM) values. From the energy plots, FWHM of GaSb/GaAs is 6.2 Ga atoms, that is, 20.79 Å.

Table 8.1 shows several parameters associated with the strain energy in 90° interfacial, periodic misfit arrays in InAs–GaAs, GaSb–GaAs, and AlSb–Si. The table includes lattice mismatch, bending and stretching force constants K_s and K_θ, lattice ratios x_f:x_s, peak energy E_{max}, FWHM, maximum bond length deviation Δl_{max}, and maximum bond angle deviation, $\Delta\theta_{max}$. The lattice mismatch is 7.2% and 7.8% for the InAs–GaAs and GaSb–GaAs systems and 13% for the AlSb–Si, which correspond to lattice ratios of 14:15, 13:14, and 7:8. The values for K_s and K_θ, which represent the amount of energy stored in the distorted bond, are rather similar for these materials: K_s ranges from 32 to 36 N/m, and K_θ ~ 6 N/m. The table lists values for E_{max}(AlSb/Si) ~ E_{max}(InAs/GaAs) ~ 230

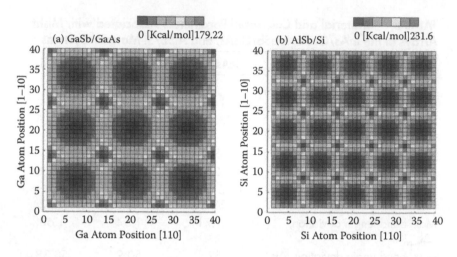

FIGURE 8.13 3D plots of total energy calculated at the (a) GaSb/GaAs and (b) AlSb/Si using molecular modeling. (From Jallipalli, A. et al. 2007. *J. Cryst. Growth*, 303: 449–455. With permission.)

(Kcal/mol), which is greater than E_{max}(GaSb/GaAs) ~ 180 (Kcal/mol). The similar values for E_{max}(AlSb/Si) ~ E_{max}(InAs/GaAs) are somewhat surprising, considering that the lattice mismatches in the two systems are quite different. This apparent inconsistency can be explained by considering the relative values of K_s and K_θ along with Δl_{max} and $\Delta \theta_{max}$, and noting that E_{max} is seemingly controlled by $\Delta \theta_{max}$; $\Delta \theta_{max}$(AlSb/Si) > $\Delta \theta_{max}$(InAs/GaAs). The FWHM value depends on the lattice constant of the substrate and the bond strength between the atoms in the substrate. For both InAs and GaSb systems, the substrate is GaAs and for AlSb, the substrate is Si. The bond strength between Si atoms will be greater compared to GaAs. For that reason, the strain energy is distributed among only 3 Si atoms compared to the other two systems, in which the strain energy is distributed among 6 Ga atoms. Figure 8.13 is a 3D plot of the total energy (stretching plus bending energies) as a function of position at both the GaSb–GaAs and AlSb–Si interfaces. The plot indicates the highly strained areas surrounding the misfits and unstrained areas between misfits using color. The color variation blue to red shows very low strain to very high strains. The peak energy value (E_{max}) for AlSb/Si is 231.6 Kcal/mol and for GaSb/GaAs, E_{max} = 179.22 Kcal/mol. The values of E_{max} in Figure 8.13 represent the total energy in both the [110] and [1-10] direction, whereas in Figure 8.12, we are considering energy only in one direction. For that reason the total energy in Figure 8.13 is twice the value of the energy in Figure 8.12. The rest of the matrix experiences a strain value that lies in between the two. Areas of yellow and green between the misfit peaks indicate residual strain between the misfits along [110] and [1-10]. The integrated strain energy relationship of these three systems follows the relationship of lattice mismatch, that is, AlSb > GaSb ~ InAs (0.44 Kcal/cm² > 0.34 Kcal/cm² ~ 0.33 Kcal/cm²). To more fully understand the implications of this data, we must perform more advanced calculations of strain propagation into the bulk material and compare with experimental x-ray diffraction measurements.

8.2.7 Growth of III-V on Si

In the monolithic III-V growth on Si substrates, there are mainly three issues to be overcome: (1) lattice mismatch, (2) thermal expansion mismatch, and (3) formation of APDs. Table 8.2 shows several parameters (lattice constants, lattice mismatch to Si, bandgap energy, and thermal expansion coefficient) in several material systems including Si, Ge, and III-Vs. As can be clearly seen, all of the III-V and Ge materials except for GaP materials (≅0.38%) have a huge lattice mismatch to Si from 4% to 13%. Growth of such lattice-mismatched materials on Si can cause the formation of 60° dislocations in them that readily thread into the material, which results in extremely high dislocation density. Another important problem here in the growth of III-Vs on Si has been the huge thermal expansion coefficient mismatch between virtually every III-V and Si. GaP's that are practically lattice matched to Si have not resulted in high-quality devices due to the fact that

TABLE 8.2 Lattice Vonstant (Lattice Mismatch to Si), Bandgap, and Thermal Expansion Coefficients for Si and Selected III-V Materials

Material	Lattice Constant (nm) (Lattice Mismatch to Si)	Bandgap (eV)	Thermal Expansion Coefficient ($\times 10^{-6}$ K^{-1})
Si	0.54307 (–%)	1.12	2.56
Ge	0.56579 (4.18%)	0.66	5.9
GaP	0.54512 (0.377%)	2.26	4.6
GaAs	0.56533 (4.10%)	1.424	6.928
GaSb	0.60959 (12.3%)	0.726	6.928
InP	0.58687 (8.07%)	1.34	6.328
InAs	0.60583 (11.5%)	0.36	5.238
AlSb	0.61355 (13.0%)	1.58	2.551

GaP has a thermal expansion coefficient that is 2.75 times that of Si, which results in a severe lattice mismatch at RT after growing materials at much higher temperature. Here, noteworthy is the fact the Si has a lower expansion coefficient than all III-Vs with the exception of AlSb, implying a net tensile strain in the material when cooled to RT. It is a known fact that mild tensile strain is often responsible for highly detrimental results such as microcracks and threading dislocations, resulting in the material quality being too poor for opto-electronic applications. Furthermore, AlSb has a smaller expansion coefficient than Si. This would result in a small compressive strain in the AlSb layer at RT, not tensile. The high bond strength of AlSb and the miniscule mismatch therefore ensure that the material is lattice matched at both growth temperatures as well as at room temperature.

Formation of APDs in the growth of AlSb on Si(001) is also indeed an inherent issue with the growth of polar III-Vs on nonpolar Si. In general, the diamond structure, which includes III-V and Si, consists of two interpenetrating face-centered cubic lattices. The two sublattices differ from each other only in the spatial orientation of the four tetrahedral bonds that connect each atom to its four nearest neighbors, which are on the other sublattice. For example, in Figure 8.14a the atoms with bond orientations indicated as A and B belong to sublattices A and B. There is no distinction between the sublattices otherwise. Both are occupied by the same atomic species. In the zinc blende structure in which both GaAs and AlSb crystallize, one of the sublattices is occupied by the group III and the other by the group V. In a crystal without antiphase disorder, the sublattice allocation is the same throughout

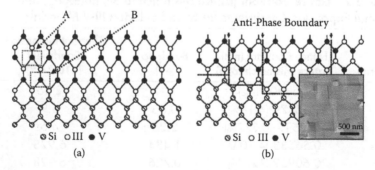

FIGURE 8.14 Schematic illustrations of (a) ideal III-V growth with step-free Si substrates and (b) APD formation on Si(100) surfaces with single steps. (From Tatebayashi. 2009. *Monolithically Integrated III-Sb-Based Laser Diodes Grown on Miscut Si Substrates.* IEEE Press 716, With permission.)

the crystal. If this allocation changes somewhere inside the crystal, the interface between the domains with opposing sublattice allocation forms a two-dimensional structural defect called anti-phase boundary (APB) and the domains themselves are called anti-phase domains (APD), shown in Figure 8.14b. Ideally, if the silicon substrate could be manufactured without a single step in it, the problem of APBs and APDs would not exist. However, since this is impossible to achieve, and the problem is further accentuated by lack of Si homoepitaxy (resulting in rougher surfaces), extensive domain formation is observed in the growth of all III-Vs (and polar) alloys on Si.

In this chapter, the growth and characterization of the IMF array between AlSb and Si substrates is investigated. As mentioned earlier, the formation of APDs in the growth of AlSb on Si (001) is an inherent issue with the growth of polar III-Vs on nonpolar Si, together with a huge lattice mismatch and thermal expansion coefficient. As shown in Figure 8.14a, the problem of antiphase boundaries (APBs) and APDs would not exist if the Si substrates could be manufactured ideally without a single step in it. However, since this is impossible to achieve, and the problem is further accentuated by lack of Si homoepitaxy (resulting in rough surfaces), extensive domain formation is observed in the growth of all III-Vs (and polar) alloys on Si. As shown in Figure 8.14b, the growth of AlSb on Si(100) results in domain formation. An inset of Figure 8.14 shows an AFM image of an AlSb–Si surface with APD formation that appears over the whole region on AlSb/Si. Although the formation of APDs can be mitigated by growing a very thick material where two domains run into each other and one of them survives and the other does not, the presence of domains could be highly detrimental to device development. Therefore, in the absence of step-free Si(001) substrates, only one established method to achieve single domain III-Vs is to use specific miscut Si (2.5° to 5°) substrates, typically characterized by a double atomic-step height.

8.2.8 Growth and Atomic-Force Microscope and Transmission Electron Microscope Analyses of the IMF Array

Samples are grown on Si(100) substrates by a solid-state MBE equipped with valved crackers for Sb and an optical pyrometer for monitoring the temperature. Prior to the growth, the surface of Si substrates is hydrogen-passivated by immersing the wafer in a diluted HF bath. The HF reacts with the SiO_2 and leaves behind a clean Si surface, with the dangling bonds passivated by hydrogen atoms. Heating the substrate to 500°C in vacuum removes the loosely bonded hydrogen. A thermal cycle at 800°C ensures the removal of the oxide remnants. This is verified by the reflection high-energy electron diffraction (RHEED), which shows a (2×2) surface reconstruction with the removal of the oxide. The substrate temperature is reduced and stabilized at 500°C for the growth of AlSb.

The RHEED pattern proceeds through two distinct phases during the initial growth. The deposition of AlSb on Si results in an interconnected chevron pattern as shown in the inset of Figure 8.15a. A 3×3 pattern is also superimposed on this pattern. This implies that the initial growth of AlSb results in the formation of islands with {111} facets and truncated on top with a (100) plane. After deposition of ~150 Å GaSb, the RHEED

FIGURE 8.15 AFM images showing surface structure after (a) 3 MLs, (b) 18 MLs, (c) 54 MLs, and (d) ≅ 500 nm of AlSb deposition on Si. Insets of (a) and (c) also show the RHEED image for the corresponding growths. (From Balakrishnan, G. et al. 2005 *Appl. Phys. Lett.*, 86: 034105-1–3. With permission.)

pattern becomes a pure (3 × 3) pattern as shown in Figure 8.15c, indicating that a planar growth mode has been initiated.

The interface and initial bulk growth are analyzed using atomic force microscopy (AFM). Figures 8.15a–8.15c show the AFM data after 3, 18, and 54 MLs of AlSb deposition. At the nominal thickness of 3 MLs, the island density is ~10^{11} QDs/cm^2 with a dot height and diameter of 1–3 and 20 nm, respectively. Figure 8.15b shows the growth at 18 MLs. The effect of this continued deposition is that the individual islands coalesce but remain crystallographic. Figure 8.15c shows the continued coalescence toward the planar growth with 54 ML deposition. The growth of >54 ML AlSb results in very smooth and high-quality surfaces as shown in Figure 8.15d. The insets of Figures 8.15a and 8.15c show the corresponding RHEED patterns at different stages (3 and 54 MLs) of the nucleation layer growth. The best RHEED patterns are obtained when the AlSb growth is terminated at 50 Å, and GaSb is grown on this surface. This effect has also been demonstrated by Akahane et al. using PL and HR-XRD studies (Akahane et al. 2004).

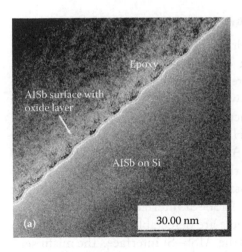

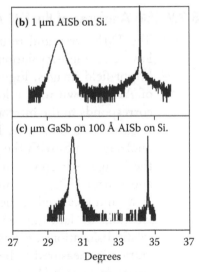

FIGURE 8.16 (a) TEM images of the undulations at the AlSb surface after 25 nm, (b) HR-XRD (004) scan of AlSb on Si, and (c) GaSb on AlSb on Si. The reduction in the full width for the same growth thicknesses indicates that the GaSb layer is successful in suppressing the undulations in the AlSb. (From Tatebayashi, J. 2009. *Monolithically Integrated III-Sb-Based Laser Diodes Grown on Miscut Si Substrates.* IEEE Press 716, With permission; Balakrishnan, G. et al. 2005. *Appl. Phys. Lett.*, 86: 034105-1–3. With permission.)

Figure 8.16a shows TEM images of the undulations at the AlSb surface after 25 nm of AlSb growth on Si. The undulations are ~10 nm wide and 1 nm high. The figure shows a layer of native oxide on the sample that occurs during atmospheric exposure. With continued growth, the surface undulations merge, and become shallower and considerably broader until they can no longer be detected by TEM after ~1 μm. However, evidence of the undulation is still visible in the RHEED pattern for AlSb layer thicknesses ~10 μm. These undulations can, however, be suppressed very effectively by keeping the AlSb nucleation layer to ~100 Å and immediately following it with a GaSb layer. The suppression of the undulations can be seen in HR-XRD studies shown in Figures 8.16b and 8.16c. Part (b) shows the XRD spectrum of AlSb grown on Si directly, and part (c) shows that of GaSb growth on a 100 Å AlSb nucleation layer. The FWHM in part (b) is ~1700 arcseconds and that in part (c) is ~320 arcseconds. Undulations in semiconductors due to misfit dislocations have been noted and modeled by other researchers (Jonsdottir 1995).

8.2.9 TEM Analyses of the IMF Array

The IMF array and resulting bulk material are studied using low- and high-resolution cross-sectional TEM (HR-XTEM) bright-field images. Figure 8.17a shows the HR-XTEM image of AlSb grown with a low defect density on Si. A bright line corresponds to the interface between the AlSb and Si substrate. Figure 8.17b shows the HR-XTEM image of the AlSb buffer nucleated on Si with the AlSb–Si interface. The bright spots in the image correspond to the misfit dislocation sites (Williams and Carter 1996). The misfits are arranged in a highly periodic array and localized at the AlSb–Si interface. No threading dislocations or dark-line defects are detectable in the bulk, and no misfit dislocations exist at the AlSb–Si interface. The misfit separation is measured to be ~34.6 Å, which is expected from the lattice mismatch ($\Delta a_0/a_0 = 13\%$), where a_0 is a lattice constant of Si and Δa_0 is a lattice mismatch between AlSb and Si. This periodicity corresponds to exactly eight AlSb lattice sites and nine Si lattice sites, as schematically shown in Figure 8.17c. Thus, every ninth Si atom has a pair of dangling bonds (one going into and the other out of the image plane) to accommodate the larger Sb atom in the next (001) plane.

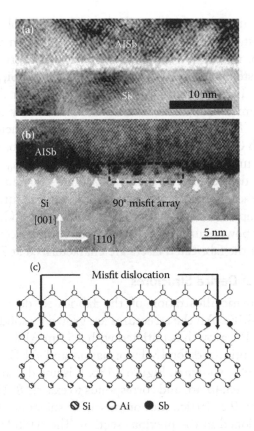

FIGURE 8.17 (a) and (b) Cross-sectional HR-TEM images of the AlSb–Si interface, showing the (110) plane with changes in crystallographic orientation. (c) Schematic illustration of the IMF interface between AlSb and Si. (From Tatebayashi, J. 2009. Monolithically Integrated III-Sb-Based Laser Diodes Grown on Miscut Si Substrates. IEEE Press 716. With permission.)

8.3 Fabrication and Device Characteristics of Electrically Injected GaSb–AlGaSb Edge Emitters on Si Substrate

8.3.1 Introduction

In this section, device characteristics of electrically injected monolithic edge-emitting broad-area lasers grown on 5° miscut Si are investigated utilizing the IMF growth mode of AlSb on Si, which has the smallest difference of thermal expansion coefficients. Much research has been devoted to the realization of III-V laser devices on silicon substrates since the mid-1980s. However, the formation of APDs as well as threading dislocation should be suppressed in order to achieve better device performance. In order to mitigate the formation of APDs, miscut Si substrates are used as an established method to achieve single domain III-Vs. The use of 5° miscut Si substrates, which is characterized by a double atomic step height, for the growth of AlSb on Si, enables simultaneous IMF formation and suppression of an APD, resulting in a dramatic suppression of dislocation density over the III-Sb growth. The active region for the laser diodes includes three 100 Å GaSb QWs separated by 200 Å $Al_{0.3}Ga_{0.7}Sb$ barriers that emit at 1.62 μm at RT, and are cladded by $Al_{0.55}Ga_{0.45}Sb$ layers. An etch pit density (EPD) characterization is carried out to evaluate the threading dislocation density of laser structures grown on Si and compare it with that of laser structures on GaAs substrates. Output power-current (L-I) and electroluminescence (EL) properties of the fabricated lasers are characterized at 77 K.

8.3.2 Growth of Device Structures

A schematic illustration of the fabricated device is shown in Figure 8.18a. All the structures are grown by solid-state MBE at 400°C. The growth is initiated with a 5 nm AlSb nucleation layer that is optimized for IMF formation and APD suppression on miscut Si, as shown in Figure 8.18b. It is noted that the misfit separation is $\cong 3.46$ nm and corresponds to exactly 8 AlSb lattice sites grown on 9 Si lattice sites on miscut Si substrates (Figure 8.18c), as mentioned in the previous section. The AlSb layer is followed by a 2 μm n-GaSb contact, a 2.3 μm $Al_{0.55}Ga_{0.45}Sb$ n-type clad, an active region, a 1.5 μm $Al_{0.55}Ga_{0.45}Sb$ p-type clad, and a highly doped 50 nm GaSb p-type contact layer. The active region comprises six-layer GaSb (10 nm) QWs separated by an $Al_{0.3}Ga_{0.7}Sb$ (10 nm) barrier cladded by 300 nm $Al_{0.3}Ga_{0.7}Sb$ waveguide layers. RT photoluminescence (PL) of the active region has an emission peak at 1.62 μm (not shown here). A conventional top-top contact processing for broad-area edge-emitting lasers is performed with different stripe widths ranging from 25 to 100 μm. The process involves an inductively coupled plasma reactive ion etching with BCl_3 into the n-GaSb contact layer, and Ti/Pt/Au metal evaporations for contact to both n- and p-GaSb. After the processing, the wafer is thinned to ~70 μm and cleaved to bar lengths of 1 mm. A thickness of ~70 μm is optimal for the better facet quality.

8.3.3 Etch-Pit Density Characterization of Laser Structures

EPD characterization of the GaSb buffer layer is carried out for evaluating the threading dislocation density of the fabricated

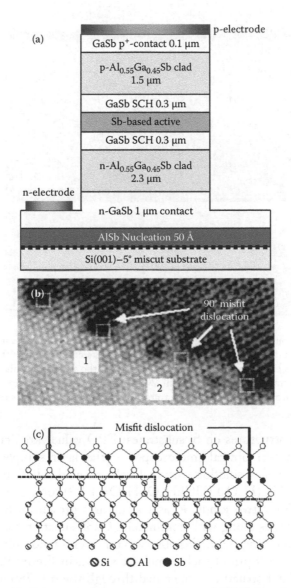

FIGURE 8.18 (a) Schematic illustration of fabricated III-Sb-based edge-emitting broad-area laser structures monolithically grown on miscut Si (001) substrates. (b) Cross-sectional TEM images of the interface between AlSb and Si. (c) Schematic illustration of the IMF interface between AlSb and miscut Si with a double atomic step. (From Tatebayashi, J. 2009. Monolithically Integrated III-Sb-Based Laser Diodes Grown on Miscut Si Substrates. IEEE Press 716. With permission. Huang, S. H., 2008. Simultaneous interfacial misfit array formation and antiphase domain suppression on miscut silicon substrate. American Institute of Physics 071102. With permission.)

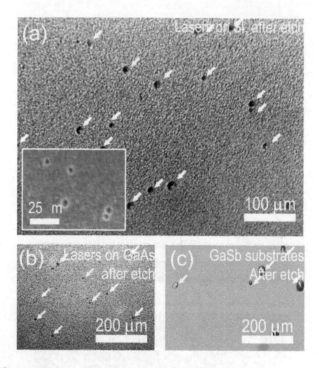

FIGURE 8.19 Surface morphology of (a) GaSb/AlSb on Si substrates, and (b) commercial GaSb substrates. Average EPDs are (a) 3.5×10^4 pits/cm², (b) 9.0×10^3 pits/cm², and (c) 2.7×10^3 pits/cm². Inset of (a) is an SEM image of the surface morphology of GaSb/AlSb on Si substrates. (From Tatebayashi, 2009. Monolithically Integrated III-Sb-Based Laser Diodes Grown on Miscut Si Substrates. IEEE Press 716. With permission.)

laser structures on Si substrates. EPD values are determined by Nomarski optical microscope on the laser samples immersed into an acid solution with hydrochloric acid (HCl) and hydrogen peroxide (H$_2$O$_2$) (HCl:H$_2$O$_2$:H$_2$O= 1:1:2) for 1 min at RT. The EPD for two kinds of laser structures grown on Si and GaAs substrates along with commercial GaSb substrates are estimated by averaging the EPDs observed on several randomly chosen 500 μm × 500 μm of each sample as shown in Figure 8.19. Both of the laser structures are etched through the n-GaSb contact layer (above ~5000 Å from the AlSb–Si interface) to evaluate the EPD at GaSb buffer layers and compare with commercial GaSb substrates. Figure 8.19a shows an EPD of ~3.5×10^4 pits/cm² from the laser samples grown on Si, which is at least 10 times smaller than the values reported previously by our group because of the use of 5° miscut Si substrates. This EPD value is approximately 4 times higher than that of the lasers on GaAs substrates (Figure 8.19b),

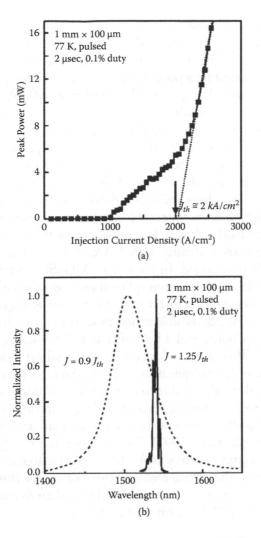

FIGURE 8.20 (a) L-I curve of the fabricated laser devices at 77 K under pulsed operation. (b) EL spectra above and below the threshold current density $J_{th} \cong 2$ kA/cm². (From Tatebayashi, J. 2009. Monolithically Integrated III-Sb-Based Laser Diodes Grown on Miscut Si Substrates. IEEE Press 716. With permission.)

and 13 times higher than that of commercial GaSb substrates (Figure 8.19c).

8.3.4 Device Characterization at 77 K

The L-I and EL spectra are shown in Figure 8.20. Lasing operation is observed at 77 K at a wavelength of 1.54 μm with a threshold current density (J_{th}) of 2 kA/cm² for a 1-mm-long device under pulsed conditions with 2 μs pulse width and a 0.1% duty cycle. A higher J_{th} compared with the same active grown on GaAs substrates (Mehta et al. 2007) is attributed to the increased mirror loss caused by a poor quality of cleaved facets, similar to that observed by other groups (Roelkens et al. 2006). This might be because it is difficult to make the facet of miscut Si substrates precisely perpendicular to the waveguide by just cleaving the substrates. Therefore, it would be necessary to polish the facet after cleaving to make a better facet quality. The maximum peak output power from the fabricated device is ~20 mW. The current-voltage (I-V) characteristics indicate a diode turn-on voltage of 0.7 V, which is consistent with a theoretical built-in potential of the laser diode. A very low resistance of 9.1 Ω and reverse bias leakage current density of 0.7 A/cm² at –5 V and 46.9 A/cm² at –15 V is obtained. In the GaSb–AlGaSb material system, the energy difference between split-off and top valence levels of GaSb exactly matches the direct bandgap photon energy, which results in a giant Auger recombination process that involves an electron, two heavy holes, and a hole of spin-orbit splitting band (CHHS process). This causes reduction of the modal gain and a substantial increase in J_{th} (Almuneau et al. 1999). An incorporation of indium into GaSb might be able to increase the compressive strain in the QW structure and suppress the Auger recombination process, which results in increase in the modal gain and reduction of J_{th} (Tatebayashi et al. 2007). Furthermore, use of QD or quantum dash structures such as InAs–InAlGaAs materials might be able to improve the device characteristics such as a threshold current density or temperature sensitivity due to their discrete sets of density of states (Balakrishnan et al. 2004).

8.4 Summary

We reviewed the fabrication, structural properties, and growth characterization of an IMF array between AlSb and Si, along with GaSb and GaAs, for the application to III-Sb-based photonic devices including edge-emitters and VCSELs monolithically grown on a Si(001) substrate. In the monolithic III-V growth on Si substrates, there are mainly three issues to be overcome: (1) lattice mismatch, (2) thermal expansion mismatch, and (3) formation of APDs. AlSb has a thermal expansion coefficient of 2.55 × 10⁻⁶/K, which is the closest to that of Si (2.59 × 10⁻⁶/K) compared with any other material systems such as GaAs, GaP, or

InP. First, the growth mechanisms of highly mismatched AlSb on Si are identified by using both AFM and TEM analyses. The initial nucleation occurs by self-assembled islands. The continuation of growth leads to coalescence of the islands followed by undulations in the AlSb. The undulations increase the surface area of the AlSb and provide strain relief. The undulating material contain the formation of IMF arrays parallel to the (100) plane; however, these do not propagate vertically as threading or screw dislocations. The AlSb layers provide a template for future monolithic integrated III-V based optoelectrical devices on Si substrates.

The use of 5° miscut Si substrates enables simultaneous IMF formation and suppression of APDs, resulting in reduction of dislocation density over the III-Sb growth and realization of electrically injected laser diodes operating at 77 K. The I-V characteristics indicate a diode turnon of 0.7 V, which is consistent with a theoretical built-in potential of the laser diode. This device is characterized by a 9.1 Ω forward resistance and a leakage current density of 0.7 A/cm^2 at -5 V. A substantial problem of the poor facet quality of Si substrates should be circumvented to obtain the CW RT operation of the GaSb QW lasers on Si substrates. Incorporation of indium into the GaSb active region can increase the compressive strain and suppress the Auger recombination process, which results in increase in the modal gain and reduction of J_{th}. The EPD characterization indicates a four times larger EPD value for the lasers on Si (3.5×10^4 pits/cm^2) than that of the lasers on GaAs (9.0×10^3 pits/cm^2), which might hinder RT lasing operation of lasers on Si. In order to achieve the RT lasing of lasers on Si, further optimization of the growth of AlSb on Si will be required to mitigate the formation of threading dislocations and APDs. Also, further analyses of the EPD characterization will be required for correlating the EPD properties with the I-V characteristics, including a leakage current density. It would be also required to polish the facet to make the facet precisely perpendicular to the waveguide in order to reduce mirror loss.

References

Adomi, K., Strite, S., Morkoc, H., Nakamura, Y., and Otsuka, N. 1991. Characterization of GaAs grown on Si epitaxial layers on GaAs substrates. *J. Appl. Phys.*, 69: 220–225.

Aindow, M., Cheng, T. T., Mason, N. J., Seong, T.-Y., and Walker, P. J. 1993. Geometry and interface structure of island and nuclei for GaSb buffer layers grown on (001) GaAs by metalorganic vapor-phase epitaxy. *J. Cryst. Growth*, 133: 168–174.

Almuneau, G., Genty, F., Wilk, A., Grech, P., Joullié, A., and Chusseau, L. 1999. GaInSb/AlGaAsSb strained quantum well semiconductor lasers for 1.55 µm operation. *Semicond. Sci. Technol.*, 14: 89–92.

Akahane, K., Yamamoto, N., Gozu, S., and Ohtani, N. 2004. Heteroepitaxial growth of GaSb on Si(001) substrates. *J. Cryst. Growth*, 264: 21–25.

Andre, C. L., Carlin, J. A., Boeckl, J. J., Wilt, D. M., Smith, M. A., Pitera, A. J., Lee, M. L., Fitzgerald, E. A., and Ringel, S. A. 2005. Investigations of high performance GaAs solar cells grown on Ge-SiGe-Si substrates. *IEEE Trans. on Electron Device*, 52: 1055–1060.

Babkevich, A. Yu, Cowley, R. A., Mason, N. J., and Stunault, A. 2000. X-ray scattering, dislocations and orthorhombic GaSb. *J. Phys. Condens. Matter*, 12: 4747–4756.

Balakrishnan, G., Huang, S. H., Rotter, T. J., Stintz, A., Dawson, L. R., Malloy, K. J., Xu, H., and Huffaker, D. L. 2004. 2.0-µm wavelength InAs quantum dashes grown on a GaAs substrate using a meta-morphic buffer layer. *Appl. Phys. Lett.*, 84: 2058–2060.

Balakrishnan, G., Huang, S., Dawson, L. R., Xin, Y.-C., Conlin, P., and Huffaker D. L. 2005. Growth mechanisms of highly mismatched AlSb on a Si substrate. *Appl. Phys. Lett.*, 86: 034105-1–3.

Balakrishnan, G., Huang, S. H., Khoshakhlagh, A., Jallipalli, A., Rotella, P., Amtout, A., Krishna, S., Haines, C. P., Dawson, L. R., and Huffaker, D. L. 2006a. Room-temperature optically-pumped GaSb quantum well based VCSEL monolithically grown on Si(100) sub-strate. *Electron. Lett.*, 42: 350–351.

Balakrishnan, G., Jallipalli, A., Rotella, P., Huang, S. H., Khoshakhlagh, A., Amtout, A., Krishna, S., Dawson, L. R., and Huffaker, D. L. 2006b. Room-temperature optically pumped (Al)GaSb vertical-cavity surface-emitting laser monolithically grown on an Si(100) substrate. *IEEE J. Sel. Top. Quant. Electron*, 12: 1636–1641.

Barbier, L., Khater, A., Salanon, B., and Lapujoulade, J. 1991. Observation of the double-step–single-step transition on a vicinal surface of Si(100). *Phys. Rev. B*, 43: 14730–14733.

Bowers, J. E., Fang, A. W., Park, H., Jones, R., Cohen, O., and Paniccia, M. J. 2007. Hybrid silicon evanescent photonic integrated circuits. *CLEO 2007*, CTuQ1.

Chang, J. C. P., Chin, T. P., and Woodall, J. M. 1996. Incoherent inter-face of InAs grown directly on GaP(001). *Appl. Phys. Lett.*, 69: 981–983.

Choi, H. K., Hull, R., Ishiwara, H., and Nemanich, R. J. 1988. Heteroepitaxy on silicon: Fundamentals, structure and devices. *Mat. Res. Soc.*, 116: Pittsburgh, PA.

Chriqui, Y., Saint-Girons, G., Bouchoule, S., Moison, J.-M., Isella, G., Von Kaenel, H., and Sagnes, I. 2003. Room temperature laser operation of strained InGaAs/GaAs QW structure monolithically grown by MOCVD on LE-PECVD Ge/Si virtual substrate. *Electron. Lett.*, 39: 1658–1659.

Deppe, D. G., Chand, N., Van der Ziel, J. P., and Zydzik, G. J. 1990. $Al_xGa_{1-x}As$-GaAs vertical-cavity surface-emitting laser grown on Si substrate. *Appl. Phys. Lett.*, 56: 740–742.

Deppe, D. G., Holonyak, Jr., N., Nam, D. W., Hsieh, K. C., Jackson, G. S., Matyi, R. J., Shichijo, H., Epler, J. E., and Chung, H. F. 1987. Room temperature continuous operation of p-n $Al_xGa_{1-x}As$-GaAs quantum well heterostructure lasers grown on Si. *Appl. Phys. Lett.*, 51: 637–639.

Deppe, D. G., Holonyak Jr., N., Hsieh, K. C., Nam, D. W., Plano, W. E., Matyi, R. J., and Shichijo, H. 1988. Dislocation reduction by impurity diffusion in epitaxial GaAs grown on Si. *Appl. Phys. Lett.*, 52: 1812–1814.

Egawa, T., Hasegawa, Y., Jimbo, T., and Umeno, M. 1994. Room-temperature pulsed operation of AlGaAs/GaAs vertical-cavity surface-emitting laser-diode on Si substrate. *IEEE Photon. Technol. Lett.*, 6: 681–683.

Egawa, T., Tada, H., Kobayashi, Y., Soga, T., Jimbo, T., and Umeno, M. 1990. Low-threshold continuous-wave room-temperature operation of $Al_xGa_{1-x}As$/GaAs single quantum well lasers grown by metalorganic chemical vapor deposition on Si substrates with SiO_2 back coating. *Appl. Phys. Lett.*, 57: 1179–1181.

Fan, J. C. C. and Poate, J. M. 1986. Heteroepitaxy on silicon. *Mat. Res. Soc.*, 67: Pittsburgh, PA.

Groenert, M. E., Leitz, C. W., Pitera, A. J., Yang, V., Lee, H., Ram, R. J., and Fitzgerald, E. A. 2003. Monolithic integration of room-temperature cw GaAs/AlGaAs lasers on Si substrates via relaxed graded GeSi buffer layers. *J. Appl. Phys.*, 93: 362–367.

Fang, A. W., Park, H., Jones, R., Cohen, O., Paniccia, M. J., and Bowers, J. E. 2006. A continuous-wave hybrid AlGaInAs–silicon evanescent laser. *IEEE Photon. Technol. Lett.*, 18: 1143–1145.

Huang, S. H., Balakrishnan, G., Khoshakhlagh, A., Jallipalli, A., Dawson, L. R., and Huffaker, D. L. 2006. Strain relief by periodic misfit arrays for low defect density GaSb on GaAs. *Appl. Phys. Lett.*, 88: 131911-1–3.

Huang, S. H., Balakrishnan, G., Mehta, M., Khoshakhlagh, A., Dawson, L. R., and Huffaker D. L. 2007. Epitaxial growth and formation of interfacial misfit array for tensile GaAs on GaSb. *Appl. Phys. Lett.*, 90: 161902-1–3.

Hwang, I., Lee, C., Kim, J.-E., and Park, H.Y. 1995. Clustering effect and residual stress in $In_xGa_{1-x}As$/GaAs strained layer grown by metalorganic chemical-vapor deposition. *Phys. Rev. B*, 51: 7894–7897.

Iga, K. 2000. Surface-emitting laser - Its birth and generation of new optoelectronics field. *IEEE J. Sel. Top. Quantum Electron.*, 6: 1201–1215.

Jallipalli, A., Balakrishnan, G., Huang, S. H., Khoshakhlagh, A., Dawson, L. R., and Huffaker, D. L. 2007. Atomistic modeling of strain distribution in self-assembled interfacial misfit dislocation (IMF) arrays in highly-mismatched III-V semiconductor materials. *J. Cryst. Growth*, 303: 449–455.

Jin-Phillipp, N. Y., Sigle, W., Black, A., Babic, D., Bowers, J. E., HuQuest, E. L., and Rühle, M. 2001. Interface of directly bonded GaAs and InP. *J. Appl. Phys.*, 89: 1017–1029.

Jonsdottir, F. 1995. Computation of equilibrium surface fluctuations in strained epitaxial-films due to interface misfit dislocation. *Modelling Simul. Mater. Sci. Eng.*, 3: 503–520

Kang, J. M., Min, S.-K., and Rocher, A. 1994a. Asymmetric tilt interface induced by 60° misfit dislocation arrays in GaSb/GaAs(001). *Appl. Phys. Lett.*, 65: 2954–2956.

Kang, J. M., Nouaoura, M., Lassabatere, L., and Rocher, A. 1994b. Accommodation of lattice mismatch and threading of dislocations in GaSb films grown at different temperature on GaAs(001). *J. Cryst. Growth*, 143: 115–123.

Kawanami, H., Sakamoto, T., Takahashi, T., Suzuki, E., and Nagai, K. 1982. Heteroepitaxial growth of GaP on a Si(100) substrate by molecular beam epitaxy. *Jpn. J. Appl. Phys. Part 2*, 21: L68–L70.

Kim, J -H., Seong, T-Y., Mason, N. J., and Walker, P. J. 1998. Morphology and defect structures of GaSb islands on GaAs grown by metalorganic vapor phase epitaxy. *J. Electron. Mater.*, 27: 466–471.

Kim, Y. H., Lee, J. Y., Noh, Y. G., Kim, M. D., Cho, S. M., Kwon, Y. J., and Oh, J. E. 2006. Growth mode and structural characterization of GaSb on Si(001) substrate: A transmission electron microscopy study. *Appl. Phys. Lett.*, 88: 241907-1–3.

Kumar, V. and Sastry, B. S. R. 2001. Thermal expansion coefficient of binary semiconductors. *Cryst. Res. Technol.*, 36: 565–569.

Kuronen, A., Kaski, K., Perondi, L. F., and Rintala, J. 2001. Atomistic modelling of interaction between dislocations and misfit interface. *Europhys. Lett.*, 55: 19–25.

Kwon, O., Boeckl, J. J., Lee, M. L., Pitera, A. J., Fitzgerald, E. A., and Ringel, S. A. 2006. Monolithic integration of AlGaInP laser diodes on SiGe/Si substrates by molecular beam epitaxy. *J. Appl. Phys.*, 100: 013103-1–7.

Lin, H. C., Chang, K. L., Pickrell, G. W., Hsieh, K. C., and Cheng, K. Y. 2002. Low temperature wafer bonding by spin on glass. *J. Vac. Sci. Technol*, B, 20: 752–754.

Mallard, R. E., Wilshaw, P. R., Mason, N. J., Walker P. J., and Booker, G. R. 1989. *Microscopy of Semiconducting Materials 1989 (Inst. Phys. Conf. Ser. 100)*. Bristol: Institute of Physics Publishing, 331–336.

Matthews, J. W. and Blakeslee, A. E. 1974. Defects in epitaxial multilayers. *J. Cryst. Growth*, 27: 118–125.

Mehta, M., Balakrishnan, G., Huang, S., Khoshakhlagh, A., Jallipalli, A., Patel, P., Kutty, M. N., Dawson, L. R., and Huffaker, D. L. 2006. GaSb quantum-well-based "buffer-free" vertical light emitting diode monolithically embedded within a GaAs cavity incorporating interfacial misfit arrays. *Appl. Phys. Lett.*, 89: 211110-1–3.

Mehta, M., Jallipalli, A., Tatebayashi, J., Kutty, M. N., Albrecht, A., Balakrishnan, G., Dawson, L. R., and Huffaker, D. L. 2007. Room-temperature operation of buffer-free GaSb-AlGaSb quantum-well diode lasers grown on a GaAs platform emitting at 1.65 µm. *IEEE Photon. Technol. Lett.*, 19: 1628–1630.

Mi, Z., Bhattacharya, P., Yang, J., and Pipe, K. P. 2005. Room-temperature self-organised $In_{0.5}Ga_{0.5}As$ quantum dot laser on silicon. *Electron. Lett.*, 41: 742–743.

Mi, Z., Yang, J., Bhattacharya, P., and Huffaker, D. L. 2006. Self-organised quantum dots as dislocation filters: The case of GaAs-based lasers on silicon. *Electron. Lett.*, 42: 121–122.

People, R. and Bean, J. C. 1985. Calculation of critical layer thickness versus lattice mismatch for Ge_xSi_{1-x}/Si strained-layer heterostrucutres. *Appl. Phys. Lett.*, 47: 322–324.

Qian, W., Skowronski, M., Kaspi, M., R., Graef, M. D., and Dravid, V. P. 1997. Nucleation of misfit and threading dislocations during epitaxial growth of GaSb on GaAs(001) substrates. *J. Appl. Phys.*, 81: 7268–7272.

Razeghi, M., Defour, M., Blondeau, R., Omnes, F., Maurel, P., Acher, O., Brillouet, F., C-Fan, J. C., and Salerno, J. 1988. First cw operation of a $Ga_{0.25}In_{0.75}As_{0.5}P_{0.5}$-InP laser on a silicon substrate. *Appl. Phys. Lett.*, 53: 2389–2390.

Rocher, A. M. 1991. Interfacial dislocation in the GaSb/GaAs (001) heterostructure. *Solid State Phenom.*, 19/20: 563–572.

Roelkens, G., Van Thourhout, D., Baets, R., Nötzel, R., and Smit, M. 2006. Laser emission and photodetection in an InP/InGaAsP layer integrated on and coupled to a Silicon-on-Insulator waveguide circuit. *Opt. Exp.*, 14: 8154–8159.

Sieg, R. M., Ringel, S. A., Ting, S. M., Samavedam, S. B., Currie, M., Langdo, T., and Fitzgerald, E.A. 1998. Toward device-quality GaAs growth by molecular beam epitaxy on offcut Ge/Si1-xGex/ Si substrates. *J. Vac. Sci. Tech. B*, 16: 1471–1474.

Sugo, M., Mori, H., Tachikawa, M., Itoh, Y., and Yamamoto, M. 1990. Room-temperature operation of an InGaAsP double-heterostructure laser emitting at 1.55 µm on a Si substrate. *Appl. Phys. Lett.*, 57: 593–595.

Tanoto, H., Yoon, S. F., Loke, W. K., Fitzgerald, E. A., Dohrman, C., Narayanan, B., and Tung, C. H. 2005. Growth of GaAs on (100) Ge and vicinal Ge surface by migration enhanced epitaxy. *Mater. Res. Soc. Symp. Proc.*, 891: 0891-EE03-17.

Tatebayashi, J., Jallipalli, A., Kutty, M. N., Huang, S. H., Balakrishnan, G., Dawson, L. R, and Huffaker, D. L. 2007. Room-temperature lasing at 1.82 μm of GaInSb/AlGaSb quantum wells grown on GaAs substrates using an interfacial misfit array. *Appl. Phys. Lett.*, 91: 141102-1–3.

Teufel, S. 2003. Adiabatic Perturbation Theory in Quantum Dynamics. Springer.

Trampert, A., Tournie, E., and Ploog, K. H. 1995. Novel plastic strain-relaxation mode in highly mismatched III-V layers induced by two-dimensional epitaxial growth. *Appl. Phys. Lett.*, 66: 2265–2267.

Van der Merwe, J. H. 2002. Misfit dislocations in epitaxy. *Metall. Mater. Trans. A*, 33: 2475–2483.

Van der Ziel, J. P., Dupuis, R. D., Logan, R. A., Mikulyak, R. M., Pinzone, C. J., and Savage, A. 1987. Low threshold pulsed and continuous laser oscillation from AlGaAs/GaAs double heterostructures grown by metalorganic chemical vapor deposition on Si substrates. *Appl. Phys. Lett.*, 50: 454–456.

Wada, H. and Kamijoh, T. 1994. Effect of heat-treatment on bonding properties in InP-to-Si direct water bonding. *Jpn. J. Appl. Phys.*, 33: 4878–4879.

Wada, H., and Kamijoh, T., 1996. Room-temperature CW operation of InGaAsP lasers on Si fabricated by wafer bonding. *IEEE Photon. Technol. Lett.*, 8: 173–175.

Wang, W. I. 1984. Molecular beam epitaxial growth and material properties of GaAs and AlGaAs on Si(100). *Appl. Phys. Lett.*, 44: 1149–1151.

Williams, D. B. and Carter, C. B. 1996. *Transmission Electron Microscopy*. New York: Kluwer Academic/Plenum Publishers.

Williams, D. B. and Carter, C. B. 1996. *Transmission Electron Microscopy*. New York: Kluwer/Plenum.

CHAPTER 9

Hybrid III-V Lasers on Silicon

Jun Yang, Zetian Mi, and Pallab Bhattacharya

Contents

9.1 Introduction

9.1.1 Need for Si Photonics

There has been a growing need for the monolithic integration of electronic and photonic components on the same chip for future high-speed and multifunctional systems. With the dense packing of active and passive elements in IC chips, metallic interconnects are fast becoming a serious bottleneck in the evolution of modern information systems. Chip-level optical interconnects are therefore required for future low-cost, high-performance, and high-productivity computing systems, due to their ultralow power consumption, low latencies, unprecedented large bandwidth, and reduced crosstalk and heat dissipation problems, compared to the conventional metal interconnects [1,2]. However, the materials and fabrication technology for conventional photonic components are generally not compatible with the existing Si infrastructure. In this regard, it is critically important to develop various active and passive photonic components, including high-performance light emitters, preferably lasers, waveguides and guided wave devices, modulators, and detectors on a Si platform that can leverage the existing manufacturing infrastructure for CMOS electronics. The development of silicon-on-insulator (SOI) substrates has further accelerated the progress of Si photonics [3]. In this chapter, we describe the recent advancements in III-V lasers and integrated optoelectronics on a Si platform, with a focus on the use of self-organized In(Ga)As/GaAs quantum dots as the laser active region.

9.1.2 Different Techniques Used for Si Photonics

In what follows, we first briefly describe the recent progress as well as the various techniques used for Si-based active and passive photonic components, including waveguides, modulators, photodetectors, and lasers.

An essential component for optical interconnects on Si-based CMOS chips is a light guide. Both single crystalline and hydrogenated amorphous Si waveguides have been developed [4–8]. While the former approach leads to reasonably low optical loss, the latter offers additional benefits in terms of lower cost and low-temperature processing. The propagation loss in Si arises primarily from waveguide surface roughness and intrinsic material absorption. Due to the very large index contrast between the core and cladding, Si nanophotonic rib waveguides can provide strong subwavelength optical confinement. With the development of microfabrication and SOI technologies, single crystalline Si waveguides can now exhibit very low optical loss ($\leq$0.5 dB/cm) in the wavelength range of 1.3–1.55 μm [4].

A Si-based modulator is required to encode the optical carrier wave with information. Various Si modulators, including Mach–Zehnder [9] and waveguide [10] modulators, have been demonstrated, which generally rely on the use of free carrier electro-optic effects in Si [11]. To overcome the large footprint (a few millimeters) associated with the conventional Mach–Zehnder modulators, nanophotonic rib waveguide-based Mach–Zehnder modulators have been developed, in which an embedded p^+-i-n^+ diode is used for charge injection [12,13]. Due to the simultaneous confinement of both carriers and photons in a very small (~0.12 μm^2) region, these devices exhibit large modulation speed (~10 Gb/s), high modulation efficiency, and low RF power consumption [13]. An alternate approach to achieve high-speed Si modulators involves the use of high-quality-factor microring resonators [14]. A modulation speed of ~18 Gb/s has recently been reported [15]. Additionally, strained Si/Ge heterostructures can exhibit a quantum-confined Stark effect (QCSE), which is the mechanism responsible for the ultrahigh-speed operation of III-V modulators [16]. An early investigation of strained Si/Ge electroabsorption modulators has employed SiGe-Si quantum wells [17]. Recently, a strong QCSE based on the direct-gap absorption has been observed in strained Ge quantum wells with SiGe barriers [18], leading potentially to a novel type of Si-based electroabsorption modulators with performance comparable to their III-V counterparts.

High-speed and highly sensitive on-chip photodetectors have also been developed [19–22]. In these devices, Ge is often incorporated as the active region for the detection of photons at 1.3–1.55 μm. Due to the relatively large (~4%) lattice mismatch between Ge and Si, both the growth and fabrication processes have to be carefully controlled in order to minimize the generation of dislocations and defects in the Ge detector active region. Relatively high efficiency (>90%) and high responsivity (~1.08 A/W) have been demonstrated in Ge photodetectors on Si [19]. To achieve both high speed and high quantum efficiency, waveguide-based Ge p-i-n photodetectors have also been developed. These detectors are characterized by a responsivity of ~0.89 A/W, dark current of ~169 nA, and optical bandwidth of over 30 GHz at 1.55 μm [22].

Si, due to its indirect bandgap, is a poor light emitter, in spite of the various techniques used to coax it to emit light. By using quantum confinement [23–25] and Raman scattering effects [26,27], an optically pumped Si Raman laser and impressive optical gain have been achieved. However, until the achievement of an electrically injected all-Si laser, the hybrid integration of a GaAs- or InP-based semiconductor laser on Si may have to be considered. In this regard, various integration strategies, including heteroepitaxial growth and wafer bonding, have been developed. As also covered in another chapter in this book, heteroepitaxial growth has been extensively investigated with the use of strained layer superlattices (SLS) [28,29], relaxed and graded SiGe layers [30], thermal cycle annealing [31], selective-area growth [32] and lateral overgrowth [33] on patterned substrates, special buffer layers including AlSb [34] and $SrTiO_3$ [35], and more recently quantum dot dislocation filters [36,37]. The wafer bonding approach generally relies on the direct die bonding of the whole III-V laser structure onto a Si wafer [38,39] or the evanescent coupling of III-V gain media onto a Si waveguide cavity [40]. Room-temperature operational, electrically injected III-V lasers on Si have been demonstrated with these techniques [28–31,36–40], and significant efforts have also been devoted to further improving their performance and reliability.

9.1.3 Direct Growth of GaAs-Based Lasers on Si

The first demonstration of electrically injected III-V semiconductor lasers on Si dates back to the 1980s [41,42]. These devices were grown directly on Si with Ge [41] or thick GaAs [42] buffer layers. Later, improved performance had been achieved with the incorporation of a $Ga(As)P/GaAs_{0.5}P_{0.5}$ [28] and InGaAs/GaAs [29, 43] SLS in the buffer layers. In what follows, we briefly review the early demonstrations of GaAs-based lasers on Si by direct growth and also discuss the fundamental challenges for achieving high-performance III-V lasers on Si using this approach.

GaAs layers grown directly on Si(001) substrates generally exhibit very large densities of dislocations and antiphase domains (APDs), due to the large lattice mismatch (~4.1%) and the polar/nonpolar surface incompatibility. Fischer et al. developed several special techniques to drastically reduce the threading dislocation

densities [43]. It was observed that, with the use of Si(001) substrates misoriented 2°–4° toward [011], misfit dislocations with
their Burgers vector lying in the plane of the substrate orientation were preferentially generated, due to the presence of surface steps. This type of dislocation can be largely confined at the
GaAs–Si misfit interface. Consequently, dislocation densities in
the device active region can be greatly reduced. The use of such
offcut Si substrates, with their surfaces featuring predominately
double-atomic-layer steps, can also significantly enhance the self-
annihilation of antiphase boundaries at the III-V and Si [44]. In
addition, migration-enhanced epitaxy, in conjunction with low-
temperature growth, was utilized to further reduce the formation of antiphase domains. To reduce dislocation densities in the
laser active region, Fischer et al. investigated the use of GaAs/
InGaAs SLS. From detailed transmission electron microscopy
(TEM) studies, it was observed that threading dislocations can be
reflected or terminated at the strained superlattice. With the use
of these special techniques, GaAs/(Al,Ga)As double heterojunc-
tions lasers were demonstrated on Si(001) substrates under pulsed
operation (0.02% duty cycle) at room temperature [29]. However,
these devices exhibited extremely large threshold current (J_{th} >
7000 A/cm²) and low output slope efficiency (~0.12 W/A) [29].
The pioneering work has greatly inspired the subsequent developments of GaAs- and InP-based lasers on Si. The special techniques described above have also been widely used in the epitaxial
growth of a host of lattice mismatched heterostructures.

9.2 Review of the Recent Developments of III-V Lasers on Si

In order to achieve high-performance and highly reliable III-V
lasers on Si, many different approaches have been developed. To
date, highly promising results have been achieved using relaxed
and graded SiGe layers [30], AlSb buffer [34], evanescent coupling [40], and self-organized quantum dots as the gain region
and dislocation filter [36,37]. In what follows, we briefly review
the present status and challenges of these approaches, as well
as the recent development of high-performance self-organized
quantum dot lasers.

Virtual Ge substrates on Si have been developed to address
issues associated with the large lattice mismatch between III-V
materials and Si [30,45–47]. In this approach, a SiGe buffer layer
is compositionally graded from pure Si to Ge, with a typically

reported grading rate of ~10% Ge/μm. Since Ge is nearly lattice matched to GaAs, the generation of threading dislocations can be greatly minimized. In addition, special growth techniques, including migration-enhanced epitaxy, low-temperature growth, and As prelayer, have been utilized to further reduce the formation of antiphase domains near the GaAs–Ge interface [45–47]. Under optimized growth conditions, GaAs grown on SiGe buffer layers on Si can exhibit relatively low threading dislocation densities (~1 × 10^6 cm^-2) and very long carrier lifetime (>10 ns) [47]. Utilizing this approach, room-temperature operational AlGaInP (λ ~ 680 nm) [47] and GaAs/AlGaAs (λ ~ 858 nm) [30] laser diodes on Si have been reported. The measured threshold current densities are in the range of ~580–1650 A/cm^2, which are largely limited by the large optical scattering loss induced by the rough surface as well as the presence of threading dislocations in the laser active region. It is further believed that the use of more defect-resistant InGaAs quantum well or dot active regions may drastically improve the laser performance. Other drawbacks of this approach include the difficulty in achieving long-wavelength (1.3–1.55 μm) emission and the use of relatively thick (~10 μm) SiGe buffer layers, which may hinder their on-chip integration with CMOS electronics and other active and passive photonic components.

Another technique to effectively relieve lattice-mismatch induced strain is to use interfacial misfit arrays, which have been observed in a range of lattice-mismatched materials systems [34,48–51]. Such interfacial misfit arrays are strongly localized at the III-V–Si interface and, through their interactions with the 90° dislocations, can greatly reduce the heterointerface strain [34]. With the use of this technique, III-Sb heterostructures grown on Si can exhibit relatively low (~1 × 10^6 cm^-2) defect densities. III-Sb-based optically pumped vertical-cavity surface-emitting lasers (VCSELs) at room temperature and edge-emitting laser diodes at 77 K have been realized on Si substrates [52,53]. The emission wavelengths of these lasers on Si are at ~1.55 μm. However, due to the presence of antiphase domains and the associated carrier non-radiative recombination, these devices exhibit a very large threshold current (>2 kA/cm^2) at 77 K. In addition, the achievement of room-temperature electrically injected III-Sb lasers on Si has not been reported.

Evanescent coupled III-V lasers on Si have also been demonstrated [40,54]. In this approach, an AlGaInAs quantum well laser is bonded to a Si strip waveguide on SOI substrates, wherein optical modes in both regions are evanescently coupled and, as a result, the laser cavity is largely defined by the Si waveguide [40]. The III-V laser heterostructure is first grown on InP substrates and then transferred onto the patterned Si wafer using a low-temperature process. Such devices are self-aligned to the underlying Si waveguides and are suitable for large-scale integration. The devices are characterized by an output power of ~1–10 mW, external quantum efficiency of ~10%, and highly temperature-sensitive operation [40,54]. Photodetectors and modulators can also be realized by optimizing the dimensions of the Si waveguides and the III-V heterostructures with this hybrid architecture. A potential drawback of this approach is the direct contact of the laser (or other devices) active region with the III-V/Si heterointerface, which poses a significant concern for the device performance and, more importantly, the device reliability.

In spite of the extensive research efforts, conventional III-V quantum well lasers on Si still exhibit relatively poor performance, limited, to a large extent, by the propagation of high densities of dislocations into the laser active region, due to the large lattice mismatch (4.1%), large thermal coefficient difference (250%), and polar/nonpolar surface incompatibility between GaAs and Si. Recently, significant progress has been made in the development of quantum dot optoelectronics [55]. Due to the three-dimensional confinement of carriers and the resulting near-discrete density of states, quantum dot lasers promise significantly improved performance and reliability, compared to the conventional quantum well lasers. Self-organized In(Ga)As quantum dot lasers grown on GaAs or InP substrates have recently demonstrated near-ideal performance characteristics, that is, ultra-low threshold current ($J_{th} \leq 20$ A/cm^2) [56,57], large output power (11.7 W) [58], temperature-invariant operation ($T_0 \cong \infty$) [59], large small signal modulation bandwidth ($f_{-3dB} \approx 24.5$ GHz) [60], and near-zero chirp and α-parameters [60–62]. Self-organized quantum dots are coherently strained, and the strain field surrounding a quantum dot can suppress, or prevent, the propagation of dislocations [36]. Therefore, lasers incorporating quantum dots as the active region grown on Si may exhibit significantly improved lifetime, in addition to high performance characteristics.

9.3 Growth of Device Quality Quantum Dots on GaAs and Si

Coherently strained and nearly defect-free quantum dots can be formed in the Stranski–Krastanow growth mode. Both molecular beam epitaxy (MBE) and metal organic chemical vapor phase deposition (MOCVD) have been utilized for the fabrication of self-organized quantum dot heterostructures. In the Stranski–Krastanow growth mode, the transition from a layer-by-layer growth to the formation of three-dimensional islands is governed by the interplay between the interface energy and strain energy. It is determined, from energy minimization considerations in a unit cell of a strained heterostructure, that the island growth mode is preferred if the lattice mismatch is ~1.8%, or larger [63]. The critical layer thickness, corresponding to the onset of island formation, is largely determined by the lattice mismatch and is ~1.7 ML and 5 ML for InAs and $In_{0.5}Ga_{0.5}As$ layers grown on (100) GaAs, respectively. Above the critical layer thickness, elastic strain relaxation occurs via the formation of coherently strained, defect-free islands. Studies on InGaAs/GaAs quantum dots further confirm that the three-dimensional islands are coherently strained and free of dislocations [64–66].

InAs/GaAs quantum dots are typically of pyramidal shape. The sizes of quantum dots, which are generally in the range of 3–10 nm in height and 10–30 nm in base width, can be controlled by varying the growth conditions, including the growth temperature, growth rate, and III/V flux ratio as well as the amount of materials deposited. A reduced growth rate and an increased growth temperature generally lead to quantum dots with larger sizes and smaller areal densities. Illustrated in Figure 9.1a is the atomic force microscopy image of an InAs quantum dot layer grown on GaAs substrate [67]. Dot densities in the range of ~1.0 × 10⁸ cm⁻² to 1.0 × 10¹¹ cm⁻² can be measured for InAs quantum dots grown on GaAs [68]. The emission wavelengths of quantum dot heterostructures are largely determined by the dot size and composition as well as the bandgap of the surrounding barrier layers. For example, the emission wavelength of self-organized InAs quantum dots can be continuously tuned from visible to ~2.0 μm by controlling the dot size and composition or using different barrier layers [69–72]. Shown in Figure 9.1b are the room-temperature photoluminescence emission spectra of InAs/InGaAs quantum dots grown on GaAs substrates, wherein dif-

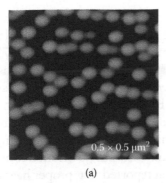

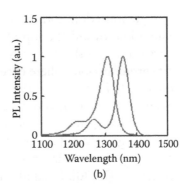

(a)

(b)

FIGURE 9.1 (a) AFM image of an uncapped InAs metamorphic quantum dot layer grown under optimized conditions. (From Z. Mi et al., *J. Crystal Growth* 301: 923–926, April 2007. Reprinted with permission from Elsevier.) (b) Room-temperature photoluminescence spectra of self-assembled InAs/InGaAs quantum dots incorporating two different In compositions in the InGaAs capping layers.

ferent emission wavelengths are achieved by altering the indium composition in the InGaAs capping layers.

In the Stranski–Krastanow growth mode, island formation is kinetically controlled based on arguments of minimization of crystal free energy by the growth front. There is therefore a distribution in the size of the islands in a given layer and also a randomness in their spatial ordering. The size distribution gives rise to an inhomogeneous broadening of optical transitions, which generally limits the maximum achievable gain of a quantum dot laser. Various techniques, including vertical coupling of dot layers [73,74] and the use of buried stressor dot layers [75–77], have been developed to reduce the transition linewidths. These techniques of "strain patterning" rely on the fact that the strain field around the quantum dots in one layer alters the adatom migration rates of subsequent dot layers such that larger and more uniform dots are formed on these layers. The best results, in terms of dot uniformity, have been obtained with a quantum dot bilayer system [75,76], in which the "stressor dots" in the first layer act as a template that influences the growth kinetics, formation, and characteristics of the second layer of "active dots." Under optimized growth conditions, bilayer InAs/GaAs quantum dot heterostructures exhibit extremely narrow spectral linewidths of 10.6 and 17.5 meV at 20 K and 300 K, respectively [78].

For practical applications, multiple quantum dot layers are generally incorporated in laser heterostructures to achieve large gain. Due to the large strain field associated with 1.3 and 1.55

μm quantum dots and the subsequent strain-driven indium atom migration toward the top of the dots, severe surface undulations develop in the subsequently grown GaAs capping layer [75,79]. At certain regions where excess strain builds up, dislocations may eventually form during the growth of GaAs barriers and subsequent InAs quantum dot layers. Recent reports suggested that the use of high growth temperature spacer layers or *in situ* annealing can greatly reduce the formation of dislocations during the growth of multiple layers of 1.3 and 1.55 μm quantum dot heterostructures [79,80]. It has also been reported that proper heat treatment can evaporate defect regions in InGaAs/GaAs quantum dot heterostructures, leading to improved crystal quality [81].

Since the demonstration of the first room-temperature operational quantum dot lasers, the growth and characterization of quantum dot heterostructures on Si substrates have also been intensively investigated [82–87]. In these studies, (001)-oriented Si substrates misoriented 2°–6° toward ⟨111⟩ or ⟨011⟩ are utilized to eliminate the formation of antiphase domains and stacking faults at the III-V and Si misfit interfaces [44,88]. Self-organized quantum dots can be directly nucleated on Si substrates or grown on suitable buffer layers on Si. In the latter case, GaAs or SiGe metamorphic buffer layers are first grown, followed by the deposition of quantum dot heterostructures. (In, Ga)As quantum dots grown on Si, with emission wavelengths ranging from ~1.0 to 1.3 μm, have been demonstrated [82–86]. Their characteristics, including size, density, uniformity, and emission wavelength, are comparable to those grown on GaAs substrates and can be precisely controlled by varying the growth conditions. However, due to the presence of large densities of nonradiative recombination centers in the surrounding GaAs layers on Si, the room-temperature photoluminescence intensity is ~10%, or less, compared to that grown on GaAs substrates. Therefore, in order to achieve quantum dot lasers on Si with performance comparable to those grown on GaAs, it is critically important to develop special techniques that can further reduce the defect densities in the laser active region.

9.4 Multilayer Self-Organized Quantum Dot Dislocation Filters

The incorporation of strained layers near the lattice-mismatched interface can bend the propagation of dislocations by enhancing their annihilation or sliding to the sample edge [43,89,90].

Incorporation of SLS in the buffer layers as dislocation filters has led to the demonstrations of III-V quantum well lasers on Si since the 1980s [28,29]. In the recent studies of high-performance self-organized In(Ga)As quantum dot lasers directly grown on Si [36,37], a thin (≤2 μm) GaAs buffer layer is first grown using MOCVD on (001)-oriented Si substrates misoriented 4° toward ⟨111⟩. The thin GaAs buffer layer exhibits dislocation densities of ~2–5 × 10⁷ cm⁻². Self-organized InAs quantum dot dislocation filters and quantum dot laser heterostructures are subsequently grown by MBE. Compared to conventional SLS, self-organized quantum dots [63,64] offer a stronger and anisotropic strain that can bend dislocations much more effectively. In addition, the localized strain region around the quantum dots can trap point defects of opposite strain. Therefore, self-organized quantum dots incorporated in the buffer layer can significantly enhance the filtering and blocking of dislocation and defects [36]. Defect-density reduction in GaN-on-sapphire material systems due to the presence of quantum dots has been observed [91]. In this section, the mechanism of quantum dot dislocation filtering, via dislocation bending by the strained islands, is theoretically analyzed, followed by the experimental characterization using TEM and photoluminescence (PL).

9.4.1 Design of Quantum Dot Dislocation Filters

A quasi-three-dimensional model based on strain relaxation is used to analyze dislocation bending by quantum dots [36]. The coherently strained, defect-free self-organized quantum dots are assumed to have a pyramidal shape. Figure 9.2 schematically shows the bending of a propagating dislocation, initially generated at the lattice mismatched interface, by the dot. In this process, a segment of misfit dislocation is generated, which glides underneath the island. Bending occurs only when the

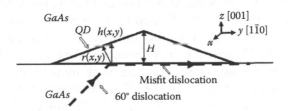

FIGURE 9.2 Cross-sectional schematic description of the bending of a 60° dislocation by a quantum dot. (From J. Yang et al., *IEEE Trans. Electron Dev.* 54(11): 2849–2855, November 2007. Reprinted with permission from IEEE.)

strain energy released due to the generation of the misfit dislocation (ΔE_{rel}) is greater than the dislocation self-energy (ΔE_{dis}) [92,93], where ΔE_{rel} and ΔE_{dis} for the dislocation of unit length are described as

$$\Delta E_{rel} = \frac{2G_{dot}(1+v)}{(1-v)} f_{eff} h b \cos\alpha \qquad (9.1)$$

$$\Delta E_{dis} = \frac{1}{2\pi} \frac{G_{buff} G_{dot}}{G_{buff} + G_{dot}} b^2 \left(\frac{1 - v\cos^2\beta}{1-v} \right) \left[\ln\left(\frac{2r}{b} \right) + 1 \right] \qquad (9.2)$$

Here, ΔE_{rel} and ΔE_{dis} are determined by the material properties, including the lattice mismatch f, Poisson ratio v, and the Young's modules G_{buff} and G_{dot} for the GaAs buffer and quantum dots, respectively. ΔE_{rel} and ΔE_{dis} are also related to the type, magnitude, and orientation of misfit dislocations, which are characterized by Burgers vector b and its orientation angles α and β with respect to the layer interface and dislocation line, respectively. Due to the island geometry and anisotropic strain distribution of quantum dots, ΔE_{rel} and ΔE_{dis} also depend on the position *(x, y)* where the dislocation encounters the island. For simplicity, the following assumptions are used: (1) the island has a uniform strain distribution with an effective lattice mismatch f_{eff}; (2) the thickness of the strain-released layer is the local height *h(x, y)* of the island; and (3) the outer cut-off radius of the dislocation strain field is the distance *r(x, y)* to the nearest facet of the island. Based on these assumptions, the bending area underlying the quantum dot, within which a propagating dislocation is bent when $\Delta E_{rel} > \Delta E_{dis}$, can be estimated using Equations 9.1 and 9.2. The bending area ratio, defined as the bending area of a single quantum dot divided by its base area, has been calculated for InAs, $In_{0.5}Ga_{0.5}As$, and $In_{0.6}Al_{0.4}As$ quantum dots. As shown in Figure 9.3a, InAs quantum dots, due to the relatively large dot size and strain field, are much more effective in bending dislocations compared to the other two types of dots. InAs dots, which have a typical base width of 20–30 nm, exhibit an effective bending-area ratio of 70%–80%. With a dot density of 2–4×10^{10} cm^{-2}, a single layer of InAs quantum dots has a bending-area ratio of ~ 10%–20%.

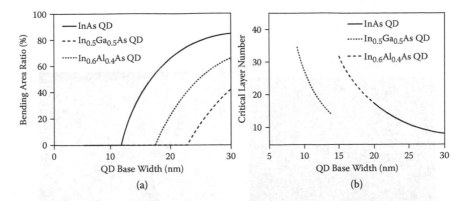

FIGURE 9.3 (a) Bending area ratio versus quantum dot base width for InAs, $In_{0.5}Ga_{0.5}As$, and $In_{0.6}Al_{0.4}As$ dots; (b) calculated critical layer number for the generation of single-kink dislocation loop for InAs, $In_{0.5}Ga_{0.5}As$, and $In_{0.6}Al_{0.4}As$ quantum dot multilayers. (From J. Yang et al., *IEEE Trans. Electron Dev.* 54(11): 2849–2855, November 2007. Reprinted with permission from IEEE.)

It is further expected that multiple, uncoupled quantum dot layers can significantly enhance the effect of dislocation bending and filtering, due to their nearly random spatial locations. However, the barrier thickness and quantum dot layer number have to be carefully optimized in order to avoid the generation of single-kink or double-kink misfit dislocation loops due to the accumulated strain. Using the excess stress model [94], one can estimate the critical number of quantum dot layers at the onset of single-kink dislocation, which depends on the depth of the buried dot layers and preferentially occurs first due to a lower energy barrier. For simplicity, the quantum dot is assumed to be equivalent to a two-dimensional uniformly strained layer with equal strain energy. Illustrated in Figure 9.3b, the critical dot layer numbers, depending on the dot size, are in the range of 10–15 for InAs dots, 20–30 for $In_{0.5}Ga_{0.5}As$ dots, and 15–35 for $In_{0.6}Al_{0.4}As$ dots, respectively, assuming that 50 nm GaAs barrier layers are used.

9.4.2 Transmission Electron Microscopy Characterization

The effect of dislocation filtering by quantum dot buffer layers has been carefully examined using cross-sectional transmission electron microscopy (XTEM) [36]. Dislocation propagation, in the 10-layer InAs quantum dot buffer region on Si, is investigated along the ⟨110⟩ axis under various diffraction conditions, $g = \langle 2\bar{2}0 \rangle$, $\langle 1\bar{1}1 \rangle$, $\langle \bar{1}11 \rangle$, $\langle 004 \rangle$, and $\langle 1\bar{1}3 \rangle$, as shown in Figure 9.4. The invisibility criterion, $g \cdot b = 0$, indicates that the

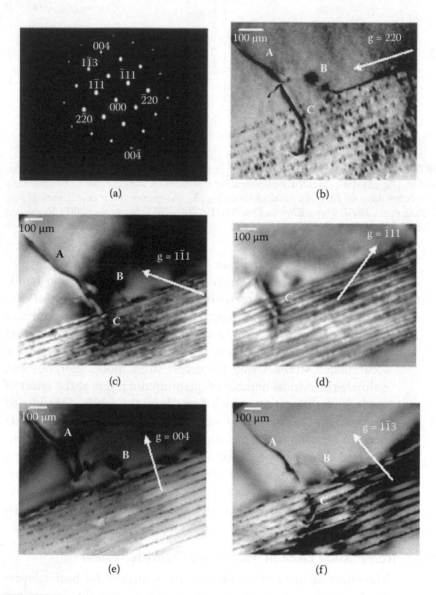

FIGURE 9.4 Electron diffraction pattern (a), and cross-sectional transmission electron microscopy image of dislocation propagation in the 10-layer InAs quantum dots buffer layer under various diffraction conditions: (b) g = $\langle 2\bar{2}0 \rangle$, (c) g = $\langle 1\bar{1}1 \rangle$, (d) g = $\langle \bar{1}11 \rangle$, (e) g = $\langle 004 \rangle$, and (f) g = $\langle \bar{1}13 \rangle$. The zone axis is $\langle 110 \rangle$. (From J. Yang et al., *IEEE Trans. Electron Dev.* 54(11): 2849–2855, November 2007. Reprinted with permission from IEEE.)

heteroepitaxy of GaAs on Si typically generates two types of threading dislocations: pure edge dislocations with Burger's vector $b = \pm\langle \bar{1}10\rangle$ (labeled C in Figure 9.4) and 60°-mixed dislocations with $b = \pm\langle 01\bar{1}\rangle$ or $\pm\langle 101\rangle$ (labeled A and B in Figure 9.4). It is observed that InAs quantum dot buffer layers can efficiently bend 60°-mixed dislocations (labeled B). More importantly, pure-edge dislocations (labeled C) can be terminated at the surface of quantum dots. Note that pure-edge dislocations cannot be blocked by conventional SLS due to the zero shear force applied by uniform 2D films [90]. Considering stronger and anisotropic stress within/surrounding the island and its interaction with the edge dislocations, the termination is probably due to the formation of either a local dislocation loop at the quantum dot surface or another dislocation with the opposite Burger's vector that annihilates the preexisting dislocation.

9.4.3 Photoluminescence Characterization

The effectiveness of such quantum dot dislocation filters has been further investigated by studying the photoluminescence emission of the active $In_{0.5}Ga_{0.5}As$ quantum dots in laser heterostructures grown on Si with the incorporation of various quantum dot dislocation filters near the GaAs–Si interface [36]. Photoluminescence spectra are recorded with an argon ion laser (~480 nm), a 0.75 m spectrometer, a lock-in amplifier, and a liquid-nitrogen-cooled Ge photodetector. Figure 9.5a depicts the room-temperature photoluminescence spectra measured from the active $In_{0.5}Ga_{0.5}As$ quantum dots with the incorporation of 10-layer InAs, $In_{0.5}Ga_{0.5}As$, and $In_{0.6}Al_{0.4}As$ quantum dot dislocation filters grown at 460°C on Si. It is seen that the active $In_{0.5}Ga_{0.5}As$ quantum dots exhibit the highest photoluminescence intensity with the use of InAs quantum dot dislocation filters. Compared to InAs quantum dots, InAlAs and InGaAs dots generally exhibit a smaller size and strain field, which may not provide enough strain to bend dislocations. This explains the much weaker photoluminescence emission from the samples grown using the InAlAs and InGaAs quantum dot buffer layers. The extremely weak and broad emission from the sample grown using InAlAs quantum dot buffer layers may also be related to the surface roughness induced by InAlAs dots. Figure 9.5b shows the photoluminescence spectra measured at 300 K from the $In_{0.5}Ga_{0.5}As$ quantum dots grown on GaAs and on Si with and without the use of InAs quantum dot dislocation filters. It is evident that, with the use of InAs quantum dot filters, both

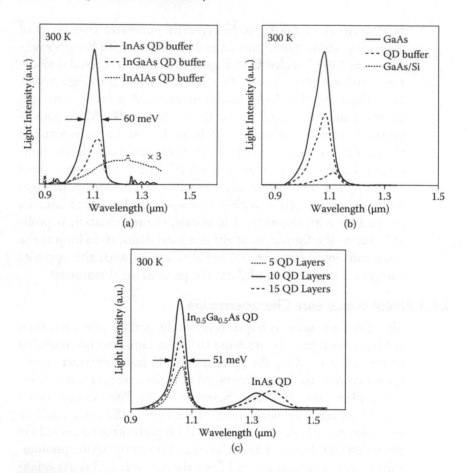

FIGURE 9.5 Photoluminescence spectra measured at 300 K from $In_{0.5}Ga_{0.5}As$ quantum dots grown on Si: (a) with different quantum dot buffer layers grown at 460°C; (b) with and without InAs quantum dot buffer layers, and on GaAs; and (c) with 5, 10, and 15-layer InAs quantum dots in the buffer layers grown at 510°C. (From J. Yang et al., *IEEE Trans. Electron Dev.* 54(11): 2849–2855, November 2007. Reprinted with permission from IEEE.)

the photoluminescence intensity and full-width-half-maximum (FWHM ~ 60 meV) of the active quantum dots are comparable to those from quantum dots grown on GaAs substrates.

Since islands with larger sizes can enhance dislocation bending, quantum dot buffers have also been grown at a relatively higher temperature of 510°C to increase the dots size. $In_{0.5}Ga_{0.5}As$ quantum dots grown on such optimized InAs quantum dot dislocation filters exhibit strong photoluminescence emission with an even narrower FWHM (~51 meV) at room temperature, shown in Figure 9.5c. Also shown in the figure are the photoluminescence spectra measured from $In_{0.5}Ga_{0.5}As$ quantum dots grown on 5- and 15-layer InAs quantum dot dislocation filters. Emission peaks (~1.3 μm) from the InAs quantum dot buffer region can also be clearly seen. These experimental results suggest that the use of an optimum number (~10) of InAs quantum dot layers can greatly reduce defect densities in III-V heterostructures grown on Si, which is consistent with the detailed TEM studies and also the theoretical analysis described earlier.

9.5 Self-Organized InGaAs/GaAs Quantum Dot Lasers on Si

In what follows, we describe the epitaxial growth, fabrication, and characterization of high-performance InGaAs/GaAs quantum dot lasers on Si substrates with the incorporation of 10-layer InAs quantum dot dislocation filters.

9.5.1 Laser Structure, Growth, and Fabrication

The schematic of a GaAs/AlGaAs separate-confinement heterostructure (SCH) $In_{0.5}Ga_{0.5}As$ quantum dot laser grown on a Si substrate is shown in Figure 9.6. To effectively suppress the formation of APD-related defects, a (001)-oriented Si substrate, misoriented 4° toward $\langle 111 \rangle$ is used in this experiment. A 2 μm GaAs buffer layer is first grown by MOCVD. Self-organized InAs quantum dot dislocation filters and the quantum dot laser heterostructures

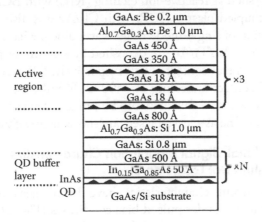

FIGURE 9.6 Schematic of self-organized $In_{0.5}Ga_{0.5}As$ quantum dot laser heterostructures grown on Si substrates with the dislocation filter consisting of *N* quantum dot layers (*N* = 0, 5, 10, and 15). (From J. Yang et al., *IEEE Trans. Electron Dev.* 54(11): 2849–2855, November 2007. Reprinted with permission from IEEE.)

are subsequently grown by solid source molecular beam epitaxy. The dislocation filter consists of 10 InAs quantum dot layers separated by 500-Å-thick GaAs layers, as described in Section 9.4.3. The GaAs spacer layers are grown at relatively high temperatures (~580°C) to minimize the generation of dislocations [80,83]. The active region consists of three coupled $In_{0.5}Ga_{0.5}As$ quantum dot layers grown at 500°C that are spaced apart by 18-Å-thick GaAs barrier layers. The p- and n-cladding layers are 1.0-μm-thick $Al_{0.7}Ga_{0.3}As$ layers grown at ~620°C, which are doped with Be and Si, respectively. In order to achieve quantum dot lasers on Si with high temperature stability, a modulation p-doped quantum dot laser is also grown. In this design, each quantum dot layer is modulation doped *p*-type using Be, averaging about 20 holes per dot. P-doping of the dots ensures that the hole ground states are filled and less injected electrons are needed for population inversion [95]. Therefore, both the gain and differential gain are increased, and the gain saturation associated with the thermal broadening of the injected holes is reduced. Significantly improved temperature stability (T_0 up to ∞) has been demonstrated in p-doped quantum dot lasers grown on GaAs [59].

Both mesa-shaped broad area and single-mode ridge waveguide lasers are fabricated using standard photolithography, wet and dry etching, and contact metallization techniques. For broad area lasers, the waveguide mesa is defined by wet etching with $H_3PO_4/H_2O_2/H_2O$-based solutions. To minimize mesa undercut, dry etching such as reactive-ion etching (RIE) with BCl_3/Ar or inductance-coupled plasma (ICP) with Cl_2/Ar was also utilized in the fabrication of single-mode ridge-waveguide devices. Pd/Zn/Pd/Au and Ni/Ge/Au/Ti/Au are deposited by e-beam evaporation as the p- and n-metal contact layers, respectively, followed by metal lift-off and contact annealing at ~400°C. Finally, the laser wafers are lapped down to a thickness of ~80 μm and cleaved along the ⟨110⟩ direction. The as-cleaved lasers are subsequently characterized.

9.5.2 DC and Small-Signal Modulation Characteristics of Quantum Dot Lasers on Si

Light–current characteristics have been measured at various temperatures under pulsed bias conditions (1% duty cycle). With the use of an optimized 10-layer InAs quantum dot buffer layer, quantum dot lasers on Si exhibit a relatively low current threshold of J_{th} = 900 A/cm² at room temperature, as shown in Figure 9.7a. For comparison, quantum dot lasers grown on Si without any

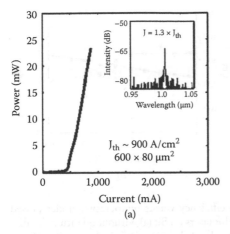 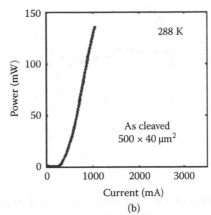

FIGURE 9.7 (a) Light–current characteristics of quantum dot lasers grown on Si and the output spectrum (inset) at room temperature with the use of 10-layer InAs QD dislocation filters. (From Z. Mi et al., *Electron. Lett.* 42(2): 121–123, Jan 2006. Reprinted with permission from IEEE.) (b) light–current characteristics of quantum dot lasers on Si showing an power output of over 140 mW. (From Z. Mi et al., *J. Vacuum Sci. Technol. B* 24(3): 1519–1522, May–June 2006. Reprinted with permission from AVS.)

quantum dot buffer layers typically exhibit a much larger threshold current ($J_{th} \geq 1500$ A/cm²) [83]. The output spectrum under an injection current of $1.3 \times J_{th}$ is shown in the inset. The lasing wavelength is at ~1.02 μm, which is shorter than the photoluminescence peak shown in Figure 9.5c, suggesting that lasing is achieved from the excited states instead of ground states due to the relatively high cavity loss. A reasonably high output power (>140 mW) can also be measured from quantum dot lasers on Si, shown in Figure 9.7b. The temperature dependence of J_{th} is usually characterized by T_0 using the empirical equation $J_{th}(T) = J_{th}(0)exp(T/T_0)$. From temperature-dependent light–current characteristics shown in Figure 9.8a, we derive a T_0 of ~103 K in the temperature range of 5°C to 95°C for quantum dot lasers on Si without using any quantum dot buffer layers. With the use of InAs quantum dot dislocation filters and the technique of p-doping, a very large value of T_0 (278 K) is obtained in the temperature range of 5°C to 85°C, as illustrated in Figure 9.8b. The output slope efficiency stays essentially constant in the temperature range of 5°C to 50°C. These attributes are highly desirable and essential for practical applications in chip-level optical interconnects.

The small signal modulation response of ridge-waveguide quantum dot lasers on Si is also measured [96]. The DC and AC

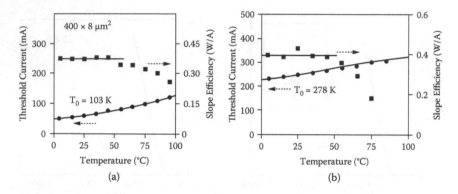

FIGURE 9.8 Threshold current and slope efficiency versus temperature under pulsed mode (1% duty cycle of 100 μs) of quantum dot lasers on Si: (a) without quantum dot dislocation filter; and (b) with 10-layer quantum dot dislocation filter and p-doping. (From J. Yang et al., *IEEE Trans. Electron Dev.* 54(11): 2849–2855, November 2007. Reprinted with permission from IEEE.)

biases applied on the laser bar, through a bias-Tee and Cascade-Microtech high-speed probe, are supplied by a pulse generator and an HP 8350B sweep oscillator. The modulated light output is recorded with a high-speed New Focus photodetector through a multimode optical fiber. The converted electrical signal is further amplified with a MITEQ low-noise amplifier and characterized with an HP 8562A electrical spectrum analyzer. The amplifier gain and loss in the microwave cables are taken into account in the measurement. The measurements are made under pulsed bias condition. Figure 9.9 shows the small signal modulation characteristics of a ridge-waveguide quantum dot laser on Si, with a –3 dB bandwidth of 5.5 GHz. It is comparable to that of conventional InGaAs quantum dot lasers grown on GaAs substrates. The attainable small-signal modulation bandwidth of SCH quantum dot lasers is limited to 5–7 GHz [97], due to the presence of wetting layer states and the associated hot carrier effect [98]. This problem can be alleviated by the special technique of tunneling injection, which injects "cold" electrons directly into the quantum dot lasing states from an adjoining quantum well [60,61].

9.6 III-V Integrated Guided-Wave Devices on Si

The monolithic integration of electrically injected lasers with guided-wave devices including waveguides and modulators on a Si platform is presented in this section. For edge-emitting

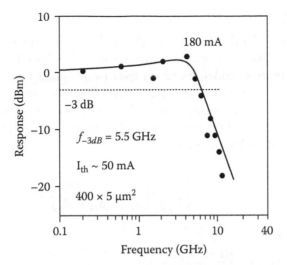

FIGURE 9.9 Small signal modulation characteristics of a ridge-waveguide quantum dot laser on Si operating under pulsed mode (1% duty cycle of 100 μs). (From Z. Mi et al., *J. Vacuum Sci. Technol. B* 24(3): 1519–1522, May–June 2006. Reprinted with permission from AVS.)

lasers, although a variety of coupling schemes such as butter-joint have been developed, groove coupling remains a simple and effective approach to achieving high-performance lasing as well as efficient coupling between the laser and guided-wave device [99]. The performance of the coupling groove depends on its facet quality and dimension accuracy. High-quality grooves, for laser cavity feedback, coupling, and electrical isolation, can be fabricated with focused-ion-beam (FIB) milling. On the other hand, the optimal dimension of grooves can be analyzed by taking into account beam quality and Gouy phase shifts [100].

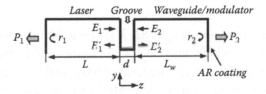

FIGURE 9.10 Schematic illustration of a laser/waveguide coupled system. (From J. Yang et al., *J. Lightwave Technol.*, 25(7): 1826–1831, July 2007. Reprinted with permission from IEEE.)

9.6.1 Model

Figure 9.10 illustrates the schematic of the coupled-cavity integrated laser/waveguide or modulator. A generalized transmission matrix model, based on laser beam quality and Gouy phase shifts, is developed to describe the reflectivity and coupling coefficients in such groove-coupled structures [100]. Using transmission matrices [99], complex amplitudes of the optical fields in the coupled cavity are related by

$$\begin{pmatrix} E_1' \\ E_2' \end{pmatrix} = \begin{pmatrix} S_{11} & S_{12} \\ S_{21} & S_{22} \end{pmatrix} \begin{pmatrix} E_1 \\ E_2 \end{pmatrix} \tag{9.3}$$

where

$$S_{11} = S_{22} = r - \frac{r(1-r^2)t(2d)}{1-r^2 t(2d)},$$

$$S_{12} = S_{21} = \frac{t(d)(1-r^2)}{1-r^2 t(2d)}$$

Here, r is the amplitude reflectivity of FIB-etched facets. Transmission functions $t(d)$ and $t(2d)$, representing phase/loss in a single pass and roundtrip in the groove with width of d, are given by

$$t(z = d, \text{ or } 2d) = \exp(ikz + i\Phi_G(z)) T(z) \tag{9.4}$$

The phase shifts include the plane wave phase shift kz ($k = 2\pi/\lambda$, λ is the wavelength in vacuum) and the Gouy phase shift

$$\Phi_G(z) = -\sum_{\xi=x,y} \left(M_\xi^2/2 \right) \tan^{-1}\left(z/z_{R\xi} \right)$$

[101]. Here, $z_{R\xi} = \pi w_{0\xi}^2 / M_\xi^2 \lambda$ is the Rayleigh range, $w_{0\xi}$ is the beam waist width, and M_ξ^2 is the beam quality factor along the x- or y-axis ($\xi = x, y$). The Gouy phase shift, an extra axial phase shift due to the transverse spatial confinement of finite beams, becomes significant when the groove width is comparable to or greater than the Rayleigh range. Note that the Rayleigh range is typically about one micrometer for edge-emitting semiconductor

lasers perpendicular to the diode junction plane [102]. The diffraction losses $T(z)$ can be estimated through the evaluation of optical filed overlap at facets after a single pass or roundtrip in the groove, namely,

$$T(z=d, \text{ or } 2d) = \prod_{\xi=x,y} \left(\frac{2w_\xi(0)w_\xi(z)}{(w_\xi(0)^2 + w_\xi(z)^2)} \right)^{1/2} \tag{9.5}$$

For simplicity, an equivalent Gaussian amplitude profile is assumed for the real beam with the beam width $w_\xi(z)$, which is defined with the intensity moments as $w_\xi(z)^2 = 4\int\xi^2 E(\xi)^2 d\xi / \int E(\xi)^2 d\xi$. Its evolution with the beam propagation is determined by the formula $w_\xi(z-z_{0\xi}) = w_{0\xi}\{1 + [(z - z_{0\xi})/z_{0\xi}]^2\}^{1/2}$ [103], where $z_{0\xi}$ is the position of beam waist along the x- or y-axis ($\xi = x, y$).

Applicable for a general beam (not limited to a Gaussian beam), this model more accurately describes the coupling behavior of an edge-emitting laser with a coupled cavity.

9.6.2 FIB Etched Facets as Cavity Mirrors and Coupling Grooves

The quantum dot lasers on Si presented in Section 9.5 use cleaved facets as cavity mirrors. However, the use of etched facets is more desirable for monolithic integration. Facet etching can be achieved using either plasma etching [99,104] or FIB milling [105,106]. FIB is more favorable due to the advantages of mask-free etching to form fine patterns, a high aspect ratio, and a smooth surface. In this experiment, the high-quality cavity mirror and coupling groove are fabricated with FEI-Nova-Nanolab FIB using gallium ions. Figure 9.11 shows the scanning electron microscopy (SEM) image of the FIB-etched facet for an InGaAs/GaAs quantum dot laser grown on Si. From a detailed analysis of the threshold and cavity loss, the facet reflectivity is estimated to be ~0.28, which is very close to the reflectivity (~0.31) for cleaved GaAs facets. As a result, the light–current (L-I) characteristics of InGaAs/GaAs quantum dot lasers with FIB-etched facets are comparable to those for cleaved lasers [100].

9.6.3 Integrated Quantum Dot Lasers and Quantum Well Electroabsorption Modulators on Si

Utilizing regrowth and FIB milling techniques, the monolithic integration of InGaAs/GaAs quantum dot lasers and

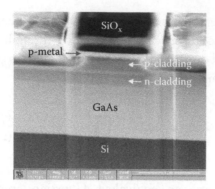

FIGURE 9.11 Scanning electron microscopy image of a focused-ion-beam etched facet of an InGaAs quantum dot laser grown on Si. (From J. Yang et al., *J. Lightwave Technol.* 25(7): 1826–1831, July 2007. Reprinted with permission from IEEE.)

QCSE-based quantum well electroabsorption modulators has been realized [107]. One advantage offered by this approach is the separate growth of laser and modulator active regions, which enables the alignment of the absorption band edge of quantum wells with respect to the lasing wavelengths of quantum dots.

The schematic of a groove-coupled laser/modulator on Si is shown in Figure 9.12a. The laser heterostructure is first grown by MBE following the procedure described in Section 9.5.1. The grown wafer is then patterned and etched to create trenches where the modulator heterostructure is subsequently regrown. The designated regions are first etched by ICP dry etching with Cl_2/Ar

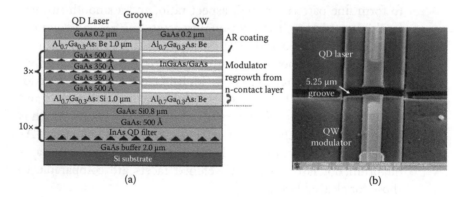

FIGURE 9.12 A groove-coupled quantum dot laser and quantum-well electroabsorption modulator on Si: (a) schematics of the device with a GaAs buffer layer and a dislocation filter consisting of 10 layers of InAs quantum dots (drawing is not to scale); (b) a scanning electron microscopy image of the device. (From J. Yang et al., *IEEE Photon. Technol. Lett.* 19(10): 747–749, May–June 2007. Reprinted with permission from IEEE.)

to provide vertical etching profiles. This step removes all of the p-contact/cladding layer, active region, and most of the n-cladding layer. Then, buffered hydrofluoric (BHF) acid solution is utilized to selectively remove the rest of the thin (~0.1–0.2 μm) n-cladding layer of $Al_{0.7}Ga_{0.3}As$. Before MBE regrowth, a 0.2-μm-thick SiO_x layer is deposited by plasma-enhanced chemical vapor deposition (PECVD) as a protection layer and then patterned to expose the trenches. Subsequently, the quantum well modulator heterostructure, consisting of seven $In_{0.2}Ga_{0.8}As/GaAs$ quantum wells sandwiched between the n- and p-$Al_{0.7}Ga_{0.3}As$ cladding layers, is grown. The quantum well and quantum dot active regions are carefully aligned during the etching/regrowth process. More importantly, the excitonic absorption edge of the quantum wells at zero bias is tuned to be ~15–20 meV higher than the lasing photon energy of the quantum dots. Due to the QCSE effect [16], the absorption edge is red-shifted with the application of an appropriate transverse bias, which results in a strong absorption of the quantum dot emission. Such optimum conditions, achieved with 85-Å-thick $In_{0.2}Ga_{0.8}As$ quantum wells with 150-Å-thick GaAs barrier layers, are characterized by PL and lasing spectra as shown in Figure 9.13. The coupled laser/modulator devices are fabricated utilizing standard photolithography, wet and dry etching, contact metallization, and cleaving techniques. A λ/4-thick Al_2O_3 antireflection coating is then deposited on the modulator output facet, which increases the transmission up to ~97%. Finally, the coupling

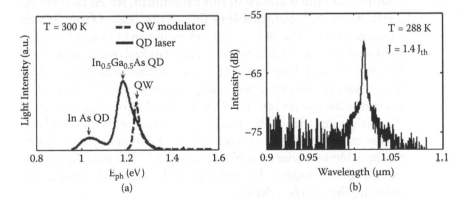

FIGURE 9.13 (a) Photoluminescence spectra measured at 300 K for $In_{0.5}Ga_{0.5}As$ quantum dots and multiple $In_{0.2}Ga_{0.8}As$ quantum wells on Si; (b) lasing spectrum for the $In_{0.5}Ga_{0.5}As$ quantum dot laser section of a coupled laser/modulator on Si. (From J. Yang et al., *IEEE Photon. Technol. Lett.* 19(10): 747–749, May–June 2007. Reprinted with permission from IEEE.)

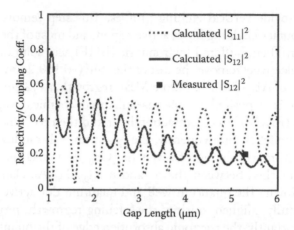

FIGURE 9.14 Calculated coupling coefficient versus groove width for a laser beam with the beam width 0.65 μm and beam quality factor $M^2 \sim 1.5$ perpendicular to the junction plane. The measured coupling coefficient is for a groove width of 5.25 μm under a laser injection current $J = 2J_{th}$. (From J. Yang et al., *IEEE Photon. Technol. Lett.* 19(10): 747–749, May–June 2007. Reprinted with permission from IEEE.)

groove is defined by FIB milling. A SEM image of the integrated device on Si is shown in Figure 9.12b. The lengths of the laser and modulator devices are 400 and 250 μm, respectively. In practice, groove widths larger than 5 μm are required in order to remove lateral grown materials formed during the 2-μm-thick regrowth process. Based on the simulation discussed in Section 9.6.1, a groove width of 5.25 μm is chosen in this experiment, which can provide a reflectivity $|S_{11}|^2$ comparable to that of the cleaved facets as well as an acceptable value of $|S_{12}|^2$ (>20%), as shown in Figure 9.14. The light–current characteristics measured from both the laser end and the coupled modulator end at zero bias are shown in Figure 9.15a. The modulated output under an injection current of $2 \times J_{th}$, is plotted versus the reverse bias in Figure 9.15b. The coupling coefficient $|S_{12}|^2$ is greater than 20%, which is consistent with the simulated results using the foregoing model. A modulation depth of ~100% can be achieved under an applied bias of –5 V. The modulation bias can be further reduced by optimizing the regrowth process and reducing the interface defects.

9.6.4 Integration of Quantum Dot Lasers with Si Waveguides on a Si Platform

Another essential component of integrated optoelectronics on a Si platform is a low-loss Si waveguide, particularly in the wavelength

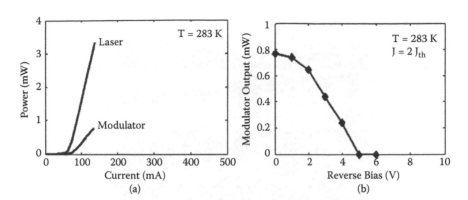

FIGURE 9.15 (a) Light–current characteristics for output from the laser end and the coupled modulator end at zero bias; (b) modulator output versus reverse bias under laser injection current $J = 2J_{th}$. The dimensions of the laser and modulator sections are $400 \times 8 \ \mu m^2$ and $250 \times 8 \mu m^2$, respectively. (From J. Yang et al., *IEEE Photon. Technol. Lett.* 19(10): 747–749, May–June 2007. Reprinted with permission from IEEE.)

range of 1.3–1.55 μm. In this respect, groove-coupled amorphous and crystalline Si waveguides with quantum dot lasers on Si [108,109] are discussed.

The use of hydrogenated amorphous Si (a:Si-H) to fabricate low-loss Si waveguides offers the advantages of low cost and low-temperature processing. It also provides other unique properties, such as desirable thermo-optic effect, tunable bandgap of 1.4–1.8 eV, and hydrogen-composition-dependent refractive index [110]. As a result, a:Si-H can exhibit acceptable absorption loss in the wavelength range of 0.95–1.15 μm (while crystalline Si has a much higher loss in this range), and certainly lower absorption loss at longer wavelengths [8]. Illustrated in Figure 9.16a, the device structure and fabrication process are similar to those of the previously described integrated laser/modulator devices, except that a SiO_x/a:Si-H/SiO_x waveguide is used. The SiO_x/a:Si-H/SiO_x waveguide is fabricated by PECVD at 380°C. The thicknesses of the core and cladding layers are carefully aligned with respect to the quantum dot laser heterostructure. Figure 9.16b shows the SEM image of a groove-coupled laser/waveguide device with lengths of 400 and 250 μm, respectively. The InGaAs/GaAs quantum dot laser on Si emits at a wavelength of 1.02 μm, as shown in the inset of Figure 9.17a. To determine the propagation loss of the PECVD a:Si-H waveguide in this wavelength range, light from a 1.05 μm Nd:glass laser is coupled into and out of the waveguide segments using single-mode fibers. The output

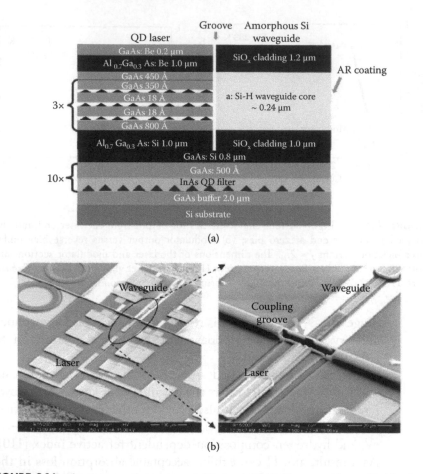

FIGURE 9.16 An integrated quantum dot laser and a:Si-H waveguide on Si: (a) schematics of the devices with a dislocation filter consisting of 10 layers of InAs quantum dots; (b) scanning electron microscopy image of an integrated device and focused-ion-beam etched coupling groove. (From J. Yang and P. Bhattacharya, *Opt. Express* 16(7): 5136–5140, March 2008. Reprinted with permission from OSA.)

power is measured for waveguide segments with varying lengths and $\lambda/4$-thick Al_2O_3 antireflection coating deposited on the facets. From these measurements, the waveguide propagation loss is estimated to be ~10 dB/cm. The propagation loss can be reduced by optimizing the fabrication process, including the optimization of H-composition in a:Si-H and the improvement of the waveguide sidewall surface. The laser–waveguide coupling is achieved through a FIB-etched groove with a width of 3.20 μm. The light–current characteristics measured under pulsed bias (1% duty cycle, 500 μs pulse) from the quantum dot laser and waveguide ends are

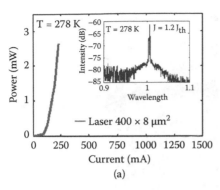 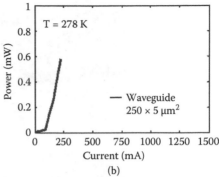

FIGURE 9.17 Light–current characteristics for output from (a) the InGaAs quantum dot laser end and (b) the coupled a:Si-H waveguide end. The lasing spectrum is shown in the inset of (a). (From J. Yang and P. Bhattacharya, *Opt. Express* 16(7): 5136–5140, March 2008. Reprinted with permission from OSA.)

shown in Figures 9.17a and 9.17b. A coupling coefficient of 22% is derived.

Crystalline Si generally exhibits smaller propagation loss in the wavelength range of 1.3–1.55 μm and high-speed electro-optic modulation property. A lift-off process has been developed to transfer Si membrane electronic devices and circuits onto foreign substrates [111,112]. This technology has been extended to achieve chip-scale integration of crystalline Si waveguides with quantum dot light sources [109]. Such transferred Si membrane waveguides can offer properties comparable to SOI-based waveguides in addition to providing more flexibility in on-chip integration. A typical process flow includes the following: (1) Si membranes are detached from commercial SOI wafers with BHF etching of both top and buried oxide layers and (2) membranes are transferred and bonded to the host substrate with the use of a special soft stamp to pick up and release with glue-free approach [111], or the use of a flip-transfer approach with SU-8 as the adhesive transfer medium [112]. In the latter case, SU-8 has to be dissolved in the solution with proper heat treatment. Specifically, the SOI wafer is first patterned with lithography and etched to have either strips or mesh holes in order to make the membrane detachment faster with higher yield. The host substrate used here is a Si wafer with MBE-grown laser heterostructures and etched trenches. PECVD SiO$_x$ is used for the bottom cladding layer, where crystalline Si membrane will be transferred and bonded. After membrane transfer, standard lithography and dry etch are performed to define the

FIGURE 9.18 A scanning electron microscopy image of chip-scale integration of a quantum dot laser heterostructure with a crystalline Si waveguide using membrane transfer. (From J. Yang et al., *Chinese Opt. Lett.* 6(10):727–731, October 2008. Reprinted with permission.)

waveguide profile for the integrated device followed by PECVD deposition of SiO_x for the top cladding layer. After contact metallization on the laser part and wafer thinning and cleaving, the device bar is deposited with the antireflection coating of $\lambda/4$-thick Al_2O_3 on the waveguide output facet. Finally, FIB is used to create coupling grooves. Figure 9.18 displays the SEM image of a transferred Si waveguide coupled with a quantum dot laser heterostructure. Efficient optical coupling is observed.

9.7 Future Prospects

For practical applications in chip-level optical interconnects, it is required that self-organized quantum dot lasers on Si can exhibit long-wavelength emission (1.3–1.55 μm) and a further reduced threshold current. Two promising approaches for achieving 1.3–1.55 μm quantum dot lasers on Si are briefly described. In the first approach, self-organized 1.3 μm InAs/GaAs quantum dot laser heterostructures are grown on Si with the use of InAs quantum dot dislocation filters. The laser heterostructure is shown in Figure 9.19a, which consists of 10 InAs quantum dot layers, with each dot layer capped by a 50 Å $In_{0.15}Ga_{0.85}As$ layer. Illustrated in Figure 9.19b, the device heterostructure exhibits strong photoluminescence emission at ~1.3 μm. Lasing has recently been

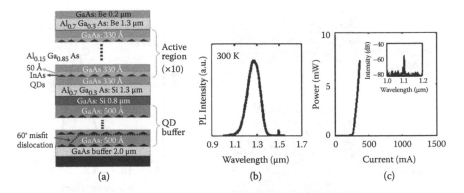

FIGURE 9.19 (a) Schematic of 1.3 μm InAs quantum dot laser heterostructure grown on Si with the incorporation of InAs quantum dots as dislocation filters. The bending of a 60° dislocation induced by the strain field of a dot layer is illustrated; (b) photoluminescence spectra measured at 300 K (linear scale in y-axis); (c) light–current characteristics and the output spectrum (inset) at room temperature. (From J. Yang et al., *Chinese Opt. Lett.* 6(10):727–731, October 2008. Reprinted with permission.)

demonstrated from the excited states of the dots, with emission wavelengths at ~1.1 μm [109], shown in Figure 9.19c. It is believed that ground state lasing can be achieved by reducing the optical cavity loss of quantum dot lasers on Si. In the second approach, metamorphic $In_{0.2}Ga_{0.8}As/In_{0.2}Al_{0.3}Ga_{0.5}As$ SCH quantum dot lasers are grown on relaxed $In_{0.2}Ga_{0.8}As$ buffer layers on a GaAs layer on Si, illustrated in Figure 9.20a. Due to the increased indium composition and reduced bandgap in the quantum dot barrier layers, emission wavelengths in the range of 1.3–1.55 μm can be readily achieved [109], as shown in Figure 9.20b. Utilizing this approach, we have recently achieved 1.5 μm quantum dot lasers on GaAs with an ultralow threshold current (~70A/cm²) [67].

In order to achieve quantum dot lasers on Si with threshold comparable to that of lasers on GaAs, it is imperative to utilize low (≤1 × 10⁶ cm⁻²) defect density buffer layers, such as relaxed and graded SiGe buffer layers [30], in conjunction with InAs quantum dot dislocation filters. Another attractive approach to achieve ultralow threshold lasers on Si is to incorporate epitaxial or colloidal quantum dots [113] in micro- and nanocavities on Si. Enhanced spontaneous emission at ~1.55 μm has been observed from Si photonic crystal microcavities embedded with PbSe or PbS quantum dots [114–116]. Recently, a 1.3 μm InAs/GaAs quantum dot photonic crystal nanocavity laser has also been demonstrated on Si substrates, which exhibits an ultralow threshold

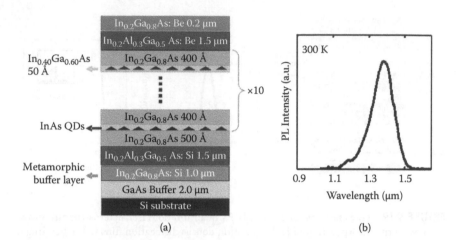

FIGURE 9.20 (a) Schematic of metamorphic InAs quantum dot laser heterostructure grown on Si with the incorporation of InAs quantum dots as dislocation filters; (b) photoluminescence spectra measured at 300 K (linear scale in *y*-axis). (From J. Yang et al., *Chinese Opt. Lett.* 6(10):727–731, October 2008. Reprinted with permission.)

pump power of ~2 µW [117]. The achievement of electrically injected micro- and nanoscale lasers on Si is being intensively investigated.

It is also important to note that, at the present time, all III–V lasers grown directly on Si involve the use of offcut Si substrates in order to minimize the formation of dislocation and APDs. There is therefore a growing concern whether offcut Si substrates are suitable for the mainstream Si electronics. Preliminary studies have been performed to compare the characteristics of metal-oxide-semiconductor field-effect transistor (MOSFET) fabricated on regular Si (100) and offcut Si substrates [118]. It was observed that the differences in some of the key parameters, such as the mobility and transconductance, between regular and offcut Si-based devices diminish when the gate length shrinks down to less than 1 µm, suggesting that offcut Si substrates may be suitable for deep-submicron CMOS transistors. Another challenge for the on-chip integration of quantum dot lasers is the processing temperature incompatibility between III–V and Si devices. In a standard CMOS process, temperatures in the range of 900°C–1000°C and 600°C–650°C are generally required for the thermal oxidation of SiO_2 gate insulators and for the LPCVD deposition of a poly-Si gate, respectively. GaAs or InP will degrade at such high temperatures. This issue can potentially be overcome by using the scheme of "laser-after-gate" or using metal gates with low-temperature-processed high-κ dielectric materials as gate insulators. Significant progress has already been made in developing such gate-insulator materials, for example, HfO_2 [119,120].

9.8 Summary

Hybrid III-V lasers on Si have been extensively investigated in the last two decades. Significant advances in the epitaxial growth, device design, and development of various dislocation reduction techniques have led to room-temperature operational, electrically injected III-V lasers on Si with emission wavelengths in the range of ~0.68–1.6 μm. With the use of self-organized InAs quantum dots as dislocation filters, InGaAs/GaAs quantum dot lasers monolithically grown on Si exhibit relatively low threshold current (J_{th} ~ 900 A/cm^2), reasonable small signal modulation bandwidth of 5.5 GHz, extremely high temperature stability (T_0 = 278 K), and reasonably high output power with output slope efficiency of 0.4 W/A in the temperature range of 5°C–50°C. Integrated quantum dot lasers and quantum well electroabsorption modulators on Si have been achieved, with a coupling coefficient of more than 20% and a modulation depth of ~100% at a reverse bias of 5 V. The monolithic integration of quantum dot lasers with both amorphous and crystalline Si waveguides, fabricated with plasma-enhanced-chemical-vapor deposition and membrane transfer, respectively, has also been demonstrated. The prospects and challenges of III-V lasers on Si for applications in future chip-level optical communications have been further discussed in this chapter.

References

1. R. Soref, The past, present, and future of silicon photonics, *IEEE J. Select. Top. Quantum Electron.* 12(6): 1678–1687, November–December 2006.
2. B. Jalali and S. Fathpour, Silicon photonics, *J. Lightw. Technol.* 24(12): 4600–4615, December 2006.
3. M. Bruel, Silicon-on-insulator material technology, *Electron. Lett.* 31(14): 1201–1202, July 1995.
4. S. Lardenois, D. Pascal, L. Vivien, E. Cassan, S. Laval, R. Orobtchouk, M. Heitzmann, N. Bouzaida, and L. Mollard, Low-loss submicrometer silicon-on-insulator rib waveguides and corner mirrors, *Opt. Lett.* 28(13): 1150–1152, July 2003.

5. Y. A. Vlasov and S. J. McNab, Losses in single-mode silicon-on-insulator strip waveguides and bends, *Opt. Express* 12(8): 1622–1631, April 2004.

6. M. Lipson, Guiding, modulating, and emitting light on silicon-challenges and opportunities, *J. Lightw. Technol.* 23(12): 4222–4238, December 2005.

7. M. J. A. de Dood, A. Polman, T. Zijlstra, and E. W. J. M. van der Drift, Amorphous silicon waveguides for microphotonics, *J. Appl. Phys.* 92(2): 649–653, July 2002.

8. A. Harke, M. Krause, and J. Mueller, Low-loss singlemode amorphous silicon waveguide, *Electron. Lett.*, 41(25): 1377–1379, December 2005.

9. G. V. Treyz, P. G. May, and J. M. Halbout, Silicon Mach-Zehnder wave-guide interferometers based on the plasma dispersion effect, *Appl. Phys. Lett.* 59(7): 771–773, August 1991.

10. G. V. Treyz, P. G. May, and J. M. Halbout, Silicon optical modulators at 1.3 μm based on free-carrier absorption, *IEEE Electron Dev. Lett.* 12(6): 276–278, June 1991.

11. R. A. Soref and B. R. Bennett, Electrooptical effects in silicon, *IEEE J. Quantum Electron.* 23(1): 123–129, January 1987.

12. A. S. Liu, R. Jones, L. Liao, D. Samara-Rubio, D. Rubin, O. Cohen, R. Nicolaescu, and M. Paniccia, A high-speed silicon optical modulator based on a metal-oxide-semiconductor capacitor, *Nature* 427(6975): 615–618, February 2004.

13. W. M. J. Green, M. J. Rooks, L. Sekaric, and Y. A. Vlasov, Ultra-compact, low RF power, 10 Gb/s silicon Mach-Zehnder modulator, *Opt. Express* 15(25): 17106–17113, December 2007.

14. Q. F. Xu, V. R. Almeida, and M. Lipson, Micrometer-scale all-optical wavelength converter on silicon, *Opt. Lett.* 30(20): 2733–2735, October 2005.

15. S. Manipatruni, Q. F. Xu, and M. Lipson, PINIP based high-speed high-extinction ratio micron-size silicon electro-optic modulator, *Opt. Express* 15(20): 13035–13042, October 2007.

16. D. A. B. Miller, D. S. Chemla, T. C. Damen, A. C. Gossard, W. Wiegmann, T. H. Wood, and C. A. Burrus, Band-edge electroabsorption in quantum well structures: the quantum-confined Stark effect, *Phys. Rev. Lett.* 53(22): 2173–2176, November 1984.

17. O. Qasaimeh, P. Bhattacharya, and E. T. Croke, SiGe-Si quantum-well electroabsorption modulators, *IEEE Photon. Technol. Lett.* 10(6): 807–809, June 1998.

18. Y. H. Kuo, Y. K. Lee, Y. S. Ge, S. Ren, J. E. Roth, T. I. Kamins, D. A. B. Miller, and J. S. Harris, Strong quantum-confined Start effect in germanium quantum-well structures on silicon, *Nature* 437(7063): 1334–1336, October 2005.

19. D. Ahn, C. Y. Hong, J. F. Liu, W. Giziewicz, M. Beals, L. C. Kimerling, J. Michel, J. Chen, and F. X. Kartner, High performance, waveguide integrated Ge photodetectors, *Opt. Express* 15(7): 3916–3921, April 2007.

20. G. Dehlinger, S. J. Koester, J. D. Schaub, J. O. Chu, Q. C. Ouyang, and A. Grill, High-speed germanium-on-SOI lateral PIN photodiodes, *IEEE Photon. Technol. Lett.* 16(11): 2547–2549, November 2004.

21. L. Vivien, M. Rouviere, J. M. Fedeli, D. Marris-Morini, J. F. Damlencourt, J. Mangeney, P. Crozat, L. EI Melhaoui, E. Cassan, X. Le Roux, D. Pascal, and S. Laval, High speed and high responsivity germanium photodetector integrated in a silicon-on-insulator microwaveguide, *Opt. Express* 15(15): 9843–9848, July 2007.

22. T. Yin, R. Cohen, M. M. Morse, G. Sarid, Y. Chetrit, D. Rubin, and M. J. Paniccia, 31GHz Ge n-i-p waveguide photodetectors on silicon-on-insulator substrate, *Opt. Express* 15(21): 13965–13971, October 2007.

23. W. H. Chang, A. T. Chou, W. Y. Chen, H. S. Chang, T. M. Hsu, Z. Pei, P. S. Chen, S. W. Lee, L. S. Lai, S. C. Lu, and M. J. Tsai, Room-temperature electroluminescence at 1.3 and 1.5 μm from Ge/Si self-assembled quantum dots, *Appl. Phys. Lett.* 83(14): 2958–2960, October 2003.

24. R. J. Walters, G. I. Bourianoff, and H. A. Atwater, Field-effect electroluminescence in silicon nanocrystals, *Nat. Mater.* 4(2): 143–146, February 2005.

25. C. M. Hessel, M. A. Summers, A. Meldrum, M. Malac, and J. G. C. Veinot, Direct patterning, conformal coating, and erbium doping of luminescent nc-Si/SiO2 thin films from solution processable hydrogen silsesquioxane, *Adv. Mater.* 19(21): 3513–3516, November 2007.

26. O. Boyraz and B. Jalali, Demonstration of a silicon Raman laser, *Opt. Express* 12(21): 5269–5273, October 2004.

27. H. S. Rong, R. Jones, A. S. Liu, O. Cohen, D. Hak, A. Fang, and M. Paniccia, A continuous-wave Raman silicon laser, *Nature* 433(7027): 725–728, February 2005.

28. S. Sakai, T. Soga, M. Takeyasu, and M. Umeno, Room-temperature laser operation of AlGaAs-GaAs double heterostructures fabricated on Si substrates by metalorganic chemical vapor-deposition, *Appl. Phys. Lett.* 48(6): 413–414, February 1986.

29. R. Fischer, W. Kopp, H. Morkoc, M. Pion, A. Specht, G. Burkhart, H. Appelman, D. Mcgougan, and R. Rice, Low threshold laser operation at room-temperature in GaAs/(Al,Ga)As structures grown directly on (100) Si, *Appl. Phys. Lett.* 48(20): 1360–1361, May 1986.

30. M. E. Groenert, C. W. Leitz, A. J. Pitera, V. Yang, H. Lee, R. J. Ram, and E. Fitzgerald, Monolithic integration of room-temperature cw GaAs/AlGaAs lasers on Si substrates via relaxed graded GeSi buffer layers, *J. Appl. Phys.* 93(1): 362–367, January 2003.

31. T. Egawa, Y. Hasegawa, T. Jimbo, and M. Umeno, Effects of dislocation and stress on characteristics of GaAs-based laser grown on Si by metalorganic chemical vapor deposition, *Jpn. J. Appl. Phys. Part 1,* 31(3): 791–797, March 1992.

32. M. Yamaguchi, M. Tachikawa, M. Sugo, S. Kondo, and Y. Itoh, Analysis for dislocation density reduction in selective area grown GaAs films on Si substrates, *Appl. Phys. Lett.* 56(1): 27–29, January 1990.

33. Y. Ujiie and T. Nishinaga, Epitaxial lateral overgrowth of GaAs on a Si substrate, *Jpn. J. Appl. Phys.* 28(3): L337–L339, March 1989.

34. G. Balakrishnan, S. Huang, L. R. Dawson, Y. C. Xin, P. Conlin, and D. L. Huffaker, Growth mechanisms of highly mismatched AlSb on a Si substrate, *Appl. Phys. Lett.* 86(3): 034105, January 2005.

35. K. Eisenbeiser, J. M. Finder, Z. Yu, J. Ramdani, J. A. Curless, J. A. Hallmark, R. Droopad, W. J. Ooms, L. Salem, S. Bradshaw, and C. D. Overgaard, Field effect transistors with $SrTiO_3$ gate dielectric on Si, *Appl. Phys. Lett.* 76(10): 1324–1326, March 2000.

36. J. Yang, P. Bhattacharya, and Z. Mi, High-performance $In_{0.5}Ga_{0.5}As/$ GaAs quantum-dot lasers on silicon with multiple-layer quantum-dot dislocation filters, *IEEE Trans. Electron Dev.* 54(11): 2849–2855, November 2007.

37. Z. Mi, J. Yang, P. Bhattacharya, and D. Huffaker, Self-organised quantum dots as dislocation filters: The case of GaAs-based lasers on silicon, *Electron. Lett.* 42(2): 121–123, January 2006.

38. H. Wada and T. Kamijoh, 1.3-μm InP-InGaAsP lasers fabricated on Si substrates by wafer bonding, *IEEE J. Select. Top. Quantum Electron.* 3(3): 937–942, June 1997.

39. K. Mori, K. Tokutome, and S. Sugou, Low-threshold pulsed operation of long-wavelength lasers on Si fabricated by direct bonding, *Electron. Lett.* 31(4): 284–285, February 1995.

40. A. W. Fang, H. Park, O. Cohen, R. Jones, M. J. Paniccia, and J. E. Bowers, Electrically pumped hybrid AlGaInAs-silicon evanescent laser, *Opt. Express* 14(20): 9203–9210, October 2006.

41. T. H. Windhorn, G. M. Metze, B. Y. Tsaur, and J. C. C. Fan, AlGaAs double-heterostructure diode-lasers fabricated on a monolithic GaAs/Si substitute, *Appl. Phys. Lett.* 45(4): 309–311, August 1984.

42. T. H. Windhorn and G. M. Metze, Room-temperature operation of GaAs/AlGaAs diode-lasers fabricated on a monolithic GaAs/Si substrate, *Appl. Phys. Lett.* 47(10): 1031–1033, November 1985.

43. R. Fischer, D. Neuman, H. Zabel, H. Morkoc, C. Choi, and N. Otsuka, Dislocation reduction in epitaxial GaAs on Si (100), *Appl. Phys. Lett.* 48(18): 1223–1225, May 1986.

44. S. F. Fang, K. Adomi, S. Iyer, H. Morkoc, H. Zabel, C. Choi, and N. Otsuka, Gallium arsenide and other compound semiconductors on silicon, *J. Appl. Phys.* 68(7): R31–R58, October 1990.

45. R. M. Sieg, S. A. Ringel, S. M. Ting, S. B. Samavedam, M. Currie, T. Langdo, and E. A. Fitzgerald, Toward device-quality GaAs growth by molecular beam epitaxy on offcut Ge/Si1-xGex/Si substrates, *J. Vacuum Sci. Technol. B* 16(3): 1471–1474, May–June 1998.

46. M. T. Currie, S. B. Samavedam, T. A. Langdo, C. W. Leitz, and E. A. Fitzgerald, Controlling threading dislocation densities in Ge on Si using graded SiGe layers and chemical-mechanical polishing, *Appl. Phys. Lett.* 72(14): 1718–1720, April 1998.

47. O. Kwon, J. J. Boeckl, M. L. Lee, A. J. Pitera, E. A. Fitzgerald, and S. A. Ringel, Monolithic integration of AlGaInP laser diodes on SiGe/Si substrates by molecular beam epitaxy, *J. Appl. Phys.* 100(1): 013103, July 2006.

48. S. H. Huang, G. Balakrishnan, A. Khoshakhlagh, L. R. Dawson, and D. L. Huffaker, Simultaneous interfacial misfit array formation and antiphase domain suppression on miscut silicon substrate, *Appl. Phys. Lett.* 93(7): 071102, August 2008.

49. W. I. Wang, Molecular-beam epitaxial-growth and material properties of GaAs and AlGaAs on Si(100), *Appl. Phys. Lett.* 44(12): 1149–1151, June1984.

50. A. Trampert, E. Tournie, and K. H. Ploog, Novel plastic strain-relaxation mode in highly mismatched III-V-layers induced by 2-Dimensional epitaxial-growth, *Appl. Phys. Lett.* 66(17): 2265–2267, April 1995.

51. Y. H. Kim, J. Y. Lee, Y. G. Noh, M. D. Kim, S. M. Cho, Y. J. Kwon, and J. E. Oh, Growth mode and structural characterization of GaSb on Si (001) substrate: A transmission electron microscopy study, *Appl. Phys. Lett.* 88(24): 241907, June 2006.

52. G. Balakrishnan, A. Jallipalli, P. Rotella, S. H. Huang, A. Khoshakhlagh, A. Amtout, S. Krishna, L. R. Dawson, and D. L. Huffaker, Room-temperature optically pumped (Al)GaSb vertical-cavity surface-emitting laser monolithically grown on an Si(100) substrate, *IEEE J. Select. Top. Quantum Electron.* 12(6): 1636–1641, November–December 2006.

53. J. Tatebayashi, A. Jallipalli, M. N. Kutty, S. Huang, K. Nunna, G. Balakrishnan, L. R. Dawson, and D. L. Huffaker, Monolithically integrated III-Sb-based laser diodes grown on miscut Si substrates, *IEEE J. Select. Top. Quantum Electron.* 15(3): 716–723, May–June 2009.

54. A. W. Fang, M. N. Sysak, B. R. Koch, R. Jones, E. Lively, Y. H. Kuo, D. Liang, O. Raday, and J. E. Bowers, Single-wavelength silicon evanescent lasers, *IEEE J. Select. Top. Quantum Electron.* 15(3): 535–544, May–June 2009.

55. P. Bhattacharya and Z. Mi, Quantum-dot optoelectronic devices, *Proc. IEEE* 95(9): 1723–1740, September 2007.

56. P. G. Eliseev, H. Li, G. T. Liu, A. Stintz, T. C. Newell, L. F. Lester, and K. J. Malloy, Ground-state emission and gain in ultralow-threshold InAs-InGaAs quantum-dot lasers, *IEEE J. Select. Top. Quantum Electron.* 7(2): 135–142, March–April 2001.

57. S. Freisem, G. Ozgur, K. Shavritranuruk, H. Chen, and D. G. Deppe, Very-low-threshold current density continuous-wave quantum-dot laser diode, *Electron. Lett.* 44(11): 679–680, May 2008.

58. R. L Sellin, C. Ribbat, D. Bimberg, F. Rinner, H. Konstanzer, M. T. Kelemen, and M. Mikulla, High-reliability MOCVD-grown quantum dot laser, *Electron. Lett.* 38(16): 883–884, August 2002.

59. S. Fathpour, Z. Mi, P. Bhattacharya, A. R. Kovsh, S. S. Mikhrin, I. L. Krestnikov, A. V. Kozhukhov, and N. N.Ledentsov, The role of Auger recombination in the temperature-dependent output characteristics (T_0 = ∞) of p-doped 1.3 µm quantum dot lasers, *Appl. Phys. Lett.* 85(22): 5164–5166, November 2004.

60. S. Fathpour, Z. Mi, and P. Bhattacharya, High-speed quantum dot lasers, *J. Phys. D–Appl. Phys.* 38(13): 2103–2111, July 2005.

61. Z. Mi, P. Bhattacharya, and S. Fathpour, High-speed 1.3 µm tunnel injection quantum-dot lasers, *Appl. Phys. Lett.* 86(15): 153109, April 2005.

62. P. K. Kondratko, S. L. Chuang, G. Walter, T. Chung, and N. Holonyak, Observations of near-zero linewidth enhancement factor in a quantum-well coupled quantum-dot laser, *Appl. Phys. Lett.* 83(23): 4818–4820, December 2003.

63. P. R. Berger, K. Chang, P. Bhattacharya, J. Singh, and K. K. Bajai, Role of strain and growth-conditions on the growth front profile of $In_xGa_{1-x}As$ on GaAs during the pseudomorphic growth regime, *Appl. Phys. Lett.* 53(8): 684–686, August 1988.

64. D. Leonard, M. Krishnamurthy, C. M. Reaves, S. P. Denbaars, and P. M. Petroff, Direct formation of quantum-sized dots from uniform coherent islands of InGaAs on GaAs surfaces, *Appl. Phys. Lett.*, 63(23): 3203–3205, December 1993.

65. A. Madhukar, Q. Xie, P. Chen, and A. Konkar, Nature of strained InAs 3-dimensional island formation and distribution on GaAs(100), *Appl. Phys. Lett.* 64(20): 2727–2729, May 1994.

66. P. D. Siverns, S. Malik, G. McPherson, D. Childs, C. Roberts, R. Murray, and B. A. Joyce, Scanning transmission-electron microscopy study of InAs/GaAs quantum dots, *Phys. Rev. B* 58(16): 10127–10130, October 1998.

67. Z. Mi, J. Yang, and P. Bhattacharya, Molecular beam epitaxial growth and characteristics of ultra-low threshold 1.45 µm metamorphic InAs quantum dot lasers on GaAs, *J. Crystal Growth* 301: 923–926, April 2007.

68. B. Alloing, C. Zinoni, V. Zwiller, L. H. Li, C. Monat, M. Gobet, G. Buchs, A. Fiore, E. Pelucchi, and E. Kapon, Growth and characterization of single quantum dots emitting at 1300 nm, *Appl. Phys. Lett.* 86(10): 101908, March 2005.

69. S. Fafard, K. Hinzer, S. Raymond, M. Dion, J. McCaffrey, Y. Feng, and S. Charbonneau, Red-emitting semiconductor quantum dot lasers, *Science* 274(5291): 1350–1353, November 1996.

70. A. E. Zhukov, A. R. Kovsh, N. A. Maleev, S. S. Mikhrin, V. M. Ustinov, A. F. Tsatsul'nikov, M. V. Maximov, B. V. Volovik, D. A. Bedarev, Y. M. Shernyakov, P. S. Kop'ev, Z. I. Alferov, N. N. Ledentsov, and D. Bimberg, Long-wavelength lasing from multiply stacked InAs/InGaAs quantum dots on GaAs substrates, *Appl. Phys. Lett.* 75(13): 1926–1928, September 1999.

71. L. Y. Karachinsky, T. Kettler, I. I. Novikov, Y. M. Shernyakov, N. Y. Gordeev, M. V. Maximov, N. V. Kryzhanovskaya, A. E. Zhukov, E. S. Semenova, A. P. Vasil'ev, V. M. Ustinov, G. Fiol, M. Kuntz, A. Lochmann, O. Schulz, L. Reissmann, K. Posilovic, A. R. Kovsh, S. S. Mikhrin, V. A. Shchukin, N. N. Ledentsov, and D. Bimberg, Metamorphic 1.5 μm-range quantum dot lasers on a GaAs substrate, *Semiconductor Sci. Technol.* 21(5): 691–696, May 2006.

72. Z. Mi and P. Bhattacharya, DC and dynamic characteristics of p-doped and tunnel injection 1.65-μm InAs quantum-dash lasers grown on InP (001), *IEEE J. Quantum Electron.* 42(11–12): 1224–1232, November–December 2006.

73. Q. H. Xie, A. Madhukar, P. Chen, and N. P. Kobayashi, Vertically self-organized InAs quantum box islands on GaAs(100), *Phys. Rev. Lett.* 75(13): 2542–2545, September 1995.

74. G. S. Solomon, J. A. Trezza, A. F. Marshall, and J. S. Harris, Vertically aligned and electronically coupled growth induced InAs islands in GaAs, *Phys. Rev. Lett.* 76(6): 952–955, February 1996.

75. E. Le Ru, P. Howe, T. S. Jones, and R. Murray, Strain-engineered InAs/GaAs quantum dots for long-wavelength emission, *Phys. Rev. B* 67(16): 165303, April 2003.

76. I. Mukhametzhanov, R. Heitz, J. Zeng, P. Chen, and A. Madhukar, Independent manipulation of density and size of stress-driven self-assembled quantum dots, *Appl. Phys. Lett.* 73(13): 1841–1943, September 1998.

77. S. Krishna, J. Sabarinathan, K. Linder, P. Bhattacharya, B. Lita, and R. Goldman, Growth of high density self-organized (In,Ga)As quantum dots with ultranarrow photoluminescence linewidths using buried In(Ga,Al)As stressor dots, *J. Vacuum Sci. Technol. B* 18(3):1502–1506, May–June 2000.

78. Z. Mi and P. Bhattacharya, Molecular-beam epitaxial growth and characteristics of highly uniform InAs/GaAs quantum dot layers, *J. Appl. Phys.* 98(2):023510, July 2005.

79. Z. Mi and P. Bhattacharya, Pseudomorphic and metamorphic quantum dot heterostructures for long-wavelength lasers on GaAs and Si (Invited paper), *IEEE J. Select. Top. Quantum Electron.* 14(4): 1171–1179, July–August 2008.

80. H. Y. Liu, I. R. Sellers, T. J. Badcock, D. J. Mowbray, M. S. Skolnick, K. M. Groom, M. Gutierrez, M. Hopkinson, J. S. Ng, J. P. R. David, and R. Beanland, Improved performance of 1.3 µm multilayer InAs quantum-dot lasers using a high-growth-temperature GaAs spacer layer, *Appl. Phys. Lett.* 85(5): 704–706, August 2004.

81. D. S. Sizov, M. V. Maksimov, A. F. Tsatsul'nikov, N. A. Cherkashin, N. V. Kryzhanovskaya, A. B. Zhukov, N. A. Maleev, S. S. Mikhrin, A. P. Vasil'ev, R. Selin, V. M. Ustinov, N. N. Ledentsov, D. Bimberg, and Z. I. Alferov, The influence of heat treatment conditions on the evaporation of defect regions in structures with InGaAs quantum dots in the GaAs matrix, *Semiconductors* 36(9): 1020–1026, September 2002.

82. K. K. Linder, J. Phillips, O. Qasaimeh, X. F. Liu, S. Krishna, P. Bhattacharya, and J. C. Jiang, Self-organized $In_{0.4}Ga_{0.6}As$ quantum-dot lasers grown on Si substrates, *Appl. Phys. Lett.* 74(10): 1355–1357, March 1999.

83. Z. Mi, P. Bhattacharya, J. Yang, and K. P. Pipe, Room-temperature self-organised $In_{0.5}Ga_{0.5}As$ quantum dot laser on silicon, *Electron. Lett.* 41(13): 742–744, June 2005.

84. L. Li, D. Guimard, M. Rajesh, and Y. Arakawa, Growth of InAs/Sb:GaAs quantum dots on silicon substrate with high density and efficient light emission in the 1.3 µm band, *Appl. Phys. Lett.* 92(26): 263105, June 2008.

85. H. Tanoto, S. F. Yoon, C. Y. Ngo, W. K. Loke, C. Dohrman, E. A. Fitzgerald, and B. Narayanan, Structural and optical properties of stacked self-assembled $Sig_{-x}Ge_x$/Si quantum dots on graded Si1-xGex/Si substrate, *Appl. Phys. Lett.* 92(21): 213115, May 2008.

86. Z. I. Kazi, T. Egawa, M. Umeno, and T. Jimbo, Growth of InxGa1-xAs quantum dots by metal-organic chemical vapor deposition on Si substrates and in $In_xGa_{1-x}As$-based lasers, *J. Appl. Phys.* 90(11): 5463–5468, December 2001.

87. B. H. Choi, C. M. Park, S. H. Song, M. H. Son, S. W. Hwang, D. Ahn, and E. K. Kim, Selective growth of InAs self-assembled quantum dots on nanopatterned SiO_2/Si substrate, *Appl. Phys. Lett.* 78(10): 1403–1405, March 2001.

88. O. Ueda, T. Soga, T. Jimbo, and M. Umeno, Direct evidence for self-annihilation of antiphase domains in GaAs/Si heterostructures, *Appl. Phys. Lett.* 55(5): 445–447, July 1989.

89. J. W. Matthews and A. E. Blakeslee, Defects in epitaxial multilayers: I. Misfit dislocations, *J. Crystal. Growth*, 27(1): 118–125, December 1974.

90. N. A. EL-Masry, J. C. Tarn and N. H. Karam, Interactions of dislocations in GaAs grown on Si substrates with InGaAs-GaAsP strained layered superlattices, *J. Appl. Phys.* 64(7): 3672–3677, October 1988.

91. D. Huang, M. A. Reshchikov, F. Yun, T. King, A. A. Baski and H. Morkoc, Defect reduction with quantum dots in GaN grown on sapphire substrates by molecular beam epitaxy, *Appl. Phys. Lett.* 80(2): 216–218, January 2002.

92. K. Tillmann and A. Forster, Critical dimensions for the formation of interfacial misfit dislocations of $In_{0.6}Ga_{0.4}As$ islands on GaAs(001), *Thin Solid Film* 368(1): 93–104, June 2000.

93. I. A. Ovid'ko, Relaxation mechanisms in strained nanoislands, *Phys. Rev. Lett.* 88(4): 046103, January 2002.

94. J. Y. Tsao and B. W. Dodson, Excess stress and the stability of strained heterostructures, *Appl. Phys. Lett.* 53(10): 848–850, September 1988.

95. O. B. Shchekin and D. G. Deppe, 1.3 µm InAs quantum dot lasers with T_0 = 161K from 0 to 80°C, *Appl. Phys. Lett.* 80(18): 3277–3279, May 2002.

96. Z. Mi, J. Yang, P. Bhattacharya, P. K. L. Chan, and K. P. Pipe, High performance self-organized InGaAs quantum dot lasers on silicon, *J. Vacuum Sci. Technol. B* 24(3): 1519–1522, May–June 2006.

97. K. K. Kamath, J. Phillips, H. Jiang, J. Singh, and P. Bhattacharya, Small-signal modulation and differential gain of single-mode self-organized $In_{0.4}Ga_{0.6}As$/GaAs quantum dot lasers, *Appl. Phys. Lett.* 70(22): 2952–2953, June 1997.

98. D. G. Deppe and D. L. Huffaker, Quantum dimensionality, entropy, and the modulation response of quantum dot lasers, *Appl. Phys. Lett.* 77(21): 3325–3327, November 2000.

99. L. A. Coldren, K. Furuya, B. I. Miller, and J. A. Rentschler, Etched mirror and groove-coupled GaInAsP/InP laser devices for integrated optics, *IEEE J. Quantum Electron.* 18(10): 1679–1688, 1982.

100. J. Yang, Z. Mi, and P. Bhattacharya, Grooved-coupled InGaAs/GaAs quantum dot laser/waveguide on silicon, *J. Lightwave Technol.* 25(7): 1826–1831, July 2007.

101. J. Yang and H. G. Winful, Generalized eikonal treatment of the Gouy phase shift, *Opt. Lett.* 31(1): 104–106, January 2006.

102. C. Ribbat, R. L. Sellin, I. Kaiander, F. Hopfer, N. N. Ledentsov, D. Bimberg, A. R. Kovsh, V. M. Ustinov, A. E. Zhukov, and M. V. Maximov, Complete suppression of filamentation and superior beam quality in quantum dot lasers, *Appl. Phys. Lett.* 82(6). 952–954, February 2003.

103. P. A. Belanger, Beam propagation and ABCD ray matrices, *Opt. Lett.* 16(4): 196–198, February 1991.

104. Y. Yuan, T. Brock, P. Bhattacharya, C. Caneau, and R. Bhat, Edge-emitting lasers with short-period semiconductor air distributed Bragg reflector mirrors, *IEEE Photon. Technol. Lett.* 9(7): 881–883, July 1997.

105. M. P. Mack, G. D. Via, A. C. Abare, M. Hansen, P. Kozodoy, S. Keller, J. S. Speck, U. K. Mishra, L. A. Coldren, and S. P. DenBaars, Improvement of GaN-based laser diode facets by FIB polishing, *Electron. Lett.* 34(13): 1315–1316, June 1998.

106. L. Bach, S. Rennon, J. P. Reithmaier, A. Forchel, J. L. Gentner, and L. Goldstein, Laterally coupled DBR laser emitting at 1.55 μm fabricated by focused ion beam lithography, *IEEE Photon. Technol. Lett.* 14(8): 1037–1039, August 2002.

107. J. Yang, P. Bhattacharya, and Z. Wu, Monolithic integration of InGaAs-GaAs quantum-dot laser and quantum-well electroabsorption modulator on silicon, *IEEE Photon. Technol. Lett.* 19(10): 747–749, May–June 2007.

108. J. Yang and P. Bhattacharya, Integration of epitaxially-grown InGaAs/GaAs quantum dot lasers with hydrogenated amorphous silicon waveguides on silicon, *Opt. Express* 16(7): 5136–5140, March 2008.

109. J. Yang, P. Bhattacharya, Z. Mi, G. X. Qin, and Z. Q. Ma, Quantum dot lasers and integrated optoelectronics on silicon platform, *Chinese Opt. Lett.* 6(10):727–731, October 2008.

110. G. Cocorullo, F. G. Della Corte, R. De Rosa, I. Rendina, A. Rubino, and E. Terzini, Amorphous silicon-based guided-wave passive and active devices for silicon integrated optoelectronics, *IEEE J. Select. Top. Quantum Electron.* 4(6): 997–1002, November–December 1998.

111. E. Menard, K. J. Lee, D.-Y. Khang, R. G. Nuzzo, and J. A. Rogers, A printable form of silicon for high performance thin film transistors on plastic substrates, *Appl. Phys. Lett.* 84(26): 5398–5400, June 2004.

112. H. Yuan and Z. Q. Ma, Microwave thin-film transistors using Si nanomembranes on flexible polymer substrate, *Appl. Phys. Lett.* 89(21): 212105, November 2006.

113. J. Xu, D. H. Cui, T. Zhu, G. Paradee, Z. Q. Liang, Q. Wang, S. Y. Xu, and A. Y. Wang, Synthesis and surface modification of PbSe/PbS core-shell nanocrystals for potential device applications, *Nanotechnology* 17(21):5428–5434, October 2006.

114. Z. Wu, Z. Mi, P. Bhattacharya, T. Zhu, and J. Xu, Enhanced spontaneous emission at 1.55 μm from colloidal PbSe quantum dots in a Si photonics crystal microcavity, *Appl. Phys. Lett.* 90(17): 171105, April 2007.

115. R. Bose, X. Yang, R. Chatterjee, J. Gao, and C. W. Wong, Weak coupling interactions of colloidal lead sulphide nanocrystals with silicon photonic crystal nanocavities near 1.55 μm at room temperature, *Appl. Phys. Lett.* 90(11), 111117, March 2007.

116. J. Yang, J. Heo, T. Zhu, J. Xu, F. Vollmer, J. Topolancik, R. Ilic, and P. Bhattacharya, Enhanced photoluminescence from embedded PbSe colloidal quantum dots in silicon-based random photonic crystal microcavities, *Appl. Phys. Lett.* 92 (26): 261110, July 2008.

117. K. Tanabe, M. Nomura, D. Guimard, S. Iwamoto, and Y. Arakawa, Room temperature continuous wave operation of InAs/GaAs quantum dot photonic crystal nanocavity laser on silicon substrate, *Opt. Express* 17(9): 7036–7042, April 2009.

118. G. Qin, H. Zhou, E. B. Ramayya, Z. Ma, and I. Knezevic, Electron mobility in scaled silicon metal-oxide-semiconductor field-effect transistors on off-axis substrates, *Appl. Phys. Lett.* 94(7): 073504, February 2009.

119. R. Chau, S. Datta, M. Doczy, B. Doyle, J. Kavalieros, and M. Metz, High-k/metal-gate stack and its MOSFET characteristics, *IEEE Electron Dev. Lett.*, 25(6): 408–410, June 2004.

120. http://www.compoundsemiconductor.net news, IBM and Intel make high-k gate breakthrough, http://compoundsemiconductor.net/cws/article/news/26922, January 29, 2007.

121. L. Cerutti, J. B. Rodriguez, and E. Tournie, GaSb-based aser, monolithically grown on silicon substrate, emitting at 1.55 μm at room temperature. *IEEE Photon. Tech. Lett.* 22(8):553–555, April 2010.

CHAPTER **10**

Three-Dimensional Integration of CMOS and Photonics

Prakash Koonath, Tejaswi Indukuri, and Bahram Jalali

Contents

10.1 Introduction

Silicon-on-insulator (SOI) materials systems have proved to be an efficient platform for the realization of both electronic and photonic devices. Recent years have seen a significant amount of research in the area of silicon photonics, resulting in the demonstration of a variety of both active and passive optical functionalities [1–16]. These technologies have demonstrated the feasibility of using silicon for the realization of integrated optical devices.

One of the most attractive features of silicon is the prospect of full integration of optical and electronic devices on the same substrate.

Transistor size continues to shrink, a trend driven by the economic benefit of having a larger number of circuits from a single silicon wafer. Ultra-large scale integration (ULSI), with the number of transistors per chip exceeding 1 billion (10^9), has increased the economic incentive to utilize silicon real estate efficiently. While today's electronic chips boast critical dimensions of 35 nm [17], the dimensions of optical waveguides have a hard lower limit of more than 200 nm, set by the optical wavelength in silicon [18,19]. It is therefore necessary to develop innovative fabrication technologies that would make the integration of optical and electronic devices viable on silicon substrates, without compromising the real-estate economics of wafer manufacturing. Furthermore, in order to utilize the foundry capabilities in silicon, it is necessary that these technologies be compatible with the well-established CMOS processing techniques.

The realization of devices on a chip in a three-dimensional (3D) fashion is an efficient way to fabricate densely integrated structures, thereby enhancing the functionality of the chip. Apart from this inherent advantage, in the realization of guided-wave structures, 3D integration offers the prospect of precise control of coupling coefficient in vertically coupled devices [20,21]. Here, the control over the critical dimension is more precise than laterally patterned structures, where the limits are set by the photolithography. Thus, complex optical circuitry with accurately controlled evanescent coupling between devices is possible by employing vertically integrated optical devices.

A novel method known as separation by implantation of oxygen (SIMOX) 3D sculpting has been developed to realize 3D integrated photonic and electronic devices on the same silicon substrate in a monolithic fashion. In this approach, devices are confined to vertically stacked layers of silicon, separated from each other by intervening layers of silicon dioxide. For example, optoelectronic integration is achieved by confining photonic circuits to buried silicon layers and fabricating transistors on the surface layer silicon of this vertical stack of silicon layers. One of the most attractive features of this technique is that the integration of photonic devices does not impact the availability of silicon real estate from the point of view of electronic circuitry as optical and electronic functionalities are separated to different layers of silicon [22]. A variety of optical devices including buried waveguides, vertically coupled microdisk resonators, add-drop multiplexers, and multipole filters have also been realized using this process of 3D integration. This chapter describes the process of SIMOX 3D sculpting and the synthesis of 3D integration of photonic and electronic devices in silicon.

10.2 SIMOX 3D Sculpting

10.2.1 Process of SIMOX 3D Sculpting

The SIMOX process involves the implantation of oxygen ions into a silicon substrate, followed by a high-temperature anneal (around 1300°C) of the substrate in order to cure the implantation damage and to effect SiO_2 formation. The thickness and depth of the buried oxide layer are, respectively, determined by the implantation dose and energy. It has been observed that in order to achieve good-quality buried oxide and to keep the defect densities in the range of $<10^5/cm^2$, the implantation dose should be in the range of 1×10^{17}–9×10^{17} ions per cm^2 with implantation energies in the range of 40–200 KeV [23]. The process is conventionally used to obtain thin silicon layers (of the order of 3000 Å) on top of a buried oxide layer of thickness of the same order of magnitude. Figure 10.1 depicts the process flow of the fabrication of vertically integrated structures using the technique of SIMOX 3D sculpting. Implantation of oxygen ions is performed on an SOI substrate that has been patterned with thermally grown oxide. The thickness of the oxide mask may be chosen suitably to decelerate the oxygen ions that penetrate into the area underneath the mask. The angled sidewall of the buried rib waveguide formed after the high-temperature anneal arises due to the lateral straggle of the implanted oxygen ions. After annealing, devices are defined on

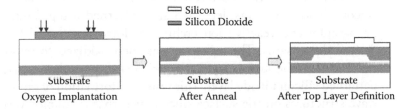

FIGURE 10.1 Process flow of SIMOX 3D sculpting. (From Koonath, P. et al. 2006. *Monolithic 3-D Silicon Photonics*. IEEE. 1796-1804; Koonath, P. et al., 2007. Multilayer 3-D photonics. *OSA*. 12686–12691; Koonath, P. et al., 2003. Sculpting 3-D nano optics in Si. AIP.4909.)

the top layer using a conventional lithography and etching process, as shown in Figure 10.1.

10.2.2 Fabrication of 3D Integrated Optical Devices

A SOI wafer (made by SOITEC Inc.) with 0.6 μm of silicon on top of a buried oxide layer of 1.0 μm thickness is oxidized and patterned using reactive ion-etching process to form oxide stripes of thickness 0.06 μm and of width 2 μm. The patterned wafer was then implanted with oxygen ions with a dose of 5×10^{17} ions per cm^2, at an energy of 150 KeV. The implanted wafers were then annealed at 1320°C for 7.5 h in an ambient of argon, with 1% oxygen, to cure the implantation damage. Figure 10.2 shows the SEM photograph of a buried rib waveguide structure that was fabricated employing the technique of SIMOX 3D sculpting, the dimensions of which are shown in the figure. It may be seen from the figure that the process Si has resulted in formation of submicron rib waveguides in the bottom silicon layer, separated from the continuous silicon layer on the top by the oxide layer formed after the oxygen implantation and subsequent anneal. The buried oxide that was formed after the implantation and anneal is found to be uniform, which is very important in achieving accurate control of evanescent coupling from the rib waveguides to the devices on the top silicon layer. It has been observed that, under nonideal conditions, the SIMOX process can result in the formation of a high density of silicon islands inside the buried oxide [24]. The optimization of the SIMOX process was successful in preventing the formation of these islands that degrade the quality of the buried oxide. Measurements based on a cutback technique reveal that these buried waveguides have propagation losses of ~4 dB/cm.

In order to fabricate optical devices on the top silicon layer, a silicon nitride layer was deposited and patterned using standard lithography and reactive ion etching to form circular disks of desired dimensions. The substrate was then oxidized to remove the silicon on the surface layer, everywhere except underneath the patterned silicon nitride layer. This results in formation of guided-wave structures on the surface silicon layer that is coupled vertically to the rib waveguides in the buried silicon layer through the oxide layer formed after the oxygen implantation. This technique has been utilized to realize a variety of vertically coupled optical devices that are described in the following sections.

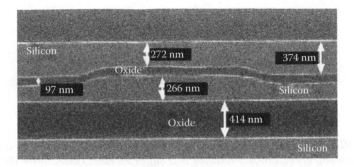

FIGURE 10.2 Cross-sectional SEM view of the buried rib waveguide formed by the process of SIMOX 3D sculpting. (From Koonath, P. et al. 2006. *Monolithic 3-D Silicon Photonics* IEEE. 1796–1804.)

10.3 Device Characteristics of 3D Integrated Optical Devices

10.3.1 Vertically Coupled Microresonators

Figure 10.3 shows the scanning electron microscopy (SEM) photographs of a microdisk resonator fabricated on the surface layer silicon, straddling the buried bus waveguide. The silicon nitride disk used as the oxidation mask is also seen in the picture on top of the microdisk. It may be noted here that the silicon dioxide that was formed after the oxidation and the buried oxide layer formed during the implantation were removed to obtain SEM photographs that clearly illustrate the structure of the device. The optical micrograph in Figure 10.4 shows the top view of the fabricated device. Figure 10.5 shows the experimental setup used to characterize the fabricated microdisk resonators. Unpolarized light from an amplified spontaneous emission (ASE) source was

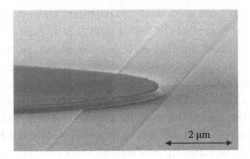

FIGURE 10.3 SEM picture of the fabricated microdisk resonators of radius 23 μm on the top silicon layer with bus waveguides underneath. (From Koonath, P. et al. 2006. *Monolithic 3-D Silicon Photonics* IEEE. 1796–1804.)

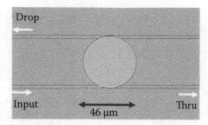

FIGURE 10.4 Top view of a microdisk resonator vertically coupled to bus waveguides. (From Koonath, P. et al. 2006. *Monolithic 3-D Silicon Photonics* IEEE. 1796–1804)

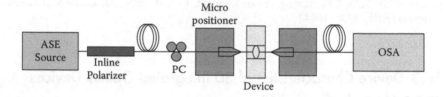

FIGURE 10.5 Experimental setup used to characterize the microdisk resonators. (From Koonath, P. et al. 2006. *Monolithic 3-D Silicon Photonics* IEEE. 1796–1804)

passed through an inline fiber polarizer (extinction ratio 40 dB) and a polarization controller that can rotate the state of polarization of the light to any desired state. This polarized light was then coupled into the bus waveguides using a tapered fiber that has a spot diameter of 2 μm. The light output from the drop ports of the filter was collected using a tapered fiber, similar to the one used at the input. The collected light was observed using an optical spectrum analyzer. This setup was used to characterize other optical devices described in this chapter as well. Figure 10.6 shows the throughput port responses of the filters characterized by launching optical power at the input port (see Figure 10.4) and observing the optical spectra at the throughput port. The spectrum displayed in Figure 10.6 is normalized with respect to the maximum transmission of the throughput port, after correcting for the spectral shape of the ASE source. Sharp resonances are observed with a free spectral range of around 5.4 nm, with the narrowest resonance observed showing a **full width at a half-maximum** (FWHM) of 1.16 nm, centered at 1564 nm. This corresponds to a loaded quality factor, Q, of around 1350 with a suppression value greater than 30 dB for the resonant wavelength at the throughput port. The measured value of Q in this device is limited by the

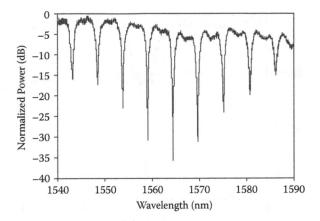

FIGURE 10.6 Throughput port transmission characteristics of the microdisk resonator. (From Koonath, P. et al. 2006. *Monolithic 3-D Silicon Photonics* IEEE. 1796–1804.)

loading of the resonator since the microdisk is separated from the bus waveguides through the intervening silicon dioxide layer of thickness of only ~100 nm.

A mathematical formalism was developed to extract the intrinsic quality factor, $Q_{intrinsic}$, of a microresonator by measuring the:

a. Loaded quality factor, Q, of the resonator in a symmetrically loaded configuration
b. Extinction ratio observed in its throughput transmission spectrum

The details of this formalism are given in [25]. From the throughput spectrum obtained in Figure 10.6, the resonator shows a loaded Q of 1350 at the resonance wavelength 1564.3 nm with an extinction ratio of 30.2 dB. Figure 10.7 shows the experimentally observed resonance at 1564.3 nm plotted along with the theoretical curve. It may be seen that there is good agreement between the experimentally observed spectrum and the theoretical plot. The estimated value of $Q_{intrinsic}$ of the device is around 4.5×10^4 at this wavelength.

10.3.2 Add-Drop Multiplexers

The process described in the previous section was used to realize add-drop multiplexers utilizing vertically coupled microdisk resonators. Figure 10.8 shows the top view of a 1×3 drop filter fabricated using this process. Disks of radii 20, 20.5, and 21

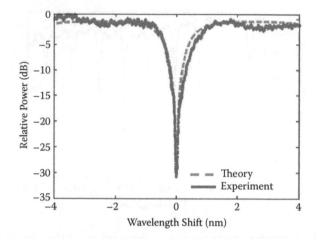

FIGURE 10.7 Comparison of the experimental data with theory for the microdisk resonance at 1564.3 nm.(From Koonath, P. et al. 2006. *Monolithic 3-D Silicon Photonics* IEEE. 1796–1804.)

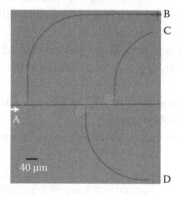

FIGURE 10.8 Top view of the fabricated 1 × 3 add-drop multiplexer using disks of radii 20, 20.5, and 21 μm. (From Koonath, P. et al. 2006. *Monolithic 3-D Silicon Photonics* IEEE. 1796–1804.)

μm were used to obtain resonators with slightly shifted resonance wavelengths for these resonators. Figure 10.9 shows the drop port responses of the filters, characterized by launching optical power at port A and observing the drop port optical spectra at ports B, C, and D. Fabricated disks show a free spectral range of around 5 nm. The adjacent channel wavelength separation between port B and port C is approximately 1.0 nm, whereas that of port B and port D is around 1.5 nm. The average value of adjacent channel crosstalk suppression, over the wavelength band of 1534–1560 nm, was

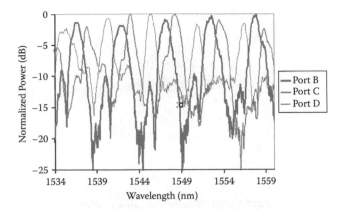

FIGURE 10.9 Drop port optical spectra of the 1 × 3 add-drop filter. (From Koonath, P. et al. 2006. *Monolithic 3-D Silicon Photonics* IEEE. 1796–1804.)

found to be 12.1 dB for channels dropped at port D. At ports B and C, these values were found to be 8.3 and 6.2 dB, respectively.

10.3.3 Multipole Filters Based on Microresonators

It has been demonstrated that cascaded resonator structures may be utilized to improve the filter characteristics of devices employing microcavities, including the flatness of the pass-band, the sharpness of the roll-off from the pass-band to the stop-band, and the suppression of adjacent channel cross-talk [26]. SIMOX 3D sculpting has been utilized to fabricate devices consisting of two microdisk resonators vertically coupled to each other through the oxide layer formed after the implantation and subsequent anneal. The resonators themselves are coupled to bus waveguides that act as the input and output ports of the device. Figure 10.10 represents a schematic of the device showing the vertically coupled microresonators and the bus waveguides with the arrows indicating the direction of flow of light through these devices.

FIGURE 10.10 Three-dimensional schematic of vertically coupled microdisk resonators and bus waveguides illustrating the flow of optical energy. (From Koonath, P. et al. 2006. *Monolithic 3-D Silicon Photonics* IEEE. 1796–1804.)

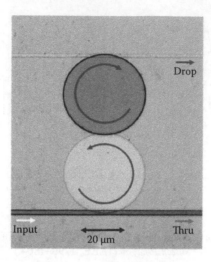

FIGURE 10.11 Top view of the vertically coupled cascaded microdisk resonators. (From Koonath, P. et al. 2006. *Monolithic 3-D Silicon Photonics* IEEE. 1796–1804.)

It may be emphasized here that waveguides as well as microdisk structure were realized on both layers of silicon coupled vertically through an intervening layer of SiO_2. These devices distinctly demonstrate the capability of the SIMOX 3D sculpting to engineer 3D structures in silicon (the intervening layer of oxide through which the coupling of light takes place is omitted in Figure 10.10 for simplicity of illustration). Figure 10.11 shows the top view of the fabricated two-stage device with cascaded microdisks of radius 20 μm coupled to bus waveguides of width 2 μm. When optical energy is introduced to the input port of the device, resonant wavelengths are transmitted to the drop port, after traversing through the cascaded disk structure. Figure 10.12 shows the drop port transmission response characterized by launching optical power at the input port and collecting the optical spectra at the drop port of the device. Fabricated disks show a free spectral range of around 5.7 nm with maximum extinction ratios of ~12 dB. A comparison of the performance of the two-stage device to that of a single-stage device, as shown in Figure 10.13, illustrates the superior characteristics of the two-stage resonator in terms of its steeper roll-off, better out-of-band rejection, and flatter pass-band. Note that the frequency scale in this figure is normalized to the 3 dB bandwidths of the individual filters. The shape factor, which is a measure of the flatness of the pass-band [27], defined as the ratio of the –1 dB bandwidth to the

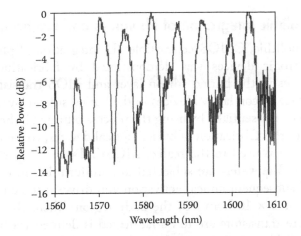

FIGURE 10.12 Drop port resonance characteristics of the cascaded device. (From Koonath, P. et al. 2006. *Monolithic 3-D Silicon Photonics* IEEE. 1796–1804.)

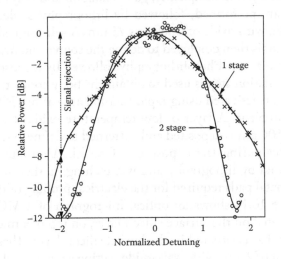

FIGURE 10.13 Comparison of the filter characteristics of the two-stage device with a single-stage device, with the frequency scale normalized to 3 dB bandwidths. (From Koonath, P. et al. 2006. *Monolithic 3-D Silicon Photonics* IEEE. 1796–1804.)

–10 dB bandwidth, is 0.47 for the two-stage device, compared to a value of 0.22 for the single-stage device. Also, at twice the value of the 3 dB bandwidth, the two-stage device shows a 4 dB improvement in the out-of-band signal rejection over the single-stage device. It may be noted that the single-stage device used for the purpose of this comparison is fabricated on the same chip as that of the two-stage device.

10.4 Monolithic Integration of Photonics and Electronics in 3D

Monolithic, CMOS-compatible 3D integration of photonic and electronic devices was demonstrated by fabricating photonic devices in the buried silicon layer and MOS transistors on the surface silicon layer. Devices in the buried silicon layer were synthesized essentially by using the process outlined in Section 10.2. After this fabrication of buried optical devices, conventional processing is used to the realize MOSFETs on the surface silicon layer. The wafers are subjected to a uniform boron implantation and subsequent anneal activation and drive-in to achieve a doping of 2×10^{17} cm^{-3} in the body region. Then, the active area where transistors are to be fabricated is defined via lithography, and field oxidation is performed everywhere except in the active region to obtain electrical isolation of the devices on the surface silicon layer. Subsequently, gate oxidation and polysilicon deposition are performed, followed by lithography to define the gate region, with oxide of thickness 25 nm. A self-aligned phosphorus implant is then employed to define the source and drain regions of the transistor. Photolithography followed by a subsequent boron implantation step is used to obtain the body contact. The dopants are then activated using rapid thermal annealing (RTA) in nitrogen ambient. A layer of low-temperature oxide (LTO) of thickness 500 nm is deposited and patterned to define vias for electrical contact during the deposition of metal. Aluminum sputtering, followed by lithography and wet etching, is then used to define the metal pads required for the electrical probing of these devices. Figure 10.14 shows an optical micrograph of a MOS transistor fabricated on the surface silicon layer, on top of a microdisk resonator that is situated in the buried silicon layer. These resonators consist of 2-µm-wide waveguides acting as input and output ports to a microdisk of radius 20 µm. The optical characterization of the microresonator filters in the buried silicon layer shows sharp resonances with a free spectral range of 5 nm and optical extinction ratios as high as 20 dB, as shown in Figure 10.15.

The electrical characteristics of these *n*-channel MOSFETs were extracted using a HP4145B semiconductor parameter analyzer. Figure 10.16 shows the drain current, I_D, versus gate voltage, V_G, characteristics demonstrating gate control in the fabricated MOSFETs. Figure 10.17 shows the drain current versus drain voltage, V_D, characteristics of the inversion channel. These transistors had a gate length of 1 µm. The devices

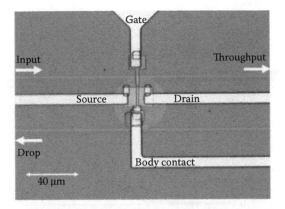

FIGURE 10.14 Optical micrograph of a MOS transistor fabricated on the surface silicon layer, on top of a microdisk resonator that is situated in the buried silicon layer. (From Indukuri, T. et al. 2006. *Three-Dimentional Integration of Metal-Oxide Semiconductor Transistor with Subterranian Photonics in Silicon.* AIP. 121108.)

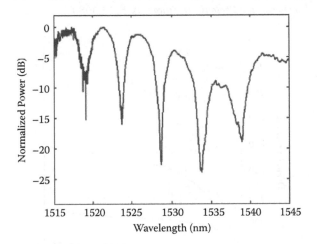

FIGURE 10.15 Thru port spectrum of the fabricated disk resonators in the buried silicon layer. (From Indukuri, T. et al. 2005. *Subterranian Silicon Photonics: Demonstration of Buried Waveguide-Coupled Resonators.* 08114. 2005.)

exhibit a threshold voltage of 2.5 V that is close to the theoretically expected value. These characteristics clearly illustrate the realization of MOS transistors that are monolithically integrated with buried optical devices.

Note that in the subthreshold region, nonzero leakage currents are observed. It has been previously observed that, in the presence of defects in the silicon wafer, there can be diffusion of dopants from source and drain regions [28]. This can result in

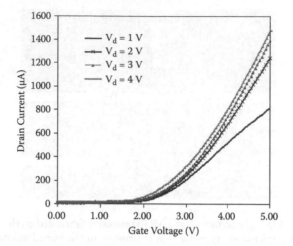

FIGURE 10.16 Gate control characteristics (I_D versus V_G) of the fabricated MOSFET. (From Indukuri, T. et al. 2006. *Three-Dimentional Integration of Metal-Oxide Semiconductor Transistor with Subterranian Photonics in Silicon.* AIP. 121108.)

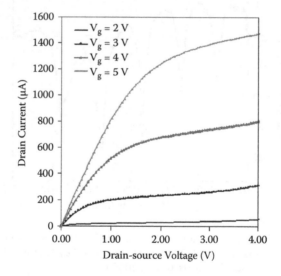

FIGURE 10.17 Drain characteristics (I_D versus V_D) of the fabricated MOSFET. (From Indukuri, T. et al. 2006. *Three-Dimentional Integration of Metal-Oxide Semiconductor Transistor with Subterranian Photonics in Silicon.* AIP. 121108.)

electrically active leakage paths from source to the drain, contributing to leakage currents in subthreshold regime. It is known that the SIMOX process can result in the creation of defects during the implantation of oxygen ions [29]. After the annealing of the wafers subsequent to oxygen implantation, defect densities on the order of 5×10^4 cm^{-2} were measured using Secco etching. It is believed that these defects are responsible for the leakage currents observed. However, the process may be optimized further to lower defect densities and obtain transistors with optimal performance characteristics. It must be emphasized, however, that a technique that results in the integration of both optical and electronic devices on the same substrate monolithically has been demonstrated. More significantly, this integration does not impact the availability of silicon real estate from the point of view of ULSI electronic circuitry as the optical and electronic functionalities are separated to two different layers of silicon.

10.5 Multilayer 3D Devices

It is desirable to extend the 3D integration approach to synthesize devices on more than two layers of silicon as it promotes another level of dense integration. During the synthesis of optical devices with MOSFETs, optical circuits had to be restricted to the subsurface silicon layer so as to dedicate the entire surface layer for electronic circuits. Thus, by extending the technology to create multilayer structures, vertically coupled optical devices can exist along with MOS transistors on the surface layer. One direct method of achieving this is through repeating the oxygen implantation and annealing steps in the SIMOX sculpting approach, by choosing the appropriate energy and dose for the oxygen ions during implantation. However, the ranges of implantation energies and doses available to create high-quality oxide layers do not lend themselves to creating multilayer structures in this straightforward fashion. This is primarily due to the limited amount of surface layer silicon that is available (<300 nm) to repeat the implantation and anneal steps. Therefore, after the first implantation and anneal, silicon needs to be epitaxially grown in order to generate multiple layers of devices.

Figure 10.18 depicts the process flow for creating multilayer SOI structures. The surface layer available after the first implantation and anneal is used as the seed layer to grow silicon epitaxially on the substrate. After the epitaxial growth, the substrate goes through another set of implantation and annealing steps, resulting in the formation of a second layer of buried devices and a surface silicon layer. Photonic or electronic devices may be defined on the surface silicon layer using a conventional lithography and etching process, resulting in the formation of three layers of 3D integrated

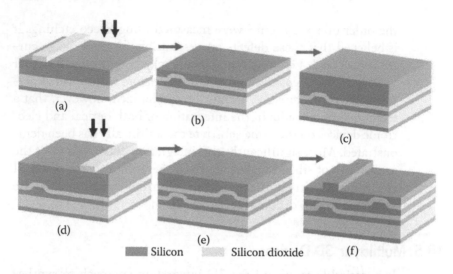

Silicon   Silicon dioxide

FIGURE 10.18 Schematic of the process flow for the fabrication of multilayer structures using SIMOX 3D sculpting. (a) Starting SOI wafer, with a semitransparent silicon oxide mask on it, is implanted with oxygen ions. (b) High-temperature anneal after the implantation results in the formation of a continuous buried oxide layer. (c) Epitaxial growth of silicon. (d) Silicon dioxide is grown thermally and patterned using photolithography to create a semitransparent oxide mask. This wafer then undergoes oxygen ion implantation as in step (a). (e) High-temperature annealing results in the realization of the second layer of subsurface waveguides separated from a surface silicon layer. (f) Photolithography and reactive ion etching performed on the surface silicon layer to create devices on the surface silicon layer. (From Koonath, P. et al. 2007. *Multilayer 3-D Photonics in Silicon*. OSA. 12686-12691. 2007.)

devices. Figure 10.19 shows the cross-sectional SEM pictures of rib waveguides realized in the three vertically stacked layers of silicon. It is very clear that the process of SIMOX 3D sculpting has successfully been employed to realize multilayer photonic structures in silicon. It may be seen from Figure 10.19b that the oxide layer that defines the second layer of buried devices is discontinuous. This is due to the fact that the amount of oxygen ion dose that entered the wafer is less than the optimum value of 5×10^{17} ions per cm^2 required for the formation of a continuous oxide layer. This can be verified by measuring the thickness of the second buried oxide layer that was formed, which is around 85 nm. For a dose of 5×10^{17} ions per cm^2, the thickness of a stoichiometric oxide layer is expected to be around 115 nm, as is measured in the case of the oxide layer formed in the first implantation step. We surmise that the difference in the dose that penetrated the wafer during the second implantation step must arise from the process variations at the commercial implantation facility where

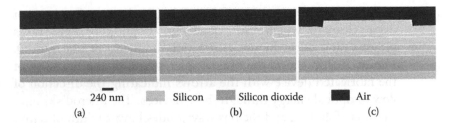

240 nm   ▨ Silicon   ▨ Silicon dioxide   ■ Air

(a)   (b)   (c)

FIGURE 10.19 Cross-sectional scanning electron microscope (SEM) pictures of devices fabricated in a multilayer structure. (a) Subsurface waveguides in the first layer of the structure after oxygen implantation and high-temperature anneal. Two layers of silicon are seen above the waveguide structure. (b) Subsurface waveguides in the second layer of the structure. A layer of silicon above and another layer of silicon below the waveguides in this layer are also seen. (c) Rib waveguides in the surface silicon realized by photolithography and etching. Two layers of silicon below this surface layer are also seen in the picture. (From Koonath, P. et al. 2007. *Multilayer 3-D Photonics in Silicon*. OSA. 12686-12691. 2007.)

the implantation was performed. By ensuring the presence of an optimum dose inside the substrate, a continuous layer of oxide can be realized. It needs to be mentioned here that, even though the oxide layer is discontinuous, finite-element-method-based simulations show that the structure supports guided optical modes.

As a vehicle for the proof of concept, an optical filter was fabricated by cascading multiple microcavities, the schematic of which is shown in Figure 10.20a. Here, the microresonators are realized in the two buried layers of silicon that are coupled to each other and to the bus waveguides fabricated on the surface silicon

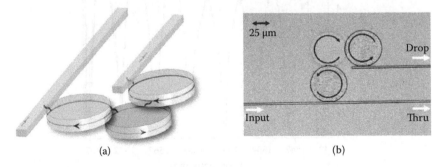

(a)   (b)

FIGURE 10.20 Three-dimensionally integrated microcavity structures in a multilayer silicon structure. (a) Schematic of the three-dimensionally coupled microcavities realized using SIMOX 3D sculpting. (b) The optical micrograph of the top view of the fabricated device where the arrows indicate the direction of flow of optical energy through the multilayer structure. (From Koonath, P. et al. 2007. *Multilayer 3-D Photonics in Silicon*. OSA. 12686-12691. 2007.)

layer through intervening oxide layers (the intervening layer of oxide through which the evanescent coupling of light takes place is omitted in Figure 10.20a for simplicity of illustration). Figure 10.20b shows the optical micrograph of the top view of the fabricated device with the arrows indicating the direction of flow of optical energy through these devices. The microdisks have a radius of 20 μm, and the bus waveguides have a width of 2 μm. When optical energy is introduced to the input port of the device, resonant wavelengths are transmitted to the drop port, after traversing vertically coupled silicon layers. The complete path of the optical field (from the input bus waveguide to the output drop port waveguide) includes five layers of silicon and four evanescent coupling stages.

The spectral characteristics of the filter at the drop port of the filter shows a free spectral range of 5.6 nm, as depicted in Figure 10.21. The extinction ratio, as measured by the ratio of the maximum power (at the resonant wavelength) to that of the minimum power (at the off-resonance wavelength), is found to have a maximum value of ~14 dB in the wavelength range of measurement. The excess insertion loss of the multistage filter structure was measured to be ~1 dB. These results validate the capability of the SIMOX 3D sculpting technique for fabricating complex 3D integrated devices.

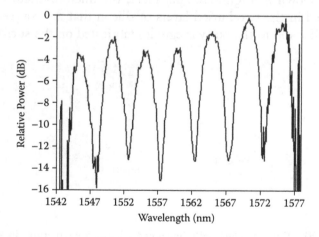

FIGURE 10.21 Spectral characteristics of the drop port of the multistage microdisk filter device. Wavelengths that are resonant with the microdisk structure travel through the multilayer structure and get collected at the drop port of the device. Nonresonant wavelengths are collected at the thru port of the device. (From Koonath, P. et al., 2007. *Multilayer 3-D Photonics in Silicon.* OSA. 12686-12691. 2007.)

10.6 Discussion and Summary

In this chapter, a novel method, SIMOX 3D sculpting, to realize 3D integrated device structures in silicon has been discussed. The method of SIMOX 3D sculpting is based on a modified version of the conventional SIMOX process used to make SOI wafers. A number of 3D integrated optical devices, including buried waveguides, vertically coupled microdisk resonators, add-drop filters, cascaded microdisk resonators, and multilayer devices, were realized. Furthermore, monolithic integration of MOS transistors with integrated optical devices was also demonstrated on this 3D integration platform. By assigning the optical and electronic functionalities to different layers of silicon, separated by an insulating oxide layer, the impact of the consumption of silicon real estate by optical devices is minimized. This is a significant advantage in the migration toward optical interconnects in silicon, as cost-effective implementation of optical interconnects requires fabrication technologies that consume the least amount of real estate on the silicon wafer. Apart from this advantage, the use of vertical coupling to inject light into resonant cavities offers precise control of the coupling coefficient in these structures. This is possible due to the fact that the critical dimension in vertically coupled devices may more precisely be controlled than laterally coupled devices. The offset between the edge of a buried waveguide and the edge of a microdisk resonator is an additional design parameter that may be used in the process developed here to engineer the coupling between the disk and the waveguide. The device characteristics of microresonant cavities depend critically on the photon lifetime in these devices. Thus, the control over the coupling coefficient acquires considerable importance in the context of designing devices based on microresonant cavities.

Acknowledgments

The authors would like to thank Koichiro Kishima for his assistance. This work was performed under the CS-WDM program funded by the MTO office of DARPA. The authors would like to thank Dr. Jag Shah of DARPA for his support.

References

1. B. Jalali, S. Yegnanarayanan, T. Yoon, T. Yoshimoto, I. Rendina, and F. Coppinger, Advances in silicon-on-insulator optoelectronics, *IEEE J. Sel. Topics Quantum Electron. (Special Issue on Silicon-Based Optoelectronics)*, vol. 4, pp. 938–47, December 1998.

2. K. Jia, W. Wang, Y. Tang, Y. Yang, J. Yang, X. Jiang, Y. Wu, M. Wang, and Y. Wang, Silicon-on-insulator-based optical demultiplexer employing turning-mirror-integrated arrayed-waveguide grating, *IEEE Photon. Technol. Lett.*, vol. 17, no. 2, pp. 378–80, February 2005.

3. T. Tsuchizawa, K. Yamada, H. Fukuda, T. Watanabe, J. Takahashi, M. Takahashi, T. Shoji, E. Tamechika, S. Itabashi, and H. Morita, Microphotonics devices based on silicon microfabrication technology, *IEEE J. Sel. Topics Quant. Electron.*, vol. 11, no. 1, pp. 232–240, January 2005.

4. A. Liu, R. Jones, L. Liao, D. Samara-Rubio, D. Rubin, O. Cohen, R. Nicolaescu, and M. Paniccia, A high-speed silicon optical modulator based on a metal–oxide–semiconductor capacitor, *Nature*, vol. 427, pp. 615–618, February 2004.

5. O. Boyraz and B. Jalali, Demonstration of a silicon Raman laser, *Opt. Express*, vol. 12, pp. 5269–5273, October 2004.

6. H. Rong, R. Jones, A. Liu, O. Cohen, D. Hak, A. Fang, and Mario Paniccia, A continuous-wave Raman silicon laser, *Nature*, vol. 433, pp. 725–728, February 2005.

7. Q. Xu, B. Schmidt, S. Pradhan, and M. Lipson, Micrometre-scale silicon electro-optic modulator, *Nature*, vol. 435, pp. 325–327, May 2005.

8. V. Raghunathan, R. Claps, D. Dimitropoulos, and B. Jalali, Wavelength conversion in silicon using Raman induced four-wave mixing, *Appl. Phys. Lett.*, vol. 84, pp. 34–36, July 2004.

9. Ö. Boyraz, P. Koonath, V. Raghunathan, and B. Jalali, *Opt. Express*, vol. 12, pp. 4094–4102, August 2004.

10. M. Borselli, K. Srinivasan, P. Barclay, and O. Painter, Rayleigh scattering, mode coupling, and optical loss in silicon microdisks, *Appl. Phys. Lett.*, vol. 85, pp. 3693–95, October 2004.

11. W. R. Headley, G. T. Reed, S. Howe, A. Liu, and M. Paniccia, Polarization-independent optical racetrack resonators using rib waveguides on silicon-on-insulator, *Appl. Phys. Lett.*, vol. 85, pp. 5523–25, December 2004.

12. P. Dumon, W. Bogaerts, V. Wiaux, J. Wouters, S. Beckx, J. V. Campenhout, D. Taillaert, B. Luyysaert, P. Bienstman, D. V. Thourhout, and R. Baets, Low-loss SOI photonic wires and ring resonators fabricated with deep UV lithography, *IEEE Photon. Technol. Lett.*, vol. 16, pp. 1328–30, May 2004.

13. V. R. Almeida, C. A. Barrios, R. R. Panepucci, and M. Lipson, All-optical control of light on a silicon chip, *Nature*, vol. 431, pp. 1081–1084, October 2004.

14. A. R. Cowan, G. W. Rieger, and J. F. Young, Nonlinear transmission of 1.5 μm pulses through single-mode silicon-on-insulator waveguide structures, *Opt. Express*, vol. 12, pp. 1611–1621, April 2004.

15. J. I. Dadap, R. L. Espinola, R. M. Osgood, Jr., S. J. McNab, and Y. A. Vlasov, Spontaneous Raman scattering in a silicon wire waveguide, *Proc. IPR*, Paper IWA4, July 2004.

16. T. K. Liang and H. K. Tsang, Pulsed-pumped silicon-on-insulator waveguide Raman amplifier, *Proc. Int. Conf. Group IV Photonics*, Paper WA4, 2004.

17. S. Tyagi et al., A 65 nm ultra low power logic platform technology using uni-axial strained silicon transistors, *IEEE IEDM Tech. Digest*, 245–247, 2005.

18. T. Tsuchizawa et al., Microphotonics devices based on silicon microfabrication technology, *IEEE J. Sel. Topics Quant. Electron.*, 11, 232–240 (2005).

19. Y. A. Vlasov and S. J. McNab Losses in single-mode silicon-on-insulator strip waveguides and bends, *Opt. Express* 21, 1622–1631 (2004).

20. S. J. Choi, K. Djordjev, J. C. Sang, P. D. Dapkus, W. Lin, G. Griffel, R. Menna, and J. Connolly, Microring resonators vertically coupled to buried heterostructure bus waveguides, *IEEE Photonics Technol. Lett.*, vol. 16, pp. 828–30, March 2004.

21. S. M. Garner, S. Lee, V. Chuyanov, A. Chen, A. Yacoubian, W. H. Steier, and L. R. Dalton, Three-dimensional integrated optics using polymers, *IEEE J. Quantum Electron.*, vol. 35, pp. 1146–1155, August 1999.

22. P. Koonath, K. Kishima, T. Indukuri, and B. Jalali, Sculpting of three-dimensional nano-optical structures in silicon, *Appl. Phys. Lett.*, vol. 83, pp. 4909–4911, December 2003.

23. M. Chen, X. Wang, J. Chen, X. Liu, Y. Dong, Y. Yu, and X. Wang, Dose-energy match for the formation of high-integrity buried oxide layers in low-dose separation-by-implantation-of-oxygen materials, *Appl. Phys. Lett,*, vol. 80, pp. 880–882, February 2002.

24. J. Jiao, B. Johnson, S. Seraphin, M. Anc, R. Dolan, and B. Cordts, Formation of Si islands in the buried oxide layers of ultra-thin SIMOX structures implanted at 65 keV, *Mater. Sci. Eng.* vol. 72, pp. 150–155, March 2000.

25. P. Koonath, T. Indukuri, and B. Jalali, Monolithic 3-D Silicon Photonics, *IEEE J. Lightwave Technol.*, vol. 24, pp. 1796–1804 (2006).

26. J. V. Hryniewicz, P. P. Absil, B. E. Little, R. A. Wilson, and P. T. Ho, Higher order filter response in coupled microring resonators, *IEEE Photonics Technol. Lett.*, vol. 12, pp. 320–322 March 2000.

27. Y. Yanagase, S. Suzuki, Y. Kokubun, and S. T. Chu, Box-like filter response and expansion of FSR by a vertically triple coupled microring resonator filter, *IEEE J. Lightwave Technol.*, vol. 20, 1525–1529 August 2002.

28. J. W. Sleight, C. Lin, and G. J. Grula, Stress induced defects and transistor leakage for shallow trench isolated SOI, *IEEE Electron Dev. Lett.*, 20, 248, (1999).

29. O. W. Holland, D. Fathy, and D. K. Sadana, Formation of ultra-thin, buried oxides in Si by O⁺ ion implantation, *Appl. Phys. Lett.*, 69, 674 (1996).

CHAPTER 11

Nonlinear Photovoltaics and Energy Harvesting

Sasan Fathpour, Kevin K. Tsia, and Bahram Jalali

Contents

11.1 The Need for Green Integrated Photonics

With its environmental, economic, and social fallout, the fast-growing global energy consumption is looming as a major challenge facing the future of human civilization. To avoid this crisis, green technologies ought to be developed for energy-hungry sectors of the economy. The solar cell technology is undoubtedly one major green technology that can target the heart of this crisis. Currently, the photovoltaic industry is growing at a rate of more than 20% annually. Despite this multibillion dollar green industry developed by the optoelectronics community, other photonic subsectors of the economy remain a big net consumer of electricity for the time being and arguably need to become more energy-efficient.

Lighting, television, and computer displays, and optical networking equipment are some photonic-based subsectors

363

consuming significant amounts of energy. For instance, lighting currently consumes 20% of the world's electricity. Solid-state lighting is a good example of a successful green photonics solution, as high-efficiency visible light-emitting diodes are steadily being deployed to replace billions of inefficient incandescent light bulbs and environment-unfriendly fluorescent lamps.

Another energy-hungry sector of the economy is information and communication technology (ICT). Three major ICT subsectors can be identified: (a) computers and peripherals, (b) data centers, and (c) telecommunication networks and devices. In 2007, the total carbon footprint of the ICT sector was 830 MtCO$_2$e (metric tons of carbon dioxide equivalent), which is about 2% of the estimated total emissions from global human activity released that year (Webb 2008). Table 11.1 summarizes the total carbon footprint of each subsector in MtCO$_2$e in 2002 and the predicted values in 2020. The 2020 predictions optimistically assume certain improvement in the energy efficiency of the components of each sector. Yet, an overall annual growth rate of 6% is predicted. Evidently, green technologies for ICT ought to be pursued.

To further highlight the case for green Internet, it is mentioned in passing that the energy consumption of Internet transmission and switching equipment is currently ~0.5% of the total electricity supply in developed countries, a figure that could rise to 1%–10% for an increased average access rate of 100 Mb/s (Tucker 2008). In the case of power-hungry data centers, their electricity bill was $7 billion (corresponding to 14 GW of power) in 2005 worldwide, 50% of which was consumed in information-technology equipments (Ferguson 2007).

Meanwhile, photonics has been increasingly playing a larger role in the ICT subsector ever since the deployment of the first long-haul fiber-optic network. Indeed, optical interconnects have been

TABLE 11.1 Carbon Footprint of the Major Sectors of the Information and Communication Technology Industry

Subsector	Footprint in 2002 (MtCO$_2$e)	Footprint in 2020 (MtCO$_2$e)	Growth Rate per annum
Computers	200	600	5%
Data centers	76	259	7%
Telecom infrastructure	133	299	5%

Source: Webb, M. 2008. SMART 2020: Enabling the low carbon economy in the information age. A report by The Climate Group, *Creative Commons.*

continuously winning over their copper counterparts for shorter and shorter links. The battle line between metal and optical interconnects is currently for rack-to-rack communications, that is, the lengths within data centers and supercomputer facilities.

Already a few private companies, some co-funded publicly, have been marketing products for such emerging applications. Luxtera Corporation (www.luxtera.com) is marketing a silicon (Si) photonics-based 4 × 10 Gb/s optical active cable for short distance communication (<4 km). The photonic–electronic microchip can be used for local area networks or Ethernet, storage area networks, and optical backplane interconnects (rack-to-rack and board-to-board). More recently in 2010, Intel Corporation has demonstrated a 50 Gb/s silicon photonics optical link. Similar solutions exist in compound semiconductors, namely, gallium arsenide (GaAs) and indium phosphate (InP). For example, Infinera Corporation has demonstrated a 10 × 10 Gb/s planar InP photonic circuits (Welch et al. 2006), and IBM has reported a 15 × 16 Gb/s GaAs-based scheme using vertical-cavity surface-emitting lasers (Schow et al. 2009). It should be emphasized that such high-speed backplane communication solutions can be also applied to supercomputing centers, since the underlying communication architecture of supercomputer racks has a lot of similarities with that of data centers. Already, IBM is implementing the aforementioned optical transceiver technology for supercomputer rack-to-rack communications (Kash 2009).

With this level of maturity in the field of electronic–photonic integrated circuits, it is timely to know whether the integration is going to make the ICT sector less or more energy efficient. Obviously, green integrated photonics is desirable to reduce the carbon footprint of the increasingly optics-dominated ICT sector of the economy. Energy efficiency of integrated photonic components will be even more crucial at shorter optical interconnects of the future, as follows.

The reader is first reminded that the battle line between optical and copper interconnects could possibly be for interchip and intrachip communications in the coming years, implying that integrated photonic components could be part of tens of millions of servers, personal computers, game consoles, fiber-to-the-home interface boxes, etc. This possibility of merging optics with consumer electronics will be a true game-changer in terms of the carbon footprint of integrated photonics. Considering that computing-related power usage is currently ~15% of the United

States' electricity consumption (Lawton 2007), energy efficiency of envisioned electronic-photonic integrated circuits will be of paramount macroeconomic relevance.

High power consumption of photonic components may also become a microeconomic issue, that is, a chip-level energy inefficiency problem. To understand this, note that the very large scale integration (VLSI) microelectronic industry is already facing an energy crisis of its own at the chip scale. The problem of heat dissipation in VLSI is so severe that it threatens to bring to a halt the continued advancement of the industry, as described by Moore's law. An acknowledgment to this threat is the recent momentous shift of the microprocessor industry away from increasing the clock speed and in favor of multicore processing in order to avoid high dynamic power of transistors at high switching rates. Evidently, it can be argued that if the photonic devices are not energy-efficient enough, they will not be heat compatible with the strict requirements of microelectronics, and the envisioned electronic–photonic integrated circuit may never be realized—at least for very large scales of integration.

11.2 Nonlinear Optical Losses in Integrated Photonics

Nonlinear losses at high optical intensities could be a major challenge in implementing green integrated photonics. This problem exists for all the three major competing (or possibly complementary) platforms that are being pursued (Si, GaAs, or InP). Let us take the example of Si. The prevailing vision for silicon photonics has been the integration of optics and electronics on the same wafer to avoid the complications needed for III-V photonics/Si electronics hybrid approaches. Stated differently, the vision has been primarily motivated by material and process compatibility of silicon photonics with the dominant microelectronics technology, namely, CMOS (complementary metal-oxide-semiconductor). As mentioned earlier, a full account of compatibility with CMOS should also consider power consumption and the interrelated heat produced by the plurality of electronic and photonic devices. Heat compatibility requires that silicon photonic devices must be able to operate on the hot VLSI chips and that their own power dissipation must be minimal.

The challenging question facing the integrated photonics community is whether replacing copper with optical interconnects will aggregate or alleviate the aforementioned energy crisis of microelectronics. From a fundamental point of view, the amount of information that can be sent through an optical channel increases with the bit rate for the same receiver sensitivity. Meanwhile, to address the large footprint of integrated silicon photonic devices, waveguide cross-sections that are as small as possible are demanded. This scaling leads to higher optical intensity for the same optical power (or the equivalent bit rate). This could lead to considerable nonlinear losses at very high bit rates of the future. Nonlinear losses could indeed dominate the linear losses at a few Tb/s aggregate bit rate, according to our first-order power budget estimates (not presented here), unless submicron waveguides with <0.5 dB/cm linear losses and subnanosecond carrier lifetimes are developed.

In addition, in the absence of second-order optical nonlinearities in Si, the operation of a wide range of silicon photonic devices is based on Kerr and Raman effects. Unfortunately, high optical intensity is a prerequisite for the onset of these third-order nonlinear effects. The inevitable high intensity in nonlinear devices, as well as in the very high-speed interconnects discussed in the previous paragraph, imposes a twofold energy efficiency crisis. First, optical power is converted to heat through the two-photon absorption (TPA) process (Figure 11.1). Second, free carriers generated by TPA must be actively removed from the waveguide core region.

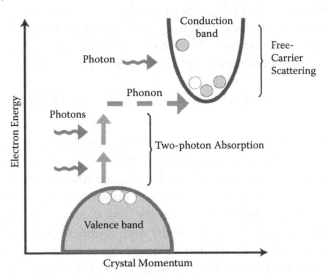

FIGURE 11.1 Two-photon absorption generates free carriers that in turn cause a significant amount of absorption. The loss of photons is the main problem facing integrated photonics at high optical intensities. (From Jalali, B. 2007. Teaching silicon new tricks. *Nat. Photonics*, 1: 193,)

Otherwise, they will accumulate, cause severe free-carrier absorption (FCA), and will prevent continuous-wave (CW) operation of the nonlinear devices. It is important to note that these predicaments of high-intensity photonic devices are not limited to silicon and, for instance, have been also observed in high-power semiconductor optical amplifiers on InP (Juodawlkis et al. 2008).

The FCA coefficient, α_{FCA}, in units of cm^{-1} at telecom wavelength λ in units of μm, is given by (Soref et al. 1986):

$$\alpha_{FCA} = \Delta\alpha_e + \Delta\alpha_h = (8.5 \times 10^{-18} \cdot \Delta N + 6.0 \times 10^{-18} \cdot \Delta P) \cdot (\lambda / 1.55)^2$$

$$(11.1)$$

where $\Delta N \approx \Delta P$ (in cm^{-3}) are the free electron and hole concentrations, respectively. In the presence of FCA and TPA, the propagation of optical intensity, $I_p(z)$, along the propagation direction in the waveguide, z, is given by the following nonlinear differential equation (Claps et al. 2004):

$$\frac{dI_p(z)}{dz} = -(\alpha + \alpha_{FCA}(z))I_p(z) - \beta I_p^2(z) \qquad (11.2)$$

where β is the TPA coefficient and α is the linear absorption coefficient of the waveguide. α_{FCA} is a function of the optical intensity I_p and bias voltage V. TPA of photons with energy E is related to the carrier generation rate G at the waveguide core:

$$G = d(\Delta N) / dt = -(1 / 2E)dI_p / dz = \beta I_p^2 / 2E \qquad (11.3)$$

The previous model can be utilized along with a drift-diffusion electronic device simulator to study the impact of nonlinear losses on the performance of very high-speed silicon optical interconnects as well as nonlinear silicon photonic devices (Fathpour et al. 2007). Figure 11.2 suggests how nonlinear losses at high-input powers can suppress the output optical power when compared to the case when only linear loss is taken into account.

At steady state, the photogeneration rate is equal to the total recombination rate, that is, $G = R = \Delta N/\tau_{eff}$, where τ_{eff} is the carrier recombination effective lifetime. Reducing τ_{eff} would therefore reduce $\Delta N \approx \Delta P$ and hence would reduce the FCA (Equation 11.1). Lifetime reduction by introducing crystal defects might be

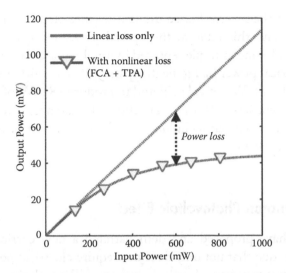

FIGURE 11.2 Reducing the cross-section of silicon waveguides for reduced footprint of the devices comes at the expense of increased nonlinear optical loss. The carrier lifetime was assumed to be 15 ns, and waveguide dimensions are identical to those in Figure 11.3. (From Jalali, B. et al. 2009. Green Silicon Photonics News by OSA. pp. 18–23.)

a solution (Liu and Tsang 2006). Effective lifetime values as low as 150 ps have been reported in submicron waveguides enhanced with argon implantation (Yamada et al. 2006). However, it is not presently clear whether such low lifetimes can be achieved without a pronounced increase of linear optical losses. With the lack of low linear loss waveguides with short carrier lifetimes, carrier sweep-out is the only demonstrated solution to date for CW operation of nonlinear silicon photonic devices (Liang and Tsang 2004; Claps et al. 2004; Rang 2005). This is typically achieved by using a reverse-biased *p-n* junction diode that straddles the waveguide (Figure 11.3). The problem with this technique, however,

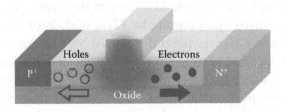

FIGURE 11.3 Schematic of a silicon rib waveguide with a *p-i-n* junction diode. In our experiments, we used devices with length of 2.2 cm, and ridge width and height of 1.5 µm and 3.0 µm, respectively. (From Jalali, B. et al. 2009. Green Silicon Photonics News by OSA. pp. 18–23.)

is that the reverse-biased diode dissipates electrical power on the chip, and this adds to the optical power dissipation caused by TPA. In one example reported by Intel researchers, about 1 W of electrical power had to be dissipated (25 V at ~40 mA) to achieve ~4 dB of CW optical gain and to produce ~8 mW of output from a Raman laser (Rong et al. 2005). In essence, a reverse-biased carrier sweep-out technique is unfortunately not heat compatible with CMOS.

11.3 Two-Photon Photovoltaic Effect

We have proposed and demonstrated a new carrier sweep-out technique that not only does not require electrical power dissipation but generates electrical power too. The technique is based on harvesting the energy of the photons lost to TPA. The core idea is that the prerequisite for carrier sweep-out from the waveguide is not negative voltage bias in a p-n junction but negative current. Thus, if the p-n junction of Figure 11.3 is biased in the fourth quadrant (current, $I < 0$ and voltage, $V > 0$) of its current–voltage (I–V) characteristics, the TPA-generated carriers are swept out by the built-in field of the junction, yet the device delivers electrical power (Figure 11.4). This novel photovoltaic effect can be perceived as the nonlinear equivalent of the conventional photovoltaic effect used in solar cells.

The reader should not forget that the main purpose of any carrier sweep-out technique is to reduce the nonlinear loss of the waveguides by decreasing the carrier lifetime. To experimentally confirm this, the lifetime values were measured from pump-probe experiments conducted on the same waveguides at different biases. Figure 11.5 depicts the change in a CW probe beam caused by a pump pulse (Tsia et al. 2006). The responses consist of a fast transient dip caused by TPA and a slow recovery associated with the recombination of TPA-generated carriers. Carrier lifetime values were extracted by fitting the slow recovery part of the curve. It is observed that the lifetime at +0.5 V (which is a fourth quadrant bias) is about 7 ns, while it is 15 ns under open-circuit conditions (Figure 11.5). Shorter lifetime gives rise to a lower free-carrier density and hence lower FCA.

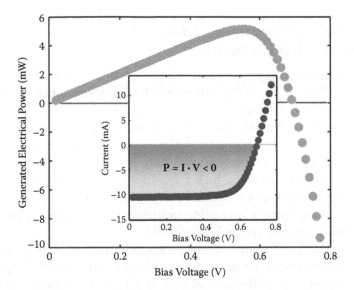

FIGURE 11.4 The measured current–voltage curve of the fabricated *p-i-n* diodes shows the photovoltaic effect in the fourth quadrant (current <0, voltage >0) where power dissipation is negative, that is, power generation (inset). The power–voltage plot clearly shows that electrical power as high as ~5.5 mW can be harvested. (From Jalali, B. et al. 2009. *Green Silicon Photonics News* by OSA. pp. 18–23.)

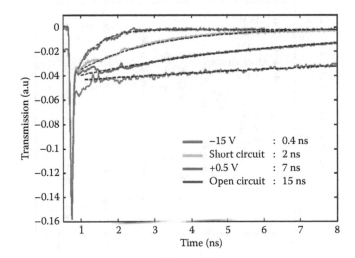

FIGURE 11.5 The temporal response of a CW signal laser to a pulsed pump laser at different biasing conditions in the *p-i-n* SOI waveguide. The fitted carrier lifetime values are shown in the legend. (From Tsia, K. K. et al. 2006. *Opt. Express.* 14:12327–12333.)

11.4 Nonlinear Photovoltaic Effect in Active Silicon Photonic Devices

A nonlinear photovoltaic effect has been successfully applied to three types of key silicon photonics devices, namely, Raman amplifiers (Fathpour et al. 2006a), optical modulators (Fathpour and Jalali 2006b), and Kerr-type wavelength converters (Tsia et al. 2006). Arguably, the technique can be considered as a universal method for achieving energy-efficient silicon nonlinear devices. As discussed earlier, high intensities may be also encountered in future high-speed silicon-based optical interconnects, and thus the same energy harvesting technique is in principle applicable to this emerging technology as well.

Figure 11.6 presents the measured on–off Raman gain at different coupled pump powers and biasing conditions. Proper mirror coatings (reflections of 80% and 10% at the input and output facets, respectively, at the pump wavelength of 1540 nm) were applied in this particular sample to increase the Raman gain. As a result of mirror coating, the optical gains presented in Figure 11.6 are higher than in our earlier work (Fathpour et al. 2006a). A maximum on–off gain of more than 6 dB is obtained at a reverse bias of 15 V. A Raman gain of ~4 dB is attained when the diode is

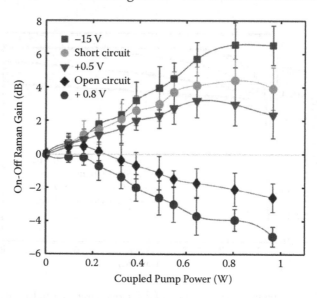

FIGURE 11.6 Measured on–off Raman gain versus coupled pump power in the devices of Figure 11.3 with mirror coatings at the facets. (From Jalali B. et al. 2009. Green Silicon Photonics News by OSA. pp. 18–23.)

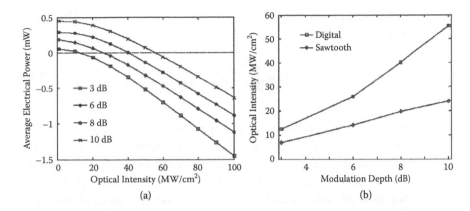

FIGURE 11.7 (a) Average on–off power dissipation versus signal optical intensity at different modulation depths when the modulator is driven with an ideal digital waveform with zero rise/fall times; (b) optical intensity required for zero power dissipation for the digital waveform ($P_{avg} = 0$) and the sawtooth waveform ($P_s = 0$). For a practical digital waveform with finite rise and fall times, the power is between these two limits. (From Fathpour, S. et al. 2006. *Appl. Phys. Lett.* 89:061109.)

short circuit (0 V), that is, zero power dissipation. More interestingly, on–off Raman gains of about 3 dB are measured when the device is forward-biased at voltages ≤ 0.7 V. The importance of this biasing regime is that power dissipation is negative (Figure 11.4). For instance, ~5 mW of electrical power is generated at a bias of +0.55 V. In contrast, more than 150 mW is dissipated at –15 V. Clearly, the two-photon photovoltaic effect offers an energy-efficient solution for nonlinear silicon photonics.

Similarly, the same technique can be applied to achieve a negative dissipation optical modulator (Fathpour and Jalali 2006b). In order to achieve maximum power efficiency, the off-state can be set at the minimum of the *P–V* characteristics, and the on-state will be determined from the desired modulation depth such that the average power consumption $P_{avg} = (P_{on} + P_{off})/2 < 0$. Figure 11.7a summarizes the trade-off between the generated electrical power and the required input optical intensity. At sufficiently large signal intensities, the attainable P_{avg} can be as high as 1 mW while maintaining a modulation depth of 8 dB. To demonstrate the dependence on the modulating data format, simulations for two extreme cases were performed. One case is a "digital" waveform corresponding to a binary signal with zero rise and fall times. The previously discussed P_{avg} corresponds to this digital case. The second case is a "sawtooth" signal (triangular waveform) in which the generated power can be conveniently calculated from the average swept power:

$$P_s = \int_{V_{on}}^{V_{off}} P dV \, / \, (V_{off} - V_{on}) \tag{11.4}$$

The generated power for a practical digital signal with finite rise and fall times lies between these two extremes. The intensities required for zero power dissipation in both cases, that is, intensities at which $P_{avg} = 0$ and $P_s = 0$, are presented in Figure 11.7b versus modulation depth. It is evident that the required intensity for $P_s = 0$ is a factor of 1.8 to 2.3 smaller than the $P_{avg} = 0$ case.

Finally, energy harvesting was demonstrated in wavelength converters based on the Kerr effect, the results of which can be found elsewhere (Tsia et al. 2006). The interesting feature of this work was the ~1.3–1.5 dB improvement in the conversion efficiency under a forward bias of +0.5 V at high pump powers as compared with the open-circuit case. It was shown that the conversion efficiency can be greatly improved by engineering the waveguide dispersion and at higher optical intensities.

11.5 Efficiency of the Two-Photon Photovoltaic Effect

Conceptually, the two-photon photovoltaic effect is a nonlinear equivalent of conventional photovoltaic effect but with certain differences. High optical intensities are required for the onset of the nonlinear effect, and two photons are involved for generating one electron–hole pair; thus, the collection efficiency cannot exceed 50%. Also, structural and geometrical differences with conventional photovoltaic devices demand alterations in the theoretical model that is typically used to describe solar cells. We have developed the theory of nonlinear photovoltaic effect elsewhere (Fathpour et al. 2007). What we would borrow in the following from that work is the mathematical expressions for some important figures of merit.

The quantum efficiency of the device, η_q, is defined as the number of collected carriers per incident photon inside the waveguide. An approximate expression for quantum efficiency at the voltage bias V_m that gives the maximum generated power is

$$\eta_q(V_m) \approx \beta L_{NL} I_{p0} \, / \, 2 \tag{11.5}$$

where E_p is the photon energy and I_{p0} denotes the coupled pump intensity at the input of the waveguide. If the coupling efficiency of light into the waveguide $\eta_{coupling}$ is known, the external quantum efficiency can be written as $\eta_{ext} = \eta_{coupling} \times \eta_q$. Finally, the external power efficiency is $\eta_P = \eta_{ext} \times qV / E_p$, which is also simply known as the "efficiency," and is approximated as

$$\eta_P \approx \frac{\eta_{copling} q \beta}{2E_p} V_m L_{NL} I_{p0} \qquad (11.6)$$

where q is the electron charge. The nonlinear photovoltaic effect is implicit in this last expression, as it is quantified in terms of a defined effective length along the length of the waveguide, that is,

$$L_{NL} \equiv \int_0^L \frac{I_p^2(z)}{I_{p0}^2} \, dz \qquad (11.7)$$

This important last quantity is the nonlinear equivalent of the interaction length commonly defined in optical fibers.

The collection efficiency η_c is the number of collected carriers per number of photons consumed by TPA. Analytical expressions for η_c are developed elsewhere (Fathpour et al. 2007). It is more important to note that the theoretical limit on this efficiency is 50%, since two photons can at best give rise to only one electron–hole pair. η_c is the appropriate efficiency for devices such as amplifiers and wavelength converters where energy harvesting is a useful by-product but not the main functionality of the device. If the two-photon photovoltaic effect is intended to be used as a photovoltaic cell, the power efficiency η_P becomes a more relevant figure of merit. As expected, β is readily recognized as the most deterministic material property for attainable power and the corresponding η_P and η_q efficiencies. The dependence of discussed efficiencies on inactively absorbed photons due to linear and free-carrier absorptions is implicit in L_{NL}. In other words, higher linear and free-carrier losses lead to smaller L_{NL}.

The collection efficiency is plotted versus voltage at different pump intensities in Figure 11.8. A good agreement between simulated and experimental results is observed. At lower pump intensities, η_c approaches the earlier mentioned theoretical limit of 50%. The value of η_c at maximum power generation bias is

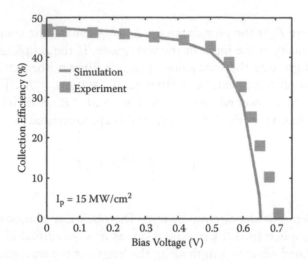

FIGURE 11.8 The collection efficiency of the two-photon photovoltaic effect at maximum power generation bias is ~40%. (From Jalali B. et al. 2009. Green Silicon Photonics News by OSA. pp. 18–23.)

extracted to be within 39%–41%, and is nearly independent of the coupled optical intensity from 5 to 150 MW/cm². Physically, this rather constant behavior is because the recombination current dominates the diffusive current at low biases and grows at approximately the same rate as the photogenerated current.

11.6 Other Applications

We would now discuss the possibility of employing the nonlinear photovoltaic effect for applications other than energy-efficient silicon photonic devices. One possible application is photovoltaic cells used in power converters and optically powered sensors (Werthen et al. 1996; Werthen et al. 2005). Energy harvesting through the two-photon photovoltaic effect could be utilized to supply the electrical power on-chip, especially for on-chip optical interconnect and optoelectronics applications. A generic schematic of such an electronic–photonic integrated circuit is shown in Figure 11.9. As an off-chip laser source is used, this could also possibly be applied to remote power delivery for physical sensors (e.g., temperature, pressure, or gas) that are installed in critical environments where it is difficult for the sensor to be powered over copper cable. One example is methane sensors in coal mines, where electrical sparks are considered extremely hazardous.

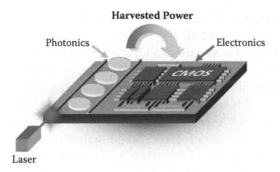

FIGURE 11.9   An electronic–photonic integrated circuit partially or fully powered by an off-chip laser source using the two-photon photovoltaic effect. (From Jalali B. et al. 2009. Green Silicon Photonics News by OSA. pp. 18–23.)

Power efficiency is the key figure of merit for the preceding applications. The structure shown in Figure 11.4 has a maximum power efficiency of ~2% attainable at $I_{p0} \sim 100$ MW/cm^2 and a device length of above 5 cm (Fathpour et al. 2007). A higher efficiency of up to 5.5% can be achieved by optimizing the device structure (Jalali et al. 2009). The earlier observations are obtained for microwaveguides with dimensions reported in the caption of Figure 11.4 but are generally valid for submicron waveguides also. Indeed, our simulations confirm that the efficiencies of the two-photon photovoltaic effect are strong functions of the optical intensity and the waveguide length, but not the waveguide cross-sectional dimensions as long as the carrier lifetime is the same (Jalali et al. 2009). This, however, may not be the case in submicron waveguides if the surface recombination is the dominant recombination process leading to ≤1 ns effective lifetime values as opposed to measured open-circuit values of 15 ns in our structures. It means that if submicron photovoltaic cells are intended, surface passivation precautions ought to be considered so that the photogenerated carriers are not recombined by the surface states but are swept out by the built-in field of the *p-n* junction.

The relatively low inherent efficiency prediction of ~5.5% in the nonlinear photovoltaic effect is partially due to the 50% limit on the collection efficiency, and more importantly due to the fact that two-photon scattering is not a very likely event as compared with interband single-photon absorption employed in solar cells. Any imaginable means that can improve β can potentially enhance the power efficiency. β is expected to be higher at shorter wavelengths, for example, 1.3 μm, as has been observed experimentally in GaAs (Hurlbut et al. 2006). In addition, FCA is lower at shorter wavelengths (according to Drude's model) that can increase L_{NL}. Therefore, the combination of the two effects should translate into higher limit efficiency of the nonlinear photovoltaic effect.

Finally, as mentioned earlier, it was also recently revealed that TPA and TPA-generated FCA have similar limiting effects on the performance of III-V based semiconductor optical amplifier output power (Juodawlkis et al. 2008). It is noteworthy that the presented nonlinear photovoltaic effect is not restricted to Si and is applicable to compound semiconductors as well. Indeed, reported values of β at 1.3 μm in InP and GaAs are 70 and 42.5 cm/GW, respectively, versus 3.3 cm/GW in Si (Tiedje et al. 2007). Meanwhile, the FCA loss of III-V materials is typically the same as in Si. Hence, higher nonlinear photovoltaic efficiencies are achievable in compound semiconductors. The maximum power efficiency versus waveguide length for Si, GaAs, and InP nonlinear photovoltaic cells operating at 1.3 μm wavelength is shown in Figure 11.10. The corresponding optical intensities at

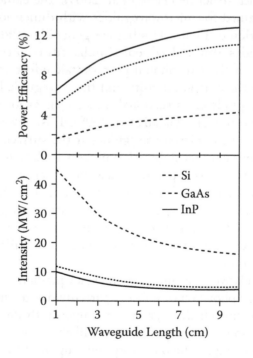

FIGURE 11.10 The maximum power efficiency and required intensities versus waveguide length for Si, GaAs, and InP at 1.3 μm.

these efficiencies are also presented. To make a reasonable comparison, it is also assumed in all cases that the *p-n* junction diodes are biased for an appropriately identical carrier lifetime of 7 ns (see Figure 11.5). The results suggest that the maximum power efficiency at 1.3 μm wavelength is ~4% for Si and is achieved at ~15 MW/cm², which should be compared to the aforementioned corresponding values of ~2% and 100 MW/cm² at 1.55 μm. More interestingly, the efficiencies for GaAs and InP can be as high as 9% and 12%, respectively, both achieved at <5 MW/cm² optical intensities. The only shortcoming of these proposed photovoltaic converters is the very long waveguide length (~10 cm) required to completely absorb the incident light. The length requirement could be considerably reduced by applying high-reflection coating at the end of the waveguides (Jalali et al. 2009).

11.7 Conclusion

Full implementation of merging integrated photonics on Si (as well as on GaAs or InP) with CMOS electronics must consider heat compatibility at very high scales of integration. A major challenge for integrated electronics–photonics is the potential aggravation of high-power dissipation of VLSI by the photonic devices present on the same chip. Semiconductors in general, and silicon in particular, have a fundamental problem—caused by two-photon absorption and free-carrier scattering—that renders the materials lossy at the high optical intensities encountered in very high-speed interconnects and nonlinear optical devices. The two-photon photovoltaic effect is a potential solution for achieving energy-efficient integrated photonic devices. The effect reduces the optical loss by free-carrier scattering and serendipitously harvests the optical energy lost to two-photon absorption.

References

Claps, R., Raghunathan, V., Dimitropoulos, V., and Jalali, B. 2004. Influence of nonlinear absorption on Raman amplification in silicon waveguides. *Opt. Express* 12:2774–2780.

Ferguson, S. 2007. Data Center Power Consumption on the Rise, Report Shows. http://www.eweek.com/article2/0,1895,2095409,00.asp

Fathpour, S., Tsia, K. K., and Jalali, B. 2006. Energy harvesting in silicon Raman amplifiers. *Appl. Phys. Lett.* 89:061109.

Fathpour, S. and Jalali, B. 2006. Energy harvesting in silicon optical modulators. *Opt. Express* 14:10795–10799.

Fathpour, S., Tsia, K. K., and Jalali, B. 2007. Two-photon photovoltaic effect in silicon. *IEEE J. Quant. Electron.* 43:1211–1217.

Hurlbut, W. C., Vodopyanov, K. L., Kuo, P. S. et al. 2006. Multi-photon absorption and nonlinear refraction of GaAs in the mid-infrared. *IEEE Conf. Lasers and Electro-Optics (CLEO)*, Paper #QThM3.

Jalali, B., Fathpour, S., and Tsia, K. K. 2009. Energy Efficiency in Silicon Photonics. In *Silicon Nanophotonics: Basic Principles, Present Status and Perspectives.* Ed. Khaiachtchev, 355–377. Singapore: World Scientific Publishing Co. Pvt. Ltd.

Juodawlkis, P. W., Plant, J. J., Donnelly, J. P. et al. 2008. Continuous-wave two-photon absorption in a Watt-class semiconductor optical amplifier. *Opt. Express* 16: 12387–12396.

Kash, J. 2009. Photonics in supercomputing: The road to exascale. *Frontiers in Optics. Frontiers in Optics 2009/Laser Science XXV.* Paper FWO2.

Lawton, G. 2007. Powering down the computing infrastructure. *Computer* 40:6–19.

Liang, T. K. and Tsang, H. K. 2004. Role of free carriers from two-photon absorption in Raman amplification in silicon-on-insulator waveguides. *Appl. Phys. Lett.* 84:2745–2747.

Liu, Y. and Tsang, H. K. 2006. Raman gain in helium ion implanted silicon waveguides. *IEEE Conf. Lasers and Electro-Optics (CLEO)*, Paper CTuU6.

Rong, H., Jones, R., Liu, A. et al. 2005. A continuous-wave Raman silicon laser. *Nature* 433:725–728.

Schow, C. L., Fuad, E. Doany, F. E., Baks, C. W. et al. 2009. A single-chip CMOS-based parallel optical transceiver capable of 240-Gb/s bidirectional data rates. *J. Lightwave Technol.* 27:915–929.

Soref, R. A. and Bennett, B. R. 1986. Kramers-Kronig analysis of E-O switching in silicon. *SPIE Integr. Opt. Circuit Eng.* 704:32–37.

Tiedje, H. F., Haugen, H. K., and Preston, J. S. 2007. Measurement of nonlinear absorption coefficients in GaAs, InP and Si by an optical pump THz probe technique. *Optics Communications* 274:187–197.

Tsia, K. K., Fathpour, S., and Jalali, B. 2006. Energy harvesting in silicon wavelength converter. *Opt. Express.* 14:12327–12333.

Tucker, R. 2008. A Green Internet. *Proc. ECOC 2008*, Brussels, Belgium.

Webb, M. 2008. SMART 2020: Enabling the Low Carbon Economy in the Information Age. A report by The Climate Group, *Creative Commons.*

Welch, D. F., Kish, F. A., Nagarajan, R. et al. 2006. The realization of large-scale photonic integrated circuits and the associated impact on fiber-optic communication systems. *J. Lightwave Technol.* 24:4674–4683.

Werthen, J. G., Andersson, A. G., and Wu, T.-C. 1996. Optically powered sensors: Are they really fiber optic sensors? *Proc. SPIE* 2872: 131–138.

Werthen, J. G., Widjaja, S. Wu, T.-C., and J. Liu 2005. Power over fiber: A review of replacing copper by fiber in critical applications. *Proc. SPIE* 5871:58710C-1–58710C-6.

Yamada, K. Fukuda, H. Watanabe, T. et al. 2006. All-optical wavelength conversion using silicon photonic wire waveguide. *IEEE Group IV Photonics Paper FB5*, 237–239.

CHAPTER **12**

Computer-Aided Design for CMOS Photonics

Attila Mekis, Daniel Kucharski,
Gianlorenzo Masini, and Thierry Pinguet

Contents

12.1 Introduction

Silicon photonics is an attractive solution for photonics integration because it allows leveraging both the functionality offered by
electronic circuits as well as the economic advantages inherent in
electronics volume production. Integrating photonic devices into

383

a complementary metal-oxide-semiconductor (CMOS) process opens up vast opportunities for fabricating diverse electro-optical modules in a cost-effective manner. Luxtera, a silicon photonics company, employed this approach and has developed what is termed a CMOS photonics technology. This technology is based on a mature silicon process modified to enable tight integration of optical components with electronic circuits on the same chip.

In the electronics industry, the superlinear growth of the number of components and integrated circuit complexity described by Moore's law (Moore 1965) caused a new discipline to be born soon after the first integrated circuit was introduced (Kilby 1964). This discipline is electronic design automation (EDA), also called *Computer Aided Design* (CAD). It encompasses and supports all levels of abstraction present in the development and design of an integrated electronic system. With today's typical integrated circuit transistor count exceeding a billion per chip, it is impossible to imagine how such complexity could be managed in the design and verification phases without the support of a suite of advanced automation tools.

It was the increasing complexity in electronics chip design that drove the need for automation, and now the same scaling is occurring in CMOS photonics chips as well. In fact, one may argue that the threshold required to build a fully automated EDA system for photonics-enabled CMOS has been crossed. Luxtera currently is in production manufacturing a 40 Gb/s optical-active cable called Blazar, shown in Figure 12.1. This sophisticated, high-performance system is capable of transmitting a bidirectional 40 Gb/s data stream over long distances up to 4 km with low latency. The

FIGURE 12.1 A 40 Gb/s active optical cable based on CMOS photonics technology.

bulk of the functionality of the transceivers at each end of the cable is contained within a CMOS photonics chip comprising a combination of electronic and optical circuits. The subject of this chapter is to describe a CMOS photonics CAD system that has enabled chip designers to achieve this tight integration.

First, we give a summary overview of the electronics design flow at different abstraction levels. Next, we introduce the building blocks of our integrated CMOS photonics technology and describe the methodology through which the photonic elements are built into the existing electronics design flow. We start by describing how the CMOS process was enabled to include optics; we detail the basic library of photonic elements; and finally we show how the automated tools of the electronic design kit were enhanced to allow for the integration of optics. At the end, we give an example of complex subsystems and a system that were designed and built in the CMOS photonics CAD environment.

12.2 The Electronic Design Flow and EDA

EDA started in the early 1970s with the introduction of the first software aimed at circuit simulation: SPICE from the University of California, Berkeley (Quarles et al. 2009) followed later on by the TCAD suite of PISCES and SUPREME from Stanford University (Dutton 2009). The first-generation EDA companies also date back to that time: they customized workstations to handle artwork for mask generation (Sangiovanni-Vincentelli 2003). Even though these tools facilitated some aspects of the design, we must remember that in those days many steps were still performed manually. For example, photomasks were produced by photographic reduction of hand-cut rubylith sheets, as was the case for the famous Intel 4004 microprocessor (Kanellos 2001). Technology, however, evolved quickly and, at the beginning of the 1980s, Mentor and Cadence were founded, and since then they have been playing a leading role in the subsequent exponential growth of the EDA sector.

One of the key reasons for the huge success of EDA is its ability to provide an integrated environment to design electronic systems from the beginning to the end. In this environment, layouts, components, circuits, subsystems, and entire chips are treated at different levels of abstraction by different tools, while keeping an overall consistency that allows one to identify the impact of a change at any level of abstraction on any other level. As an example, it is possible to modify a small layout detail of a large chip and verify the impact at the system level. Conversely, in the digital realm, it is possible to automatically generate the layout that best implements complex logic functionality, potentially described through a text language such as VHDL or Verilog, given a set of technology constraints.

The electronic design flow starts with the definition of the architecture implementing a system that satisfies a set of specifications. Frequently, a system simulation at this abstraction level is done by using behavioral models to verify feasibility and provide rough performance estimates. In the following step, the single subsystems are described in terms of components and interconnects in a *schematic view* (schematic capture) for analog and simple digital designs, while a *Register Transfer Level* (RTL) description is created for complex digital subcircuits. In both cases, a thorough simulation is performed at this level to verify compliance with specifications for temperature, supply voltage, and process corners. Several analog simulators are available today for this task (HSPICE, SPECTRE, MICROSIM, MICA, etc.) based on advanced transistor models such as BSIM3 (Hu 2009). Model parameters are usually provided by the semiconductor foundry and include *corners*, which are the limiting cases of statistical distributions. For transistors, it is common to refer to "cold" or "hot" corners for devices, respectively, showing lower or higher than typical drain current and, consequently, lower or higher speed. Digital design verification, in addition to functionality, includes also a *timing closure* check ensuring, at this level, that gate delays do not cause timing issues.

Once the schematic or RTL design has been verified, the layout can be generated. For high-performance analog design, this activity is still mostly manual, but for digital circuits, automatic place-and-route tools are available. During the layout phase, help comes to the designer from the *Design Rule Check* (DRC) tools that flag violations of the foundry design rules. As an example, a DRC check can identify and graphically show the presence and position of two metal lines running too close to one another. Modern DRC tools are not limited to flagging rule violations but also report situations that, while not violating specific rules, could cause low yield in production. The simplest case is represented by local and global *pattern density* on specific layers (metals, active silicon, and gate polysilicon) that can affect yield because of sensitivity to chemical–mechanical polishing (CMP) steps. These types of checks, usually referred to as *Design For Manufacturing* (DFM), are becoming increasingly more important in advanced technology nodes where the feature size is approaching the limit of lithography tools and yield can vary greatly based on layout choices.

Given the complexity of modern circuits, it is of the utmost importance to ensure that the layout represent exactly what was captured in the schematic. This is done automatically by the *layout versus schematic* (LVS) tool. LVS recognizes and extracts devices and connectivity from layout. For instance, a polysilicon stripe overlapping a region of active silicon is recognized as a metal-oxide-semiconductor (MOS) transistor. If the well implant in that region is n-type, it must be a p-MOS transistor. Once devices and terminals are extracted, a layout *netlist* is compiled and compared to the schematic netlist. An error message is issued if a connectivity or sizing mismatch is detected. Together with DRC, LVS tools are an invaluable help to the modern chip designer.

Once DRC errors are eliminated and the layout is found to match the schematic, a final simulation can be performed on a netlist including *parasitic components*. These components are extracted from the layout using models for the substrate and the electrical interconnects, and may include shunt capacitances to the substrate or to other lines, series resistances on long metal lines, inductances, and other elements. Time closure is also re-verified. At this point, the artworks can be released for mask production with a good probability of first-pass success, which is very important given the extremely high cost of masksets in advanced CMOS technologies combined with aggressive time-to-market targets in competitive integrated circuits (IC) industries.

12.3 CMOS Photonics Process Technology

Luxtera's CMOS photonics process technology, named LuxG, is based on a standard, high-volume 130 nm SOI-CMOS process run by Freescale Semiconductor, called hip7_soi. A large number of popular chips, such as the G4 PowerPC® processor, ubiquitous in embedded applications, as well as in mainstream Apple computers, have been designed and fabricated using this technology. The process benefits from the high performance of the hip7_soi transistors and dense metal and dielectric stack (or back-end), allowing analog and digital data processing at high speeds in excess of 10 Gb/s with good substrate isolation, thanks to the SOI wafer.

A modern CMOS technology, such as hip7_soi, has a complex flow with typically more than 40 lithographic steps interleaved with a variety of other fabrication steps, such as oxidations, implants, etches, film deposition, polishing, and annealing. While the step sequence is generally established by the type of principal structure to be fabricated, such as a MOS transistor, a key role in assembling and fine-tuning the fabrication flow is played by the thermal budget required by each step and by the maximum temperature the wafer can withstand at each through the flow. The latter tends to decrease throughout the process as soon as more temperature-sensitive films, such as salicides or low-k dielectrics, are grown or deposited. Other critical factors to be considered during integration are contamination and topography. One example of a potential contaminant interaction is copper, which is widely used in modern back-end metallization, it is also a fast diffuser and device lifetime killer when introduced in silicon. Topographical considerations are also increasingly important for high-resolution lithography that requires an exceptionally flat surface.

Modifications to the hip7_soi flow that enable photonics can be grouped in two categories: those introduced to define the photonics connectivity layer (including the input/output (I/O) as well as passive and active optical devices) and the germanium module. The first group starts with the adoption of an atypical SOI wafer: the very thin overlayer (~100 nm) used in modern technologies does not provide enough headroom for low-loss optical-mode propagation. Therefore, a thicker overlayer wafer is chosen. In addition, the buried oxide (BOX) is also thickened to reduce losses due to radiation losses to the substrate. These seemingly minor changes have several consequences impacting both processing and transistor performance. Process steps must be modified, mainly due to a different thermal conductivity of the new wafer. Transistor models must also be updated owing to a different vertical junction profile of the source/drain implants and to the partial depletion of the transistor body.

The photonics connectivity layer, including the I/O elements and the body of the active devices, is defined by adding an additional trench isolation step. Unlike the standard SOI step that fully removes the Si and reaching the BOX, this step etches only part of the Si overlay film. This allows for electrical contact and carrier injection in optoelectronic devices while keeping the optical mode away from the high-loss contact regions. All Si belonging to the photonics connectivity layer is protected from salicidation by a patterned dielectric layer. This is needed because of the very high optical losses occurring in salicided Si. In addition to the standard well implants, a number of custom implant steps are added to define the lateral p-n junctions employed in the active optoelectronic components.

The Ge module is inserted much later in the process, just before the contact module, mainly because of thermal constraints. Ge cannot withstand the high-temperature anneals used during the gate module and the activation of the source and drain implants that occur near 1000°C. Consequently, Ge epitaxy must be delayed until after these steps. At the same time, it is rather convenient to use the same contact module for transistors, active optoelectronic devices, including Ge-based devices. Also, the low-k back-end dielectric films are not able to withstand the relatively high temperatures required during Ge epitaxy (~400°C). Ge epitaxy, therefore, is inserted after the Si salicidation and poly processing steps.

A lithography step is used to open windows in the dielectric stack protecting the Si photonics connectivity layer in the region where Ge is to be deposited. The wafer is then wet-cleaned and loaded in the epitaxial growth tool where an additional, *in situ*, cleaning occurs followed by selective epitaxial growth. A nitride dielectric layer is finally deposited and patterned to seal the Ge islands and protect the film from the chemicals used in the following lithography steps that are too aggressive for Ge. The process is completed by junction formation using an implant of phosphorus and boron and activated by rapid thermal anneal. At this point, the standard process resumes with the contact module followed by the back-end. Figure 12.2 shows a simplified flowchart of the LuxG process, where the departures from the original hip7_soi process are marked in red.

12.4 Photonic Device Libraries

In addition to the process, the technology includes design libraries that contain parameterized layout cells and simulation models for standard electronic components, such as transistors, resistors, inductors, capacitors, and transmission lines. It also includes an optical device library with components at a high maturity level in terms of model accuracy, variability, and reliability (Mekis et al. 2008). In this section, we give examples of the optical library

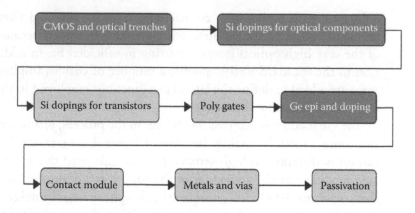

FIGURE 12.2 LuxG process flow. The steps used to integrate photonic components are colored in red.

components that fall into four categories: optical interconnects, optical interfaces, light modulators, and light detectors.

LuxG optical interconnects come in two flavors: narrow single-mode waveguides (SMW) and wide multimode waveguides (MMW). The SMW has higher propagation loss than the MMW and is used for interfacing to other standard optical elements and for redirecting light with bends. To propagate light over long distances in straight lines, the waveguide width is adiabatically tapered to an MMW, reducing mode interaction with the etched Si surface and, consequently, loss. As mentioned in the previous section, the waveguides are of the rib type in which the Si is only partially etched to obtain lateral confinement. Vertical confinement comes from the BOX and the dielectric stack deposited on top of the Si to prevent salicidation. Waveguide combiners and splitters, such as directional couplers with different tap ratios and Y-junctions, are also available in the optical library.

Light I/O from the CMOS chip is among the most critical functions provided in the LuxG device library. The use of near-normal coupling enables wafer-scale testing in which optical devices can be sorted before die singulation in exactly the same way as electronic components are. This revolutionary approach in integrated optics allows fast and inexpensive screening of dies and wafers enabling high-volume production of CMOS photonics-integrated circuits. The library I/O elements (grating couplers) are periodic grating structures designed to redirect light and efficiently focus a relatively large 10 μm beam emitted by an optical fiber into a submicron Si waveguide. Two types of grating couplers are available in the device library: single-polarization and polarization-splitting versions. The former is used whenever the polarization of the incoming beam is linear and known, such as when light from the laser source is coupled into the chip or when modulated light from the chip is coupled into an optical fiber. The latter is used when polarization is arbitrary, such as when receiving an optical signal from a single-mode optical fiber. In this case, light polarization is decomposed into two orthogonal components, and each is coupled to a separate waveguide.

Phase modulation of an optical signal traveling in an Si waveguide can be accomplished by varying temperature, taking advantage of the relatively high thermo-optic coefficient of Si, or by electronic means via light interaction with free carriers (Soref and Bennet 1987). The former, while more efficient and less lossy, is also much slower, and it does not suit the requirements of high-speed transmission. Efficient interaction of an optical mode with free carriers can be realized by modulating the depletion region of a p-n or p-i-n junction straddling a narrow Si waveguide. Our high-speed modulator uses a p-n junction allowing signals in excess of 10 Gb/s to be impressed on the phase of the traveling optical mode. Amplitude modulation is accomplished by combining two differentially phase-modulated signals using a Mach–Zehnder interferometer (MZI). For lower-speed applications, such as biasing the MZI at the quadrature point, forward bias p-i-n devices can be used to provide higher-phase modulation efficiency.

Due to the almost complete transparency of Si at the wavelengths used in optical communications (1.3 and 1.55 μm), an alternative material must be used to ensure efficient photodetection. Ge is an attractive possibility because of its chemical and processing compatibility with Si (Masini et al. 2001). Moreover, thanks to its high-absorption coefficient near 1.55 μm, a distance of only a few microns is required to fully absorb light in Ge. Combined with submicron ridge waveguide cross-section, the short absorption length allows the design of extremely compact waveguide detectors with high efficiency, relatively low dark current, and extremely low capacitance. The latter is crucial in designing ultra-high sensitivity high-speed receivers and is one of the key benefits of integration.

12.5 CMOS Photonics Design Deck

Adapting the layout and simulation tools as well as the concepts of DRC and LVS to monolithically integrated photonic devices presents some particularly challenging problems that we will discuss in this section.

12.5.1 Photonic Devices in a CMOS Design Environment

The basic optical building blocks typically have much more complex layouts than their electronic counterparts. While, for instance, a transistor can be drawn with just a handful of rectangles, a basic optical component, such as a grating coupler, can contain dozens or sometimes even thousands of distinct curved elements on multiple layers (Figure 12.3). Such complexity has necessitated the development of scripts to automate creation of the layouts for the optical library elements. Since Cadence's built-in scripting language, Skill, can become cumbersome for drawing these nonorthogonal shapes, we have opted for writing parameterized layout generation code in Matlab and developing a seamless interface from Matlab to Cadence's layout tool, Virtuoso. This approach has facilitated rapid development and optimization of optical elements in the LuxG technology.

Analogously to the structure of the electronics libraries, each photonic element cell possesses a layout, a symbol, and a corresponding simulation model that may consist of multiple layers of hierarchy. Figure 12.4 shows an example of the layout and schematic of a high-speed phase-modulator diode. The schematic of the diode has four ports for connectivity: two optical ports for

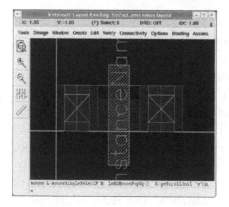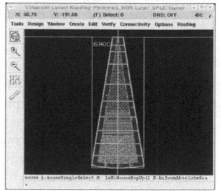

FIGURE 12.3 Layouts of a basic electronic building block (transistor, left) and of a basic optical building block (grating coupler, right).

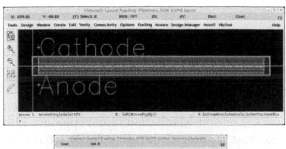

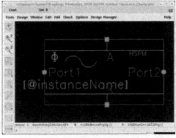

FIGURE 12.4 Layout (top) and symbol (bottom) for a high-speed phase modulator.

interfacing with waveguides (denoted by *Port1* and *Port2*) and two electrical ports, the anode and cathode of the diode (denoted by *A* and *K*). The optical and the electronic interfaces are both present in a single schematic that allows tight integration with the electronic blocks.

The optical portions of the layouts are primarily defined in layers that are specific to optics, such as the optical trench etch, the Ge islands, and the optical implants. However, compatibility with CMOS processes requires a careful consideration of process interactions and has necessitated the addition of other layers in each optical device layout. As an example, a waveguide must be protected from salicidation steps; therefore, it will require an overlay of the salicide-blocking mask layer. The optical mode propagating in the waveguide also must be far enough from any metals that could cause absorption and scattering losses. Therefore, metal interconnects can cross optical interconnects only if they are at one of the higher metal layers. As a last example, a grating coupler must be able to transfer light through the entire dielectric stack with low loss, which implies that any metal above such a structure must be blocked. Such requirements have become part of the design rule set for the CMOS photonic deck and include DRC rules as described in the next subsection.

12.5.2 DRC

From a layout perspective, most optical devices rely on nonstandard orthogonal shapes. The base Freescale Semiconductor 130 nm process uses a strict 15 nm design grid, and all electronic devices are oriented along the *x* and *y* axes in a "Manhattan" geometry, with a few elements laid out with lines along 45° angles. Off-grid and non-45° or non-Manhattan layouts produce DRC errors.

In contrast, optical elements typically require smooth curves to operate efficiently. Figure 12.5 shows cross-sectional scanning electron microscopy (XSEM) photos of some typical optical devices that are outside of the standard CMOS norm. The picture shows a waveguide bend, approximately 500 nm wide, laid out on a 15 nm grid. The sides of the bend would look very jagged, and in silicon this would create an unacceptable scattering loss. To obtain satisfactory performance, we had to design our waveguide and other optical devices on a 1 nm grid, essentially matching the capabilities of the mask writing laser tool.

While these alterations may seem innocuous, they have major repercussions throughout the design flow. One reason is that most CAD tools designed to handle layouts have been optimized for Manhattan or 45° shapes. In addition, running a standard DRC deck on a photonic design produces thousands of spurious errors due to these off-grid and non-45° geometries. Simply filtering these errors may cause one to overlook a real problem or a mistake that could cause functional or yield issues at the chip level. The solution we employed was to alter the electronics design deck to

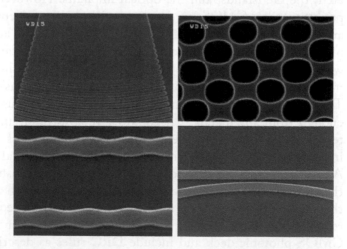

FIGURE 12.5 Examples of optical structures outside the CMOS "Manhattan" norm.

flag and automatically separate real errors from the false ones, and to supplement the design deck with optics-specific design rules. Beyond this, the DRC deck implements traditional rules such as minimum width, space, area, enclosures for the layers specific to optics such as the optical trench and Ge modules, and optical implants. All of these are complicated by the presence of nonstandard shapes, but this can be circumvented by snapping all non-critical layers to the standard 15 nm grid, while the layers critical to photonics remain on the finer grid.

In addition, photonics-specific layout rules are added to the DRC deck, such as the ones mentioned in the previous subsection. These rules typically cover two different areas. Some ensure that there are no unwanted interactions between optical and electrical devices. For instance, custom optical implants should not overlap with transistors, and transistor implants should not overlap with optical devices. Other device-specific rules have to be defined to ensure that overall integration is not affected. Design rules regarding blocking metal interconnects above grating couplers fall into this category. These rules are explicitly defined and checked for in the photonic DRC deck.

Additionally, while technically not in the realm of DRC, the design kit needs to be able to deal with vastly different densities of features. For instance, in a transceiver chip, the density of optical devices is rather low compared to the electronics blocks. Since process control usually relies on density equalization techniques, the average density of features over some areas is constrained within certain limits. This necessitates that optical features be added to the list of layers that need density equalization through a tiling algorithm, which is the action of placing dummy features in empty areas to compensate for low density. A separate deck is usually used for this purpose.

12.5.3 LVS

As discussed previously, the goal of LVS is to verify connectivity between electrical elements. The goal of an optoelectronic LVS must also include connectivity checks between optical elements, as well as optical and electrical domains. The integration of optical devices presents two major challenges for LVS (Figure 12.6). One challenge is the recognition and extraction of optical devices. Traditionally, the extraction of electrical devices has been performed by the use of Booleans on combinations of layers; for instance, a poly stripe overlapping a region of active is a transistor.

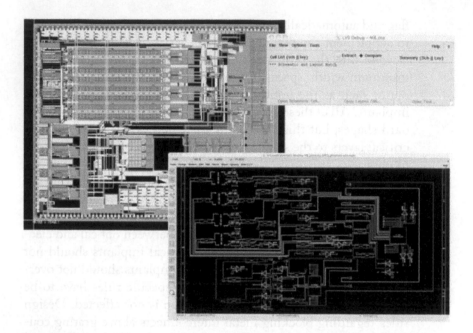

FIGURE 12.6 Example of integrated LVS in a CMOS Photonics technology.

In the case of photonic devices, especially passive devices, which are all simply different geometries of the same layer, it is impossible to adhere to this approach. One of the ways to achieve this is to add specific recognition utility layers and labels that are not present in the final mask layers to uniquely recognize different device types. The LVS deck uses a combination of overlap and labeling methods.

The second challenge is that the optical connectivity layer, comprising the waveguides, is drawn on the same layer as other optical devices. In a CMOS process, metal interconnects are usually separate from device layers, making it much easier to determine connectivity. However, some technologies do use devices built with metal layers, such as fringe capacitors, inductors, or transmission lines, creating similar issues to the ones discussed here. A typical solution is to identify and extract devices, remove them from the layer in question, and what remains is the connectivity layer.

Optical parasitic extraction is also an important feature that should match the capabilities of electrical parasitic extraction. Results from the extraction can be fed back to the simulation to determine post-layout block performance. Examples of optical parasitic extraction are computing waveguide loss, reflections, nonlinear effects, and signal skew due to optical path differences.

12.5.4 Simulating Optical Systems

Many optical communication systems, particularly of the hybrid variety, are designed using subcomponents from multiple sources, usually specialized companies focusing on one or just a handful of technologies. Optoelectronic device suppliers offer a selection of lasers, photodetectors, lithium-niobate modulators, and other components fabricated in their proprietary manufacturing processes, often based on exotic compound semiconductors. On the other side of the coin are fabless IC companies providing specialized driver and receiver circuits in a variety of more or less mainstream CMOS, SiGe, InP, and GaAs technologies. Optical module companies select these components based on published data sheets and usually validate their suitability based on measurements and characterization of a limited number of samples. This is followed by a prototyping phase where practical compatibility issues may be discovered, particularly at optoelectronic interfaces where components from different suppliers connect to each other.

A high degree of specialization can lead to excellence at the individual component level, but it can hinder the design of optimized systems. Specialized companies possess an in-depth knowledge of their product and volumes of data about performance, variability, and reliability, but such data is seldom passed to customers in its entirety. It is likely to be highly condensed to fill a data sheet with a handful of key metrics, which the manufacturer deems sufficient for procuring design wins, while minimizing the risk of revealing proprietary information to its competitors. Performance nonidealities may be omitted, and only come to the forefront during prototyping, which slows down the design cycle by forcing designers back to the drawing board.

An example of this approach is an optical module manufacturer pairing a vertical-cavity surface-emitting laser (VCSEL) with an off-the-shelf driver IC. Other than the obvious speed requirement, the selection is usually done based on output current of the driver relative to threshold current, slope efficiency, and series resistance of the VCSEL. All these parameters come from data sheets. Secondary effects, such as dynamic output impedance variation of the driver or relaxation oscillation of the laser, may be missed or not taken into consideration due to lack of data and models. Consequently, this approach is likely to produce a prototype that is more or less functional, but will often fall short of performance targets because of excessive deterministic jitter or intersymbol interference (ISI) arising from the previously mentioned nonidealities. Further experimentation may be needed, but even so this empirical approach is unlikely to cover the full range of component variability or ageing. This makes accurate yield predictions difficult, particularly in systems pushing technology boundaries. In principle, it is possible for an optical module company to engage IC and component supplies into closer collaboration, but in practice, thin margins and the constant search for more cost-effective second sources make module manufacturers reluctant to commit to suppliers and make suppliers reluctant to share data for fear of giving up their competitive advantage.

Recent progress in silicon photonics technology presents a unique opportunity not only for monolithic integration of communication systems but also for integrating and streamlining the entire design flow from product concept all the way to final verification and statistical yield prediction. The vertical integration it affords can naturally bridge the gap between the bottom-up approach of component designers and the top-down mindset of system architects. It starts with optoelectronic devices, such as modulators and photodetectors, being codesigned and integrated with optimized electronic circuits. The next step is to bring all the models into a common simulation environment and to provide a means of converting signals between optical and electrical domains, enabling simulation of complete optoelectronic systems.

The Cadence design environment has become a de facto standard for analog and mixed signal IC design. It provides a wide variety of circuit simulation tools, an analog hardware description language Verilog-A, support for synthesizable digital blocks, as well as scripting and analysis automation languages SKILL and OCEAN. This makes the Cadence design environment a convenient platform for implementing integrated optoelectronic design flow. Based on IC industry demand, Cadence process design kits (PDKs) are commonly provided by semiconductor foundries. These PDKs include models for electronic devices. Commercially viable silicon photonics processes are derived from such standard semiconductor technologies. They may include additional fabrication steps specific to optics and optoelectronic (OE) devices, but as a general rule, they try to avoid altering characteristics of core electronic devices to retain the advantage of rapid integration of existing IP cores. Consequently, a commercial PDK is a logical starting point for silicon photonics modeling and design flow enhancements.

The first question that needs to be answered is how to represent optical signals in simulation environments that evolved around only two electrical quantities, namely, current and voltage. Light, as a form of electromagnetic energy, is characterized by extremely high frequencies on the order of hundreds of terahertz for typical telecommunication wavelengths of 850–1510 nm. However, serial data throughput, and hence the electronic circuit speed in such communication systems, rarely exceeds 10–40 Gb/s, so their high frequency content is at least three orders of magnitude lower than the associated optical carrier. Attempting to simulate optical signals at the actual optical carrier frequency would necessitate subfemtosecond time steps, and barring a sudden leap in computing technology, the simulations could take years to complete. Clearly, a more efficient solution is needed. It is helpful to note that the information in such communication systems is carried either by variations in optical-power intensity (the envelope) or variations in phase of the optical carrier. These data transmission schemes are known as amplitude modulation (AM) and phase modulation (PM), respectively. In either case, the frequency associated with optical carrier modulation is identical to that of electronic circuits, and the knowledge of the instantaneous state of the terahertz optical carrier is irrelevant for data transmission. The solution becomes clear. Optical signals can be represented by an array of two fundamental quantities: (1) amplitude of the optical carrier envelope and (2) optical carrier phase. They can be supplemented with additional quantities as needed. For example, wavelength division multiplexing (WDM) systems require wavelength information as their multiplexing and demultiplexing structures have strong wavelength dependence. Some provisions for polarization diversity and multimode propagation may also be needed in some cases. But since monolithic silicon photonics systems tend to be single-wavelength, single-mode, polarization-maintaining solutions, envelope and phase may suffice.

12.5.5 Behavioral Models Using Verilog-A

Now that we have established the quantities that comprise an optical signal, the second question is how to implement models for optical and OE blocks. In case of electronic devices, de facto industry standards exist, such as BSIM for MOSFET transistors, which provide a framework for model fitting. Proliferation of such models is the consequence of commoditization of mainstream semiconductor technology and relative homogeneity in device design. These are hallmarks of a mature technology. On the other hand, silicon photonics is a young technology, characterized by a diversity of solutions and a rapid rate of innovation. A handful of companies are already on the verge of large-scale commercialization, each taking a somewhat different design approach. Years may pass before uncontested winners emerge on the device design front that would justify creating a standardized modeling framework. Fortunately, this is only a minor obstacle. Design environments such as Cadence can be readily adapted for this purpose. Specifically, Verilog-A, a hardware description language developed for behavioral modeling of analog circuits, is a good candidate for addressing this need. Verilog-A code can be simulated alongside electronic device models, and it allows defining equation-based models with conditional branches and iterative loops, using relatively simple syntax. Compared to data sheets, such models are a more reliable way for device designers to communicate with circuit and system designers. Their complexity can be enhanced more easily than the standardized models used for electronics. Also, they can be branched into different tiers of complexity, which can then be selected via Cadence's Hierarchy Editor to allow a wide range of trade-offs between simulation speed and accuracy, as appropriate for each level of abstraction and design maturity stage.

Let us look at some Verilog-A examples. In addition to physical devices, auxiliary blocks may be used to facilitate simulation setup, such as a behavioral optical source in Example 1. It converts electrical signals VmodOptPower and VmodPhasePower to an optical output consisting of OutOptPower and OutOptPhase. The code consists of a *module* statement specifying the name and I/O list of the block. Next, direction and discipline (electrical, opticalPower, and opticalPhase) of each I/O are defined. A discipline is a Verilog-A construct that groups related potential and flow quantities, referred to as *natures*. For example, voltage and current are, respectively, the potential and the flow natures of electrical discipline. Customized optical disciplines and natures define how the signal behaves when multiple nets or nodes are connected together, specify access function names, and assure that appropriate units are displayed when probing signals (optical phase in radians and optical envelope in watts). Finally, the behavior of the block is contained within the *analog* statement. In

EXAMPLE 1: BEHAVIORAL OPTICAL SOURCE MODEL IN VERILOG-A

```
module OpticalSource(VmodOptPower, VmodOptPhase,
OutOptPower, OutOptPhase);

    input VmodOptPower, VmodOptPhase;
    output OutOptPower, OutOptPhase;
    electrical VmodOptPower, VmodOptPhase;
    opticalPower OutOptPower;
    opticalPhase OutOptPhase;

    analog begin
            if(V(VmodOptPower) > 0)
OptPower(OutOptPower) <+ V(VmodOptPower);
            else OptPower(OutOptPower) <+ 0;
            OptPhase(OutOptPhase) <+ V(VmodOptPhase);
    end

endmodule
```

this case, a phase- and/or amplitude-modulated optical signal is created based on two time-varying voltage inputs. For a detailed explanation of the Verilog-A syntax, please refer to the Verilog-A reference manual available from Cadence Design Systems.

Next, a waveguide model is presented in Example 2. This is a passive optical component, and it has only optical input and output. Its behavioral model implements insertion loss scaled based on length and loss_dB_per_mm parameters. It also adds propagation delay based on length and a predefined constant P_C_in_WG representing the speed of light in the waveguide. These parameters are also defined in a Cadence simulation environment as CDF parameters that allow designers to access and modify them via the component's symbol view, without the need to edit the code.

Example 3 implements a simple photodetector. It converts the incident optical power InOptPower into photocurrent based on constant responsivity of 0.8 A/W, which is added to a dark current proportional to the reverse bias voltage. The phase of the optical signal is disregarded. Note that this is a purely behavioral model, and it does not accurately represent diode impedance,

EXAMPLE 2: OPTICAL WAVEGUIDE MODEL IN VERILOG-A.

```verilog
module Waveguide(InOptPower, InOptPhase, OutOptPower,
OutOptPhase,);

    input InOptPhase, InOptPower;
    output OutOptPhase, OutOptPower;
    opticalPower InOptPower, OutOptPower;
    opticalPhase InOptPhase, OutOptPhase;

    parameter real length = 100 from; // length in
um
    parameter real loss_dB_per_mm = 0.02; //
insertion loss in dB/mm

    real loss_scale, propagation_delay;

        analog begin

            @(initial_step) begin
*
                loss_scale = pow(10,-loss_
dB_per_mm*length/(10*1k));
                propagation_delay = length
1u/`P_C_in_WG;
            end

    OptPower(OutOptPower) <+
absdelay(OptPower(InOptPower) * loss_scale,
propagation_delay);
    OptPhase(OutOptPhase) <+
absdelay(OptPhase(InOptPhase), propagation_delay);

    end

endmodule
```

EXAMPLE 3: BEHAVIORAL PHOTODETECTOR MODEL IN VERILOG-A.

```verilog
module PhotoDetector(A, C, InOptPower, InOptPhase);

    input InOptPhase, InOptPower;
    inout A, C;
    electrical A, C;
    opticalPower InOptPower;
    opticalPhase InOptPhase;

    parameter real R_dark = 100k; // dark current
equivalent R [Ohms]
    parameter real responsivity = 0.8; //
responsivity [A/W]

        analog begin

        I(C,A) <+ (V(C,A) / R_dark) + responsivity
* OptPower(InOptPower);

            @(cross(V(C,A) + 0.1, -1)) begin
                $strobe("\n***************");
                $strobe(  "* WARNING:
*");
                $strobe(  "* Photodetector
in forward bias! *");
                $strobe(  "* Entering
invalid model region! *");
                $strobe("\n***************");
            end

            @(cross(V(C,A) + 0.1, +1)) begin
                $strobe("\n***************");
                $strobe(  "* NOTICE:
*");
                $strobe(  "* Photodetector
in reverse bias    *");
                $strobe(  "* Returning to
valid model region *");
                $strobe("\n***************");
            end

        end

endmodule
```

current–voltage (I–V) and capacitance–voltage (C–V) characteristics, noise, or many other physical aspects of behavior. This level of detail is sufficient for many system-level simulations, trading accuracy for simulation time; however, it would not be used by a circuit designer working on a high-speed transimpedance amplifier (TIA). In practice, the photodetector can be branched into multiple models with progressively higher levels of accuracy selectable for simulation with Cadence's Hierarchy Editor. Because this model is grossly inaccurate in forward bias, the code generates run-time warnings when the photodetector enters or leaves this invalid model region. Generating such messages is advisable in places where behavioral models fail to approximate real device behavior, and it allows identifying design mistakes early without the computational overhead of simulating complete, physics-based models at every stage of a design.

Further accuracy refinement of a diode-based OE component is illustrated by variable optical attenuator (VOA) in Example 4. It changes the optical insertion loss based on the forward diode current. But this time, the I–V characteristics are represented by Shockley diode equation that covers both forward and reverse regions of operation, eliminating the need for run-time warnings. The diode impedance model still lacks C–V characteristics (or any capacitance), so to improve transient and small-signal simulation accuracy its electro-optical bandwidth is limited to 5 GHz using a single-pole Laplace filter.

Finally, our last example shows how to implement an OE model without sacrificing any accuracy on the electronic side. The case in point is a phase modulator model shown in Figure 12.7. It consists

EXAMPLE 4: VARIABLE OPTICAL ATTENUATOR MODEL IN VERILOG-A.

```
module VOA(InOptPower, InOptPhase, OutOptPower,
OutOptPhase, A, C);

    input InOptPhase, InOptPower, A, C;
    output OutOptPhase, OutOptPower;
    electrical A, C;
    opticalPower OutOptPower, OutOptPower;
    opticalPhase OutOptPhase, OutOptPhase;
```

```
    parameter real length = 100; // length [um]
    parameter real loss_dB_per_mm = 0.3; // fixed
insertion loss [dB/mm]
    parameter real EO_bandwidth = 5G; // electro-
optical bandwidth [Hz]
    parameter real Is = 1.5e-12; // diode Is [A]
    parameter real n = 1.2; // diode ideality factor
    parameter real Rs = 2.0; // diode series
resistance [Ohms]

    real filter_pole_array[0:1], loss_scale, Rs_
scale, Is_scale, propagation_delay, dynamic_attn_dB;

        analog begin

        @(initial_step) begin
            filter_pole_array[0] = -`M_TWO_PI *
EO_bandwidth;
            filter_pole_array[1] = 0;
            loss_scaled = pow(10,-loss_dB_per_
mm*length/(10*1k));
            Rs_scaled = Rs / (length/100);
            Is_scaled = Is * (length/100);
            propagation_delay = length * 1u /
`P_C_in_WG;

        // Electrical model: diode equation
        I(A,C) <+ Is_scaled * (exp((V(A,C)-
I(A,C)*Rs_scaled)/(n*$vt)) - 1);

        // Optical Model: apply low-pass filtering
and delay dynamic_attn_dB = laplace_np(3.0 *
pow(1k*I(A,C), 0.8), {1}, filter_pole_array);
            OptPower(OutOptPower) <+
absdelay(OptPower(InOptPower) * loss_scaled,
propagation_delay) * absdelay(pow(10,-dynamic_attn_
dB/10), propagation_delay/2);
            OptPhase(OutOptPhase) <+
absdelay(OptPhase(OutOptPhase), propagation_delay);
    end

endmodule
```

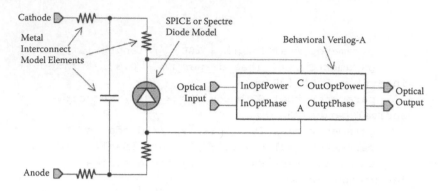

FIGURE 12.7 Two-level model of a phase modulator.

of two levels of hierarchy. The top level comprises a physics-based diode model fitted to a standard SPICE or Spectre format. This includes both I–V and C–V diode characteristics in all operating regions. In addition, metal interconnects leading to the diode terminals are represented with a resistor–capacitor (RC) network. The optical aspects of the device are modeled by the Verilog-A block whose code is shown in Example 5. The Verilog-A block samples the voltage at the diode terminals to generate a dynamic optical phase shift based on an equation. Due to its accurate representation of diode impedance, this model can be used by a circuit designer implementing a multi-Gb/s optical MZI modulator, a ring resonator, or other building block reliant on high-speed optical phase shifters.

The preceding examples provide only a rudimentary glimpse at one possible method of implementing optoelectronic simulations in an environment developed for IC design. This particular approach leverages flexibility of Cadence's Verilog-A to model and simulate blocks and functions of arbitrary complexity. Our intention within the scope of this chapter is not to provide a complete solution but to outline a path that the reader may choose to follow and refine to fit his or her specific simulation objectives. For example, the modeling framework presented here can be extended to include statistical data. Furthermore, Verilog-A can be used to automate block-level and system-level simulations to extract and summarize specification compliance information, and many other aspects of optical and electrical performance across process corners, temperature, ageing, and environmental conditions.

At this stage, the possibilities for extending the presented approach are very broad, and we encourage our readers to

EXAMPLE 5: PHASE MODULATOR MODEL IN VERILOG-A.

```
module PhaseMod(InOptPower, InOptPhase, OutOptPower,
OutOptPhase, A, C);

    input InOptPhase, InOptPower;
    output OutOptPhase, OutOptPower;
    inout A, C;
    electrical A, C;
    opticalPower OutOptPower, InOptPower;
    opticalPhase OutOptPhase, InOptPhase;

    parameter real length = 100; // lenght [um]
    parameter real loss_dB_per_mm = 0.1; // insertion
loss [dB/mm]

    real dynamic_phase, loss_scale, propagation_
delay;

            analog begin

                @(initial_step) begin
                        loss_scale = pow(10,-loss_
dB_per_mm*length/(10*1k)));
                        propagation_delay =
length*1u/`P_C_in_WG;
            end

                // calculate voltage dependent
phase shift
                dynamic_phase = 9.0 * pow(V(C,A),
0.6);
            OptPower(OutOptPower) <+
absdelay(OptPower(InOptPower) * loss_scale,
propagation_delay);
            OptPhase(OutOptPhase) <+
absdelay(OptPhase(InOptPhase), propagation_delay) +
absdelay(dynamic_phase, propagation_delay/2);

            end

endmodule
```

experiment and share their ideas with the community. Eventually, as the silicon photonics technology matures, prepackaged modeling and simulation solutions will be introduced by semiconductor foundries and design automation companies. But until then, those wishing to stay at the forefront of the technology will have to rely on their ingenuity.

12.6 Designing Optoelectronic Subsystems and Systems

The LuxG process was developed to enable high performance and design complexity in optoelectronics. The advantages brought by combining optics and electronics in a single chip can overcome penalties incurred by some of the components because of the constraints imposed by technology integration. In this section, this concept is briefly illustrated using three subsystems realized in this technology: a high-speed optical receiver and transmitter, the integrated control system, and, finally, a 40 Gb/s transceiver chip.

12.6.1 High-Speed Integrated Receiver

The high-speed optical receiver has a high-impedance architecture (Personick 1973), shown schematically in Figure 12.8. While this topology simplifies the design when compared to the more common transimpedance architecture, also shown in Figure 12.8, it is seldom used in practice because of the stringent input capacitance requirements that are difficult to meet in hybrid systems using surface-illuminated detectors. The process, owing to the extremely low capacitance of the Ge detector (up to 3 orders of magnitude lower than an equivalent high-speed III-V photodetector) and to the integrated approach minimizing interconnect

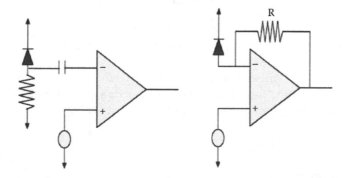

FIGURE 12.8 High-impedance (left) and transimpedance (right) architectures for an integrated optical receiver.

parasitics, not only enables the high-impedance topology but also allows to push the electrical sensitivity well beyond typical reported values for III-V-based receivers, down to –24 dBm at 10 Gb/s. The subsystem has a higher performance even though the field-effect transistors used in the amplifier are underperforming bipolar devices typically used in high-performance hybrid TIAs in terms of noise, gain, and bandwidth.

12.6.2 High-Speed Integrated Transmitter

Due to the relatively low phase modulation efficiency in electro-optic modulators, typically at least a few millimeters of length are required to achieve a phase shift close to $\pi/2$, which is needed for a sufficient extinction ratio in a differentially driven MZI modulator. In high-speed applications near 10 Gb/s, this length requires the use of a traveling-wave configuration for the modulator in which the electrical signal propagates via a transmission line. The velocity of the transmission line is matched to that of the optical mode, and the line is periodically tapped to subsequent sections of the modulator. This approach is illustrated in Figure 12.9.

In order to minimize the amplitude of the signal launched in the transmission line, a distributed amplification scheme can be used in which a low-power signal traveling in a transmission line is periodically tapped and locally amplified to drive each section

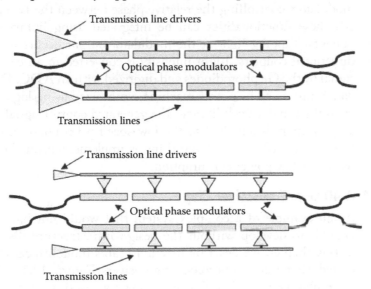

FIGURE 12.9 MZI modulator architecture with a transmission line driver (top) and with a distributed driver (bottom).

of the phase modulator. This scheme allows optimizing the transmission line for loss and velocity matching with the optical mode, independently of the driving requirements of the phase modulator diodes, which in turn are taken care of by the local amplifiers. In the specific case of our design, adopting the distributed driver architecture resulted in a dramatic reduction in power dissipation (Narasimha et al. 2007).

12.6.3 Integrated Control

A number of control functions are required in high-speed optical transceiver modules, including stabilization of the bias point of the MZI modulator, regulation and temperature compensation of laser output, and detection of valid data. All these functions require an algorithm that can be implemented through a digital state machine, a microcontroller program, or using analog signal processing. The first approach is appealing because of its parallelism as a single-state machine can be built for each function in each transceiver channel, and it can run independently of the others.

The control system of the MZI modulator bias is designed to keep the modulator close to the quadrature point. In order to do that, the system requires reading the average light intensity from both arms of the modulator output, processing that information, and generating a control signal to actuate the low-speed phase modulators controlling the relative phase between the two arms. All these functionalities can be integrated using library components. Tapping light at the modulator output occurs through directional couplers. Light intensity is converted to an electrical signal by the Ge photodiodes and digitized with an ADC. Digital data is processed by a state machine synthesized using logic gates from the standard cell library. The generated control signal, after amplification, is supplied to the low-speed p-i-n phase modulators in the MZI arms. Everything is implemented on the chip with very low power consumption.

12.6.4 A 40 Gb/s Transceiver Chip

As an example of a complex system that would not have been possible to develop without the design infrastructure discussed in this chapter, we describe here a 40 Gb/s monolithically integrated single-chip photonics transceiver. Figure 12.10 shows a die photograph. This chip integrates four optical transmitters and receivers with all corresponding high-speed and auxiliary control circuitry. It also includes Ge photodiodes for high-speed

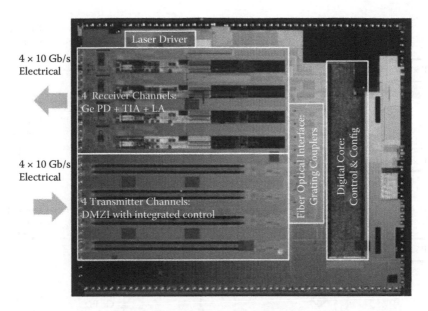

FIGURE 12.10 A monolithically integrated CMOS photonics transceiver chip.

detection as well as monitoring of MZI and laser outputs. It also contains digital electronics to manage the interface to the outside world. All optical components, with the exception of a single distributed feedback (DFB) laser, are monolithically integrated. The transmitters and the receivers operate at 10 Gb/s per channel to provide an aggregate data throughput of 40 Gb/s. The clear advantage of this technology is that assembly and packaging complexity is dramatically reduced due to the low number of components to be packaged.

Figure 12.11 shows transmitter architecture. The module uses a single external continuous-wave laser coupled to the chip through a grating coupler and split four ways to supply light to all transmit channels. The modulation is performed by four distributed-driver MZI modulators. Grating couplers are used to couple the modulated light from each transmitter channel to a standard single-mode fiber (SMF). Each of the modulators is dynamically controlled to set the desired quadrature operating point by a combination of monitoring Ge photodiodes that measure the output power of both modulated MZI output phases, low-speed phase modulators in each arm of the MZI to compensate for static phase offset, and a digital control loop. The chip does not need to be actively cooled, as the control loop ensures accurate biasing across both environmental and process variations.

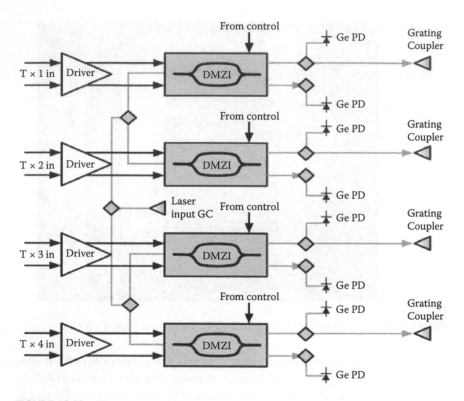

FIGURE 12.11 Transmitter architecture.

On the receive side, the fiber interface is provided by polarization-splitting grating couplers that split the light into two output waveguides based on the polarization of the incoming signal. Within each receiver channel the two waveguides are connected to the two inputs of a high-speed Ge photodiode, whose photocurrent is then converted to voltage and fed to a series of voltage amplifiers to obtain a signal swing sufficiently large to be transmitted outside the chip (Figure 12.12).

Figure 12.13 shows the electrical eye diagram for the full optical link. The eye is fully open and error-free for data transmitted through an optical transmitter and received on a second, identical chip through an optical receiver.

12.7 Summary

In this chapter, we have described our approach to a fully integrated CMOS photonics CAD. We have illustrated one method that allows a tight integration of photonic devices into a CMOS

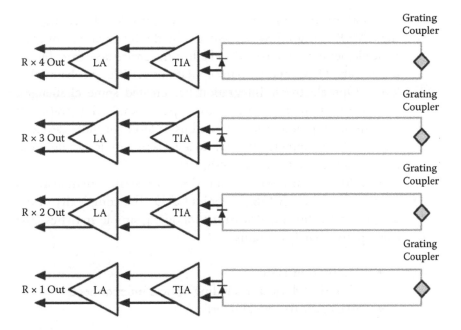

FIGURE 12.12 Receiver architecture.

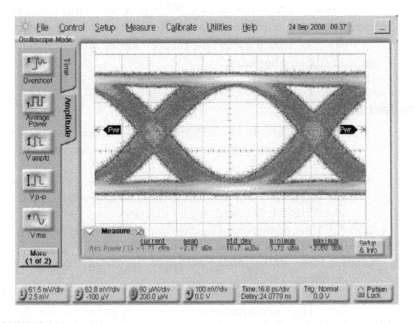

FIGURE 12.13 10 Gb/s eye diagram for complete system (electrical to electrical through optical link).

flow both on the process and on the design side. The changes to the CMOS design flow necessitated by the introduction of photonic elements are nontrivial and affect the entire flow from libraries, layouts, DRC, and LVS to simulations at different abstraction levels. Optoelectronic integration has created some challenges, but the evolution of CAD tools to support complex integration of system-on-chip design has allowed us to leverage these developments to bring optoelectronics into a well-known design flow and make it accessible to system designers.

The 40 Gb/s transceiver chip described earlier demonstrates the power of the CMOS photonics CAD. The complexity of the design can be illustrated with the following simple statistics. The chip approximately contains

- 135,500 transistors
- 310 optical devices, excluding optical interconnects
- 550 electrical–optical interfaces

Without an infrastructure such as the one described in this chapter, there would have been little chance to design and fabricate successfully a functional transceiver chip of this complexity. The integration of photonics into a CMOS process requires that automated design capabilities be available to the system designers, so they can take advantage of new functionalities offered by innovations in silicon photonics.

References

Dutton, R. 2009. Stanford TCAD. http://www-tcad.stanford.edu http://bwrc.eecs.berkeley.edu/Classes/icbook/SPICE (accessed August 10, 2009).

Hu, C. 2009. Berkeley BSIM3. http://www-device.eecs.berkeley.edu/~bsim3/intro.html http://bwrc.eecs.berkeley.edu/Classes/icbook/SPICE (accessed August 10, 2009).

Kilby, J. S. 1964. Minituarized Electronic Circuits, U.S. patent 3,138,743, June 3, 1964.

Kanellos, M. 2001, Intel's accidental revolution. *CNET News*, November 14.

Masini, G., Colace, L., Assanto, G., Luan, H.-C., and Kimerling, L. C. 2001. High performance p-i-n Ge on Si photodetectors for the near infrared: From model to demonstration. *IEEE Trans. Electron Devices*, 48: 1092–1096.

Mekis, A., Abdalla, S., Analui, B., et al. 2008. Monolithic Integration of Photonic and Electronic Circuits in a CMOS Process. *Proc. SPIE. Optoelectronic Integrated Circuits* X: 6897, 6987OL-1-14.

Moore, G. E. 1965. Cramming more components onto integrated circuits. *Electronics* 38(8).

Narasimha, A., Analui, B., Liang, Y., Sleboda, T. J., and Gunn, C. 2007. A fully integrated 4×10 Gp/s DWDM optoelectronic transceiver in a standard 0.13 μm CMOS SOI. *IEEE ISSCC Dig. Tech. Papers*: 42-586, 244–245.

Personick, S. D. 1973. Receiver design for digital fiber optic communication systems. *Bell Syst. Tech. J.*, 52: 843–874.

Quarles, T., Pederson, D., Newton, R., Sangiovanni-Vincentelli, A., and Wayne, C. 2009. Berkeley SPICE. http://bwrc.eecs.berkeley.edu/Classes/icbook/SPICE (accessed August 10, 2009).

Sangiovanni-Vincentelli, A. 2003. The Tides of EDA, *IEEE Design and Test of Computers*, November–December: 59–75.

Soref, R. and Bennet, B. 1987. Electrooptical effects in silicon. *IEEE J. Quantum Electron.*, 23: 123–129.

Index